AF892983

Dissimilar Metal Joining

Dheerendra Kumar Dwivedi

Dissimilar Metal Joining

Springer

Dheerendra Kumar Dwivedi
Department of Mechanical and Industrial Engineering
Indian Institute of Technology Roorkee
Roorkee, India

With Contribution bys
Anup S. Kulkarni
Department of Mechanical and Industrial Engineering
Indian Institute of Technology Roorkee
Roorkee, India

Pankaj Kaushik
Department of Mechanical and Industrial Engineering
Indian Institute of Technology Roorkee
Roorkee, India

ISBN 978-981-99-1896-6 ISBN 978-981-99-1897-3 (eBook)
https://doi.org/10.1007/978-981-99-1897-3

© The Editor(s) (if applicable) and The Author(s), under exclusive license to Springer Nature Singapore Pte Ltd. 2023
This work is subject to copyright. All rights are solely and exclusively licensed by the Publisher, whether the whole or part of the material is concerned, specifically the rights of translation, reprinting, reuse of illustrations, recitation, broadcasting, reproduction on microfilms or in any other physical way, and transmission or information storage and retrieval, electronic adaptation, computer software, or by similar or dissimilar methodology now known or hereafter developed.
The use of general descriptive names, registered names, trademarks, service marks, etc. in this publication does not imply, even in the absence of a specific statement, that such names are exempt from the relevant protective laws and regulations and therefore free for general use.
The publisher, the authors, and the editors are safe to assume that the advice and information in this book are believed to be true and accurate at the date of publication. Neither the publisher nor the authors or the editors give a warranty, expressed or implied, with respect to the material contained herein or for any errors or omissions that may have been made. The publisher remains neutral with regard to jurisdictional claims in published maps and institutional affiliations.

This Springer imprint is published by the registered company Springer Nature Singapore Pte Ltd.
The registered company address is: 152 Beach Road, #21-01/04 Gateway East, Singapore 189721, Singapore

I would like to dedicate this work to my father Late Shri Mahabir Prasad Dwivedi and Mother Late Shrimati Shanti Dwivedi.

Preface

The application of dissimilar metal joining is continuously growing to develop energy-efficient, lightweight and reliable systems, especially in the energy and automotive industries. Accordingly, research and development and training of human resources needed to effectively deal with the challenges of dissimilar metal joining. The book entitled *Dissimilar Metal Joining* provides a fundamental understanding of issues related to dissimilar metal joining, technological solutions and approaches used for developing sound dissimilar metal joints. Since dissimilar metal joining involves two or more metals that behave differently under the influence of heat and deformation, joining approach (fusion, deformation, chemical interaction, etc.) significantly affects the microstructure, mechanical and corrosion properties, residual stress and distortion tendency of dissimilar metal joints. This book focuses the importance of metallurgical transformation in dissimilar metal joining by fusion (arc, laser) welding, solid joining, resistance welding and adhesive joining. These aspects have been presented considering interest of all those pursuing research and development at the post-graduate and doctoral research in the area of dissimilar metal joining. Researchers will find it useful in many ways: (a) understanding the fundamental issues of dissimilar metal joining by fusion and solid-state joining processes, and factors that affect the soundness, and mechanical performance of the joints, (b) the significance of heat and deformation in dissimilar metal joining and their effect on mechanical, and metallurgical properties of joints and (c) structure–property relationship for dissimilar metal joints. Practicing engineers and managers dealing with the challenges of dissimilar metal joining will find very handy information on what can be done to ensure the quality of joints.

This book is based on the following four: (a) fundamental understanding of the author on the subject matter, (b) findings of R&D activities of post-graduate and doctoral research students under author's supervision, (c) recent research publications and (d) literature available in public domain in form of books, handbooks and web.

Roorkee, India

Dheerendra Kumar Dwivedi

About This Book

Dissimilar metal joining is considered difficult mainly due to difference in chemical, physical, mechanical and thermal properties of the parent metals to be joined. Fundamental principles and approaches unique to dissimilar metal joining related to fusion joining (arc, resistance, radiation-based welding process) and solid-state joining (friction stir welding, ultrasonic welding, explosive welding and adhesive joining) have been presented. Further, difficulties encountered during fusion/solid-state joining of dissimilar metals have been described. Methodologies to overcome issues related to dissimilar metal joining by fusion welding (GTAW, A-GTAW and its variant), resistance spot welding and solid-state joining (friction stir welding and diffusion bonding) have been included. Advances on dissimilar metal joining using both fusion and solid-state joining processes have been included. Manuscript has been developed based on our fundamental understanding, finding of experimental studies of research group at welding research laboratory. Main highlighting points of the book on *Dissimilar Metal Joining* making it unique include:

(a) Use of detailed coloured schematics to demonstrate the effect of heat, and temperature gradient on structure, properties and formation of different zones
(b) Integrating different aspects of dissimilar metal joining like heat flow, thermal expansion and contract, metallurgical transformation and weldability has been covered to develop a comprehensive understanding in graduate and doctoral research students.
(c) Inclusion of advances in the area of dissimilar metal joining by fusion and solid-state joining technologies.

Contents

About the Author

Dr. Dheerendra Kumar Dwivedi is a Professor in the Department of Mechanical and Industrial Engineering at the Indian Institute of Technology (IIT) Roorkee. He is the recipient of Indian and American National Awards, namely the Binani Gold Medal Award by IIM (2001), the W. H. Hobart Memorial Award by AWS (2022), and the INSA Teachers Award by INSA (2022), and has been recognized amongst the top 2% Global Scientists in Materials Domain by Elsevier-Stanford (2020–2022). He has been involved in teaching, research and development, and industrial consultancy for the last 28 years broadly in the area of manufacturing technologies in general and solid state joining and fusion welding processes in particular. He supervised 17 Ph.D. thesis and over 50 M.Tech. Dissertations. He has published in over 134 research papers in peer-reviewed SCI and SCIE-indexed international journals with h-factor 43 and i-10 index 104. He has been granted two Indian patents in the area of dissimilar steel joining. He has undertaken nine bilateral international collaborative research projects with reputed universities, namely Chemnitz University, Germany, University of Coimbra, Portugal, University of Uberlandia, Brazil, University of Zacatecas, Mexico and Physical Technical Institute, Belarus. Dr. Dwivedi has undertaken more than 24 research projects in the area of weld-bonding, friction stir welding, flux-assisted gas tungsten arc welding, laser cladding, and friction stir processing-Bronzes from funding agencies such as DST, BRNS, ARDB, CSIR etc.

Chapter 1
Fundamentals of Dissimilar Metal Joining

This chapter presents the need for dissimilar metal joining, types of dissimilar metal joints, issues encountered during dissimilar metal joining, the role of physical, mechanical, chemical and dimensional proportion of parent metals in dissimilar metal joining, strategies for developing dissimilar metal joint and performance of dissimilar metal joint.

1.1 Need

The need to design and manufacture components, products and systems made of different metals depends on many important factors such as functional/technological requirements as per severity of service conditions, economical compulsions to cut down cost to increase competitiveness and ease of fabrication (Sharma and Dwivedi 2017; Shankar and Dwivedi 2017). These factors predominantly push the growth of dissimilar metal combinations for manufacturing goods in many sectors such as transport (bicycle, cars, aerospace), energy (windmills, thermal power plants), chemical (food processing, petro-chemical) and electronics. However, the primary reason for the application of dissimilar metal combinations in a particular sector varies significantly. For example, in the transport sector, different metal combinations are primarily used to reduce the dead weight of the entire system while satisfying other functional and technological requirements (specific strength, impact and fatigue resistance) and keeping the cost within control (Kulkarni et al. 2018). In thermal power (coal, nuclear) plants, different subsystems like superheaters, reheaters, economizers, boilers, etc. experience different service temperatures (300–750 °C). Accordingly, different types of steel (ferritic steels such as P22, P91 and P92 to austenitic stainless steels such as AISI 304, 316) are used in thermal power plants as per the need for resistance to thermal softening, oxidation and metallurgical instability while keeping cost in control. Similarly, in chemical industries, various subsystems like pumps, pipelines, nozzles and mixers handling fluids of different severities in

© The Author(s), under exclusive license to Springer Nature Singapore Pte Ltd. 2023
D. K. Dwivedi, *Dissimilar Metal Joining*,
https://doi.org/10.1007/978-981-99-1897-3_1

the course of production tend to attack metallic components in various ways. To enhance the life of such components, therefore, it is required to use different metallic combinations considering the need for resistance to general corrosion, stress corrosion, pitting and crevice corrosion to fluid transported (Kumarni et al. 2018; Kulkarni et al. 2019; Sharma and Dwivedi 2019a, b).

Since the service conditions experienced by the metallic components and systems differ very widely, the technological requirements for desired performance and life may also vary in terms of requirements of high yield strength, toughness, ambient and high-temperature hardness, resistance to oxidation, corrosion, fatigue, thermal fatigue stress corrosion, electrical and thermal conductivity/insulation, density, appearance, ease of manufacturing, etc. For example, the application of low thermal conductivity and lightweight plastic or polymer matrix composite for the one part while high strength, impact and fatigue resistance for another part and both may be needed to join for the proper functionality of machine/system. The joining of different material combinations like plastic and metal, steel and aluminium, steel and copper, stainless steel and titanium are a few common examples (Shankar and Dwivedi 2017).

1.2 Types of Dissimilar Metal Joints

Broadly, the dissimilar metal joining is categorized as (a) conventional joint between dissimilar parent metals forming an integral part of the assembly to take the load and withstand as per the needs of the service conditions and (b) transition joint made to facilitate the connection between parent metals having significantly different physical, mechanical and chemical characteristics (Fig. 1.1). In the first case of conventional dissimilar metal joining, a joint is developed between two parent metals with or without filler metal/interlayer during fusion or solid-state joining especially when the difference in properties of both the parent metals is not much to deteriorate the performance of the joint, for example, joining between different grades of ferritic steels (mild steel and alloy steel), aluminium alloys (Al–Cu alloy and Al–Si–Mg alloys), etc.

The choice of filler metal/interlayer in such a case becomes important primarily to reduce the mismatch in properties of parent metals. The filler metal/interlayer used to develop (a) the weld metal of intermediate characteristics through controlled dilution and (b) a butter layer in case of fusion welding to isolate one of the parent metals, which is more sensitive to weld thermal cycle or dilution degrading the weld metal characteristics.

The transition joints are developed when parent metals (to be joined) possess significantly different characteristics, namely composition, melting point and thermal expansion coefficient leading to increased difficulty in the development of sound joints besides the increased possibility of reduced performance of joints during service (Fig. 1.2). To develop a transition joint, an additional metallic member having intermediate characteristics is brought in between the two parent metals. For example,

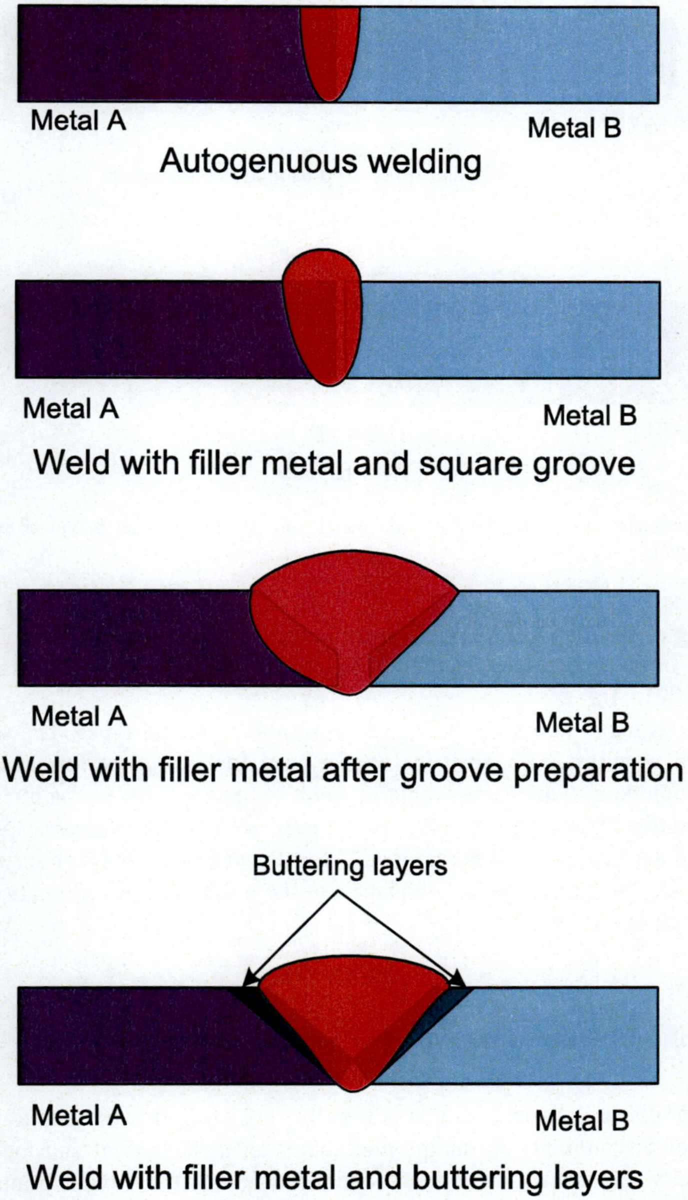

Fig. 1.1 Schematic showing different ways of developing dissimilar metal joints

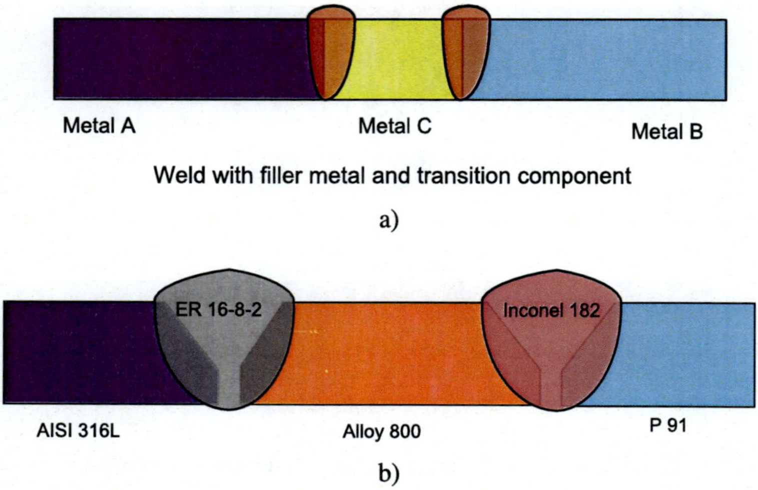

Fig. 1.2 Schematic of tri-metallic dissimilar metal joint **a** approach and **b** typical example of tri-metallic weld joint

the joining of ferritic steel (alloy steel, P22, P91, P22) with austenitic steel (AISI 304, 316) in thermal power plant components (Kulkarni et al. 2020; Sharma and Dwivedi 2021a, c).

Dissimilar metal joints can be structural or non-structural types. The structural joints have to withstand against the service load for desired performance of the system/component, while non-structural joints primarily help in maintaining the relative position of the parent metals being joined therefore level of stress experienced by these joints during service is very low. These do not transfer the high service load. For example, joints made at the foundation of the machines, brazed joints between shank and cutting tool tips, etc.

1.3 Issues in Dissimilar Metal Joining

The ease of dissimilar metal joining is usually very low, primarily due to the high tendency of discontinuity formation and chemical, metallurgical and mechanical heterogeneity across the joint. The discontinuities in dissimilar metal joints (DMJ) develop mainly due to the differences in various physical, mechanical and chemical characteristics of parent metals. These discontinuities may appear in various forms such as unmixed zone (UMZ), segregation of alloying elements at joint interface, hard and brittle intermetallic compounds, the asymmetric joint between parent metals, skewed interfacial deformation, residual stress and distortion, large variation in mechanical and corrosion properties of the joint and heat-affected zone (HAZ) on both sides of joint (Kaushik and Dwivedi 2021a, 2022a; Kulkarni et al. 2021; Sharma

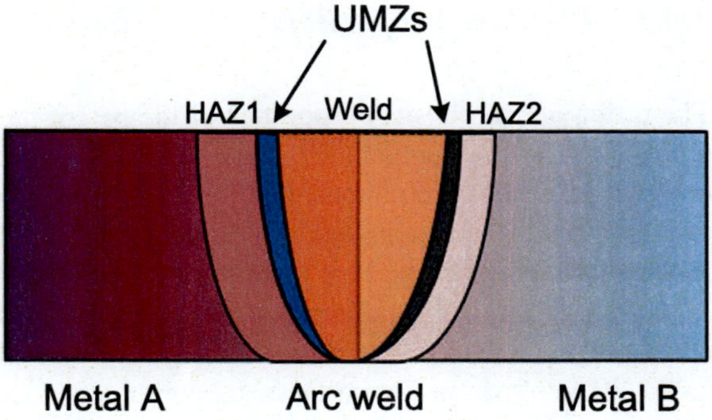

Fig. 1.3 Schematic showing different zones formed in a typical fusion weld of dissimilar metal joints

and Dwivedi 2021b). The dissimilar metal joints typically form five to seven zones (PM1-HAZ1-UMZ1-WZ-UMZ2-HAZ2-PM2) having different microstructures and mechanical properties in dissimilar metal joints (Fig. 1.3). These factors, discontinuities and zones mostly degrade the (mechanical and corrosion) performance of dissimilar metal joints. The common discontinuities in DMJ include cracks in the weld and HAZ, hard and brittle zone formation, softening and weakening, stress raisers.

Moreover, the extent of heterogeneity in the properties of dissimilar metal joints is determined by not just chemical composition, but also by parent metal conditions and strengthening mechanisms based on which parent metals are designed. The parent metals strengthened by grain refinement, work hardening, precipitation hardening and transformation hardening are affected more by mechanical and thermal stresses related to dissimilar metal joining than those strengthened by solid solution strengthening and dispersion hardening. The weld thermal cycle in general softens the autogenous weld joint and HAZ if the parent metals are strengthened by grain refinement, work hardening, precipitation hardening, and on the contrary, the hardening of the joint and HAZ is noticed in the transformation hardening parent metals like steel and cast iron (Kulkarni et al. 2021; Sharma and Dwivedi 2021b). Moreover, the HAZ of Q and T steel is softened while the weld is hardened by the weld thermal cycle of fusion welding. The effect of weld thermal cycle on solid solution and dispersion of hardened metals is marginal.

1.4 Parent Metal Characteristics and Dissimilar Metal Joining

The parent metals of dissimilar combinations may have partially or completely different physical, mechanical, chemical and dimensional characteristics. The extent of difference in these characteristics determines the ease of joining, and efforts needed for developing quality joints, and accordingly, the performance of such dissimilar metal joints during the service.

1.4.1 *Physical Properties*

The difference in physical properties of the parent metals, namely melting temperature, solidification temperature range, thermal conductivity and thermal expansion coefficient, needs careful consideration to avoid discontinuities in dissimilar metal joints especially by fusion welding using comparatively low power density welding processes leading to high heat input (Table 1.1).

A large difference in melting temperature (> 150–200 °C), solidification temperature range (> 50 °C), thermal conductivity and thermal expansion coefficient (> 20–30%) of parent dissimilar metal combination results in the following:

(a) Asymmetric weld shifting towards one side of base metal due to skewed heat distribution,
(b) Different widths of heat-affected zones and so variations in mechanical properties,
(c) Different thermal expansion and contraction of parent metals (due to weld thermal imposed during welding) leading to non-uniform residual stress and so distortion tendency
(d) Tendency of liquation and solidification cracking of weld and HAZ.

A large difference in melting point and thermal conductivity of base metals like Al and Fe makes the control of molten metal of a weld pool difficult because one parent metal may remain in red-hot condition while the other one after melting may start flowing here and there (Fig. 1.4). Similarly, a significant difference in the thermal expansion coefficient of parent metals increases the cracking and distortion tendency of the weldment primarily due to increased differential expansion and contraction setting in high residual stress (Fig. 1.5) (Kulkarni et al. 2021).

Table 1.1 Physical properties of common metal

Metal	Melting temperature (K)	Boiling temperature (K)	Density ($kg\ m^{-3}$)	Thermal conductivity ($W\ m^{-1}\ K^{-1}$)	Specific heat capacity ($J\ kg^{-1}\ K^{-1}$)	Thermal expansion coefficient ($10^6\ K^{-1}$)
Fe	1809	3133	7870	78	456	12.1
Al	933	2793	2700	238	917	23.5
Cu	1356	2833	8930	397	386	17.0
Ni	1728	3188	8900	89	452	13.3
Ti	1940	3558	4500	22	528	8.9
Zn	693	1184	7140	120	394	31.0
Mo	2888	4883	10,220	137	251	5.1
W	3673	5828	19,300	174	138	4.5
Zr	2125	4673	6490	23	289	5.9
Nb	2740	5013	8600	54	268	7.2

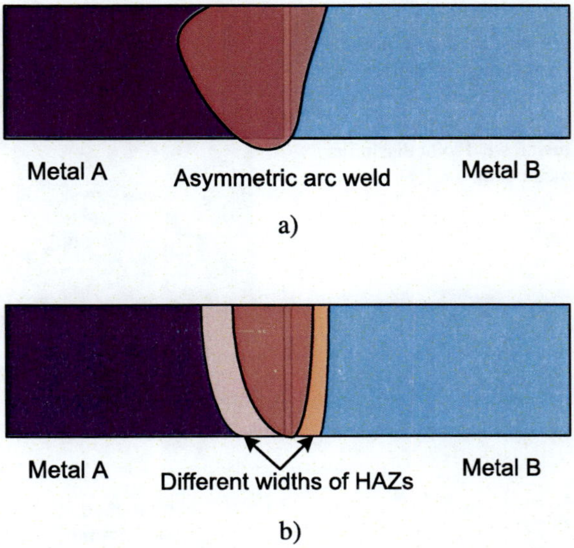

Fig. 1.4 Schematic showing issues of dissimilar metal fusion welding **a** asymmetric weld and **b** different widths of heat-affected zones

1.4.2 Mechanical Properties

The influence of the difference in mechanical properties of the parent metals namely yield strength, ductility, toughness, work hardening and thermal softening, etc. on the ease of the joining depends on the approach applied such as fusion welding or solid-state joining (Fig. 1.6). The yield strength and ductility of parent metals play a more important role in solid-state joining (except diffusion bonding) relying on macro/micro-scale deformation than fusion welding processes. Both the yield strength and ductility of a parent metal determine the extent/ease of deformation needed during solid-state joining to facilitate the metallurgical bonding. Additionally, thermal softening and work hardening due to plastic deformation and heat generation, if any, during joining can also affect the flowability of metal at the joint interface. Thermal softening during metal joining, in general, reduces yield strength and increases ductility due to recovery and recrystallization with the rise in temperature as per the weld thermal cycle. Work hardening on the other hand increases hardness and strength besides reducing the ductility. Therefore, the thermal softening and work hardening behaviour of a metal both affect the ease of joining by solid-state joining processes. Low yield strength and high ductility metals allow easy deformation during solid-state joining to develop a joint. The difference in mechanical properties during dissimilar metal joining, therefore, results in asymmetric deformation of parent metals at the joint interface, which in turn may appear in the form of limited interfacial metallurgical bonding and so poor joint strength (Kaushik and Dwivedi 2021b, 2022b).

The role of mechanical properties (yield strength, ductility and toughness) of parent metals in the case of fusion welding is primarily limited to residual stress

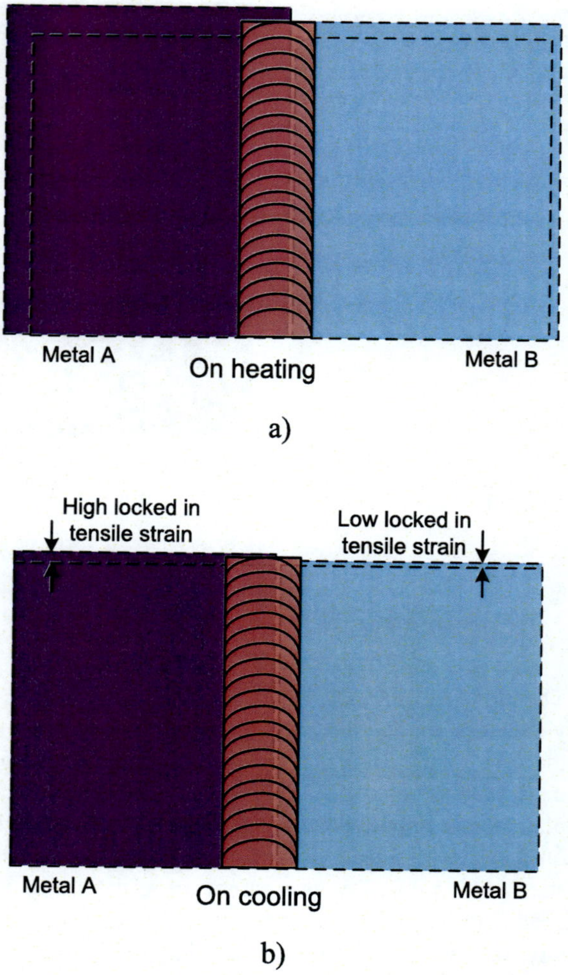

Fig. 1.5 Schematic showing different thermal expansion and contraction behaviours of parent metals during **a** heating and **b** cooling of fusion weld joint dissimilar metal combinations

development and cracking tendency. An increase in yield strength and reduction in ductility in general increases residual stress and cracking tendency of weld metal and heat-affected zone. Further, the tolerance to cracking of metallic joints due to residual tensile stress (especially in the presence of stress raiser) increases with ductility and toughness (Kaushik and Dwivedi 2021b; Kaushik and Dwivedi 2020).

1.4.3 Chemical Composition

The metallurgical characteristics of parent metals significantly depend on chemical composition besides thermal and mechanical history. Parent metals having

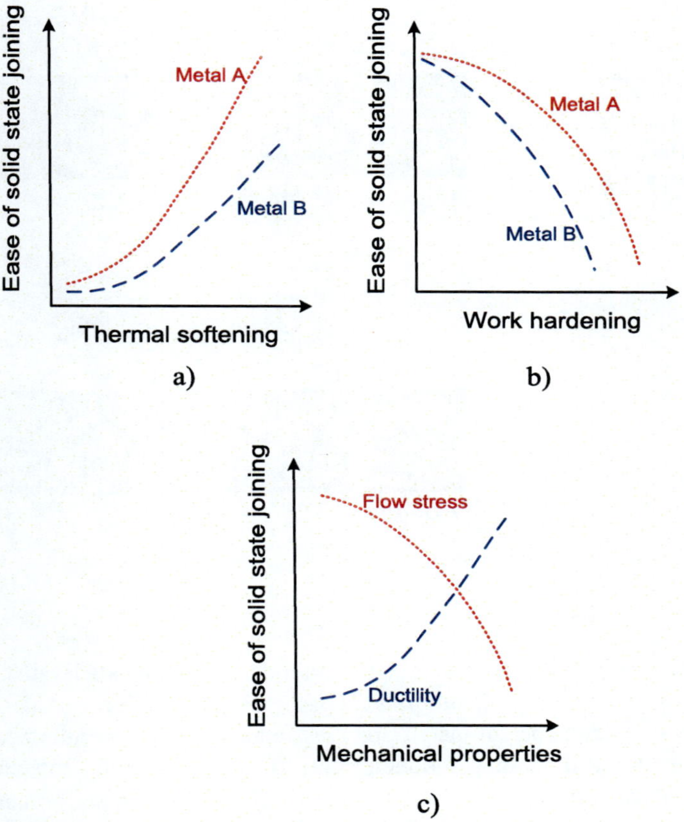

Fig. 1.6 Schematic showing the ease of solid-state joining as a function of **a** thermal softening, **b** work hardening and **c** mechanical properties

completely different components and crystal structures exhibit metallurgical incompatibility, which in turn makes it difficult to produce a sound joint with good mechanical properties for dissimilar metal joining. The ease of joining of dissimilar metals by solid-state or fusion welding depends on the mutual solubility of elements in parent metals of the given dissimilar combination in liquid and solid state. Accordingly, dissimilar parent metals having good mutual solid solubility, in general, offer good weldability than those showing limited or nil mutual solid solubility due to the increased tendency of formation of intermetallic compounds (IMCs) and inclusions. The IMCs and inclusions are usually weak and degrade the fatigue, impact and tensile properties of the joint. Further, the fusion welding of dissimilar metals can cause significant compositional heterogeneities in the form of localized segregation and the formation of the unmixed zone and banded structures (Fig. 1.7a).

Apart from the metallurgical aspects, the affinity of parent metals with atmospheric gases at elevated temperatures imposed during joining also affect the ease of dissimilar metal joining. Few reactive metals (Al, Mg, Cr, Ti, stainless steel,

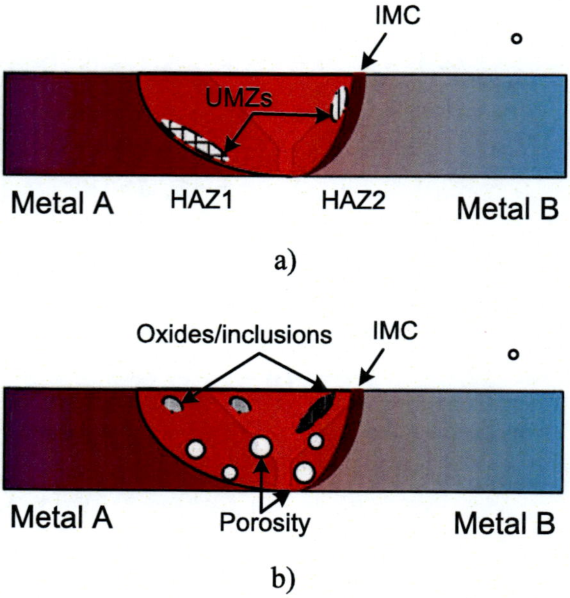

Fig. 1.7 Schematic showing the common issues **a** unmixed zone formation and **b** inclusions, pores and IMCs encountered in the dissimilar metal weld joint due to chemical and metallurgical incompatibility

etc.) rapidly form refractory oxides and nitrides during joining at elevated temperatures due to interactions with atmospheric gases (Fig. 1.7b). These oxides interfere with the fusion of the parent metals and may remain at the joint interface causing limited metallurgical bonding and inclusions. Therefore, suitable protection of the parent metals applied during joining to enhance the quality and performance of the dissimilar metal joints. The protection can be provided in the form of vacuum and inert/inactive gas shielding during the joining.

1.4.4 Dimension Characteristics

Section thickness of parent metals in dissimilar combinations is probably the most important dimensional parameter in dissimilar metal joining as it significantly determines the suitability of the joining process. Therefore, section size (as per choice of joining process) affects heat input, weld thermal cycle, thermal expansion/contraction behaviour, residual and distortion tendency, the stress state of the metal and tolerance to discontinuities. Too high/low section thickness reduces the ease of joining due to different reasons. The dissimilar parent metals having different section thicknesses further complicate the joining due to difficulties related to control over fusion and plastic deformation. However, few joining processes like ultrasonic welding and electromagnetic field-assisted solid-state joining allow the development of lap joints of parent metals of the different thicknesses (Fig. 1.8). Even in the case of dissimilar metal joining, the joining of dissimilar parent metal combinations of very thin

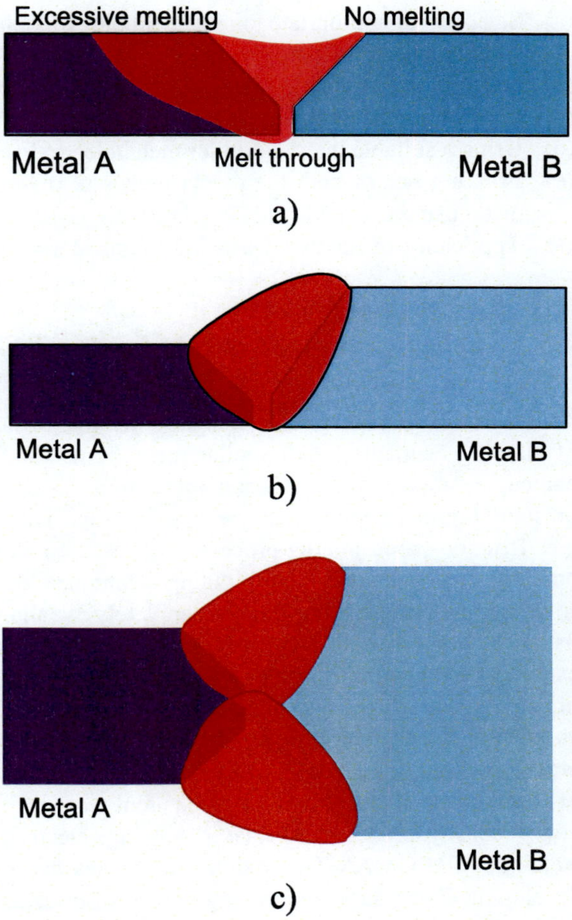

Fig. 1.8 Schematics related to dissimilar metal joining of parent metals having different thicknesses

sheets using fusion welding frequently imposes the issues related to distortion and misalignment of the sheets during joining itself due to poor stiffness.

1.5 Approaches to Address Issues of Dissimilar Metal Joining

Depending upon the performance requirement, the nature of dissimilarities (in terms of physical, mechanical, metallurgical, chemical and dimensional characteristics) and section thickness of parent metals to be joined, a suitable strategy/procedure for dissimilar metal joining is developed. The strategy for controlling the issues of dissimilar metal joining primarily includes:

(a) Selection of appropriate joining technology like fusion welding, brazing, braze welding and solid-state joining
(b) Need of using transition joint (for tri-metallic joints),
(c) Controlling the dilution
(d) Using a suitable interlayer/filler metal/electrode as per the joining process
(e) Isolating one or both the parent metals of dissimilar combination from weld metal using a suitable cladding/buttering layer,
(f) Application of suitable thermal/mechanical treatment of dissimilar metal joints.

1.5.1 Selection of Appropriate Joining Technology

The extent of metallurgical incompatibility (based on liquid/solid-state solubility of alloying elements) of dissimilar metal combinations primarily determines the choice of suitable joining technology from fusion welding, solid-state joining or solid–liquid joining technique like brazing. The solid-state joining technologies (friction-based joining processes, ultrasonic and explosive welding, diffusion bonding, etc.) and solid–liquid joining techniques like brazing, braze welding and soldering are preferred for joining completely metallurgically incompatible dissimilar metal combinations (Fig. 1.9). The dissimilar metal combinations wherein the extent of incompatibility in terms of physical and chemical properties are somewhat limited (conversely largely compatible), and the joining can be achieved using fusion-based approaches like controlled dilution, use of high power density fusion welding processes, buttering layer, functionally graded weld, transition joint, filler/electrode of intermediate characteristics of two parent metals (Fig. 1.10). Application of high power density processes like laser welding, electron beam welding, plasma arc welding, pulse GMA/GTA welding, etc. lowers the net heat input requirement for developing dissimilar metal joining than low power density welding processes like shielded metal arc welding, submerged arc welding, etc. (Sharma and Dwivedi 2019a; Sun and Ion 1995).

1.5.2 Using Transition/Tri-metallic Joint

In case the dissimilar metal combination having a difference in physical, mechanical and chemical properties to such an extent that the service performance of even a sound (direct dissimilar metal) joint is badly compromised, then unique approaches like tri-metallic joint and functionally graded weld joints should be applied. For example, the dissimilar metal joints (DMJs) have a large difference in thermal expansion coefficient, composition and electro-negativity of parent metals when subjected to cyclic heating and cooling in a corrosion environment, the premature failure of such dissimilar metal joints is caused by thermal fatigue, corrosion and stress corrosion cracking. Therefore, a common and simple approach to enhance the service performance of

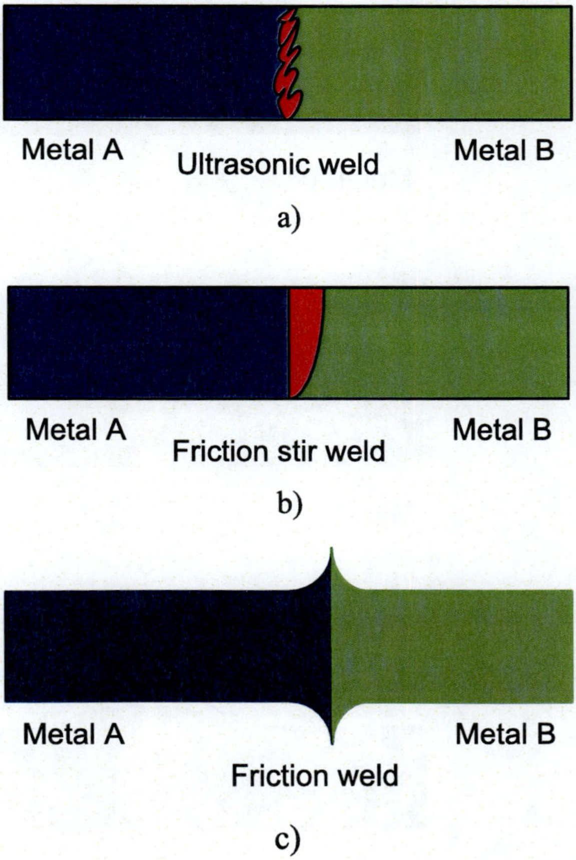

Fig. 1.9 Schematic showing approaches of solid-state joining for the development of dissimilar metal joints using **a** ultrasonic welding, **b** friction stir welding and **c** rotary friction welding

such dissimilar metal joints is to use a small segment (same cross section) of third metal (C) incorporated in between dissimilar parent metals (A and B) and joined using suitable joining technology (Fig. 1.11). The third metal (C) acts as a bridge between two dissimilar parent metals and should have desired intermediate characteristics (in terms of composition, strength, toughness, thermal expansion coefficient, etc.) as per the technological requirement for improvement of the service performance.

1.5.3 Dilution

The dilution in dissimilar metal joining by fusion welding process is very critical as it leads to the intermixing of two dissimilar parent metals to form a weld. Therefore, weld metal composition, microstructure, mechanical and corrosion properties of dissimilar metal joints developed by fusion welding significantly depend on dilution. Depending upon the metallurgical incompatibilities of two parent metals to form

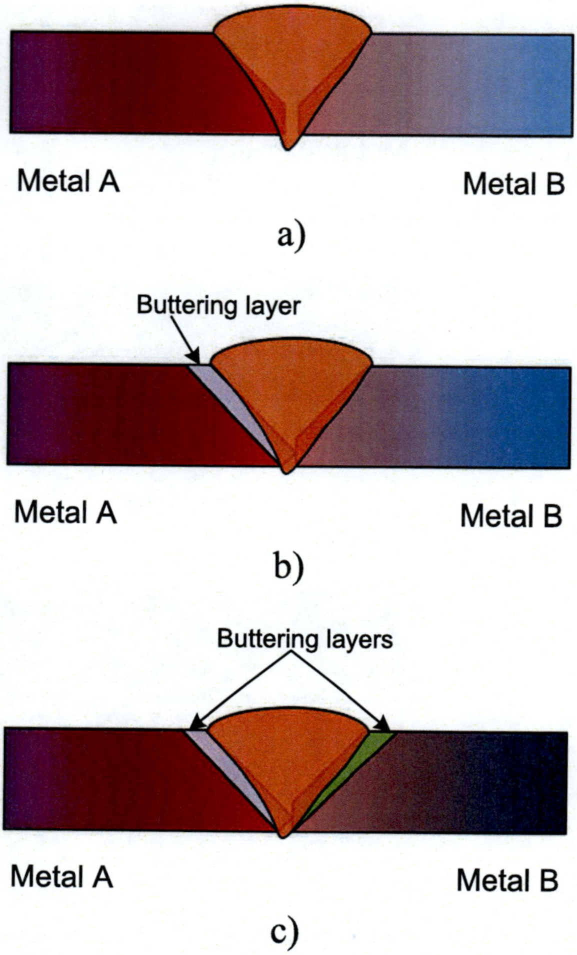

Fig. 1.10 Schematic showing fusion welding approaches for joining of parent metal with increasing in terms of physical, chemical, mechanical and metallurgical dissimilarities: **a** low, **b** medium and **c** high

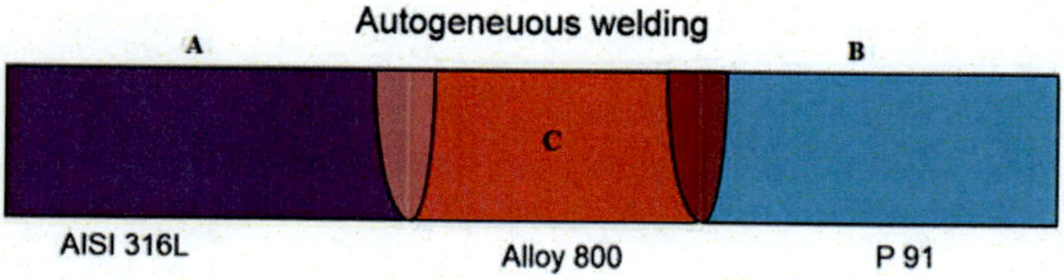

Fig. 1.11 Schematic of tri-metallic fusion weld joint

undesirable unmixed zone formation, unfavourable metallurgical transformation and intermetallic compound formation, the dilution needs to be adjusted/regulated by suitable electrodes/filler metal, buttering layer and reduced heat input using high power density process, etc. Schaeffler diagram helps a lot in choosing electrode/filler of suitable composition to adjust the weld metal composition as per the need to realize desired microstructure of weld metal while considering estimated dilution (%) during fusion welding of different types/grades of stainless steels (Sun and Ion 1995).

The composition in the fusion zone depends on the extent to which filler metal diluted ($D\%$) with the parent metal. The dilution established with knowledge of the composition of the parent metal (CP), the filler metal (CF) and the weld metal (CW):

$$\mathrm{CW} = \mathrm{CF} + (\mathrm{CP} - \mathrm{CF})(D/100)$$

This equation predicts that the composition varies linearly from 0 to 100% under the assumption that good mixing has been achieved throughout. The concentration gradients in dissimilar welds have been previously shown to exist over regions of hundreds of microns. This needs chemical composition analysis using spectroscopy. There are other ways to calculate the dilution using the geometry of the weld bead of dissimilar metal weld joints as shown in Fig. 1.12.

The dilution may be good or bad depending upon the chemical and metallurgical compatibilities of dissimilar metals joined by fusion welding and the tendency to form favourable/unfavourable metallurgical structures. For example, high dilution during welding of mild steel and cast iron increases the cracking tendency due to embrittlement of the weld metal caused by martensitic transformation, therefore, attempts made to reduce dilution. On the contrary, high dilution may be good if favouring the formation of the desired microstructure and joint characteristics.

1.5.4 Electrode, Filler Metal and Interlayer

The direct and autogenous joining of two dissimilar metals using fusion-based joining processes is considered to be more difficult than solid-state joining processes primarily due to metallurgical incompatibilities, the tendency to form unfavourable IMCs at the joint interface/weld, formation of hard and brittle microstructures. Therefore, a suitable interlayer, electrode and filler metal introduced between two dissimilar parent metals expected to act as a mediator/bridge for developing a weld metal or joint interface having more acceptable and favourable metallurgical characteristics including IMC. Such electrode, filler metals and interlayers should have compatibility with both the parent metals of dissimilar combinations (Fig. 1.13). For example, Ni electrode/filler/interlayer is preferred for joining of Al and steel, Ti and steel, Cu and steel, stainless steel and mild/alloy steel and steel and cast iron combinations. Low yield strength and high ductility filler metals and electrodes compatible with both the parent metals help in reducing residual stress, distortion and even cracking tendency.

Fig. 1.12 Schematic showing different ways to calculate the dilution based on weld bead geometry of the dissimilar metal weld joints

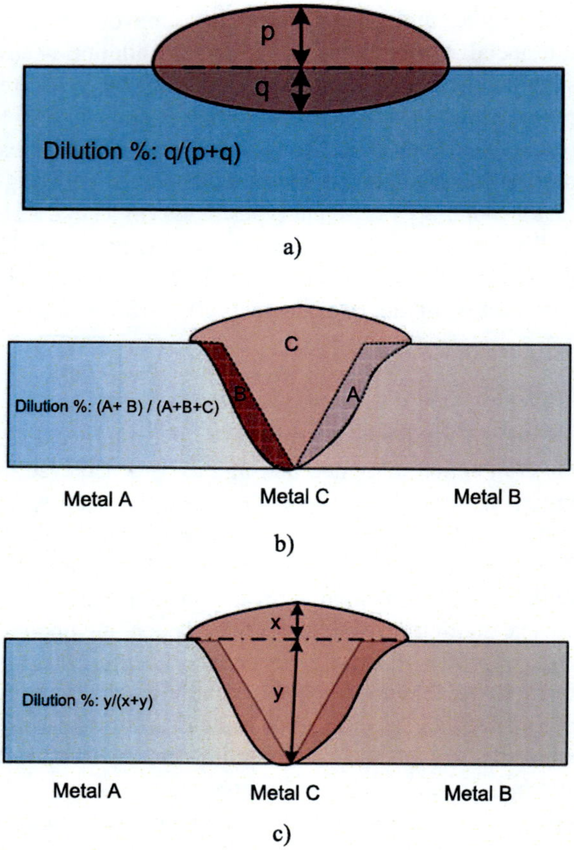

Fig. 1.13 Schematic showing dissimilar metal joint developed using suitable filler/interlayer in **a** fusion welding and **b** spot welding

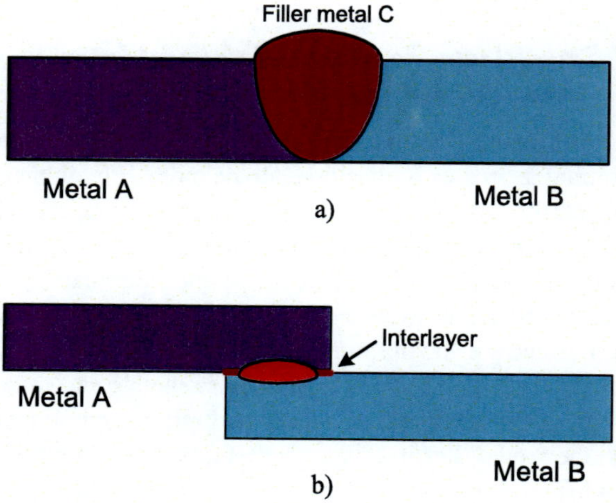

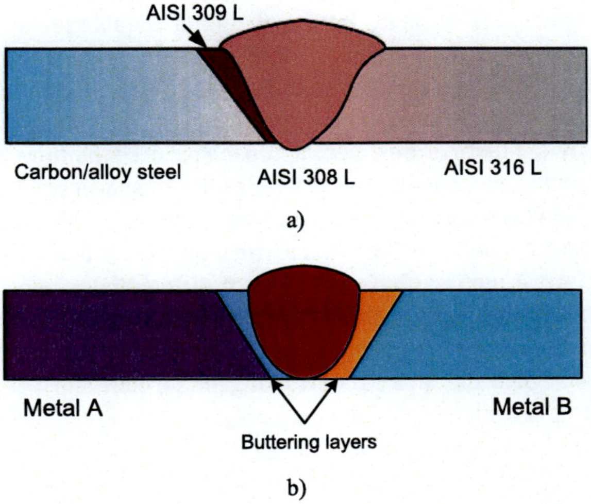

Fig. 1.14 Schematic showing the application of buttering/cladding layer on **a** just one side parent metal and **b** both side parent metals depending upon the need of isolating the parent metal from the weld metal

1.5.5 Buttering and Cladding

The approach of buttering and cladding the parent metals of dissimilar metal joining is based on the idea of isolating one or both of the parent metals from the weld metal during fusion welding as per the need to develop a weld joint of desired characteristics (Fig. 1.14). Thickness and composition of the buttering layer are the two important technical parameters selected suitably to minimize the undesirable effect of dilution (on weld metal properties) from any one or both the base metals during fusion welding. The thickness of the butter / clad layer depends on the possibility of a fusion of faying surfaces of underlying the parent metal as per the heat input of the welding process. A thin butter layer would be fine for low heat input fusion welding processes. Electrode/filler metal for buttering/cladding must be compatible (in terms of physical, chemical and metallurgical characteristics) with the respective base metals of dissimilar metal combinations. Thereafter, the gap between parent metals of dissimilar metal combinations is filled by fusion welding using suitable filler metal/electrode as per the service requirement of the application. The buttering reduces the extent of mismatch between base metals of dissimilar metal combinations and weld metal by reducing residual stress, embrittlement and cracking tendency of the weld metal besides controlling undesirable diffusion of elements across the weld metal/joint interface.

1.5.6 Thermal and Mechanical Treatment

Thermal and mechanical treatments like stress relieving, tempering and ultrasonic vibratory treatment of dissimilar metal joints are primarily designed to reduce the residual stress apart from realizing favourable metallurgical structure and mechanical properties. Thermal treatment in the form of different types of post weld heat treatment designed by considering the physical metallurgy of parent metals of dissimilar metal combinations. The heating of dissimilar metal joints for stress relieving by the thermal treatment approach reduces the flow stress of base metals and weld joints, which in turn helps to relax the locked-in strain to reduce the residual stresses. However, thermal treatment may induce an additional new set of residual stresses also due to the differential thermal expansion behaviour of dissimilar metals. Therefore, a thorough consideration of differential expansion and contraction is needed while designing thermal treatment for relieving residual stresses. Mechanical treatments like overloading and ultrasonic vibratory treatment are therefore preferred, designed and applied considering the stiffness and weight for reducing residual strain in dissimilar metal joint assembly.

1.6 Performance Parameters of Dissimilar Metal Joints

Considering the complicacies occurring due to the difference in physical, chemical and mechanical properties of dissimilar parent metals in the development of joints irrespective of joining technologies applied, the performance of dissimilar metal joint is evaluated differently than that of similar metal joints. However, the deterioration in-service performance of dissimilar metal joints of a given dissimilar metal combination mainly depends on the approach of joining namely fusion, solid-state joining and solid–liquid joining process.

In general, an increase in heat input during fusion welding reduces the performance of dissimilar metal joints due to wider HAZ, higher dilution possibilities, large weld zone, coarser grain structure in the weld and HAZ. Additionally, hardening and softening of weld and HAZ as per metal system, compositional heterogeneity in the form of selective depletion/segregation of few alloying elements in the weld and HAZ also degrade the joint performance. Therefore, mechanical and corrosion performance evaluation of dissimilar metal joints are done as per the technological requirement of service. The metallurgical characteristics of dissimilar metal joints determine the mechanical and corrosion performance; therefore, variation in alloying element and microstructure across the dissimilar metal joints (PM1-HAZ1-UMZ1-WZ-UMZ2-HAZ2-PM2) are studied using suitable techniques like stereoscopy, optical emission spectroscopy (OES), electron probe micro-analyser (EPMA), optical and scanning electron microscopy (SEM). The composition analysis of weld zone, parent metals and electrode/filler metals is used to evaluate the dilution (%). Further, the hardness profile of dissimilar metal joints gives enough idea about the weak zone and degree of hardening / softening of the joint. Similarly, general exposure of dissimilar metal joints in a corrosion environment shows relative corrosion sensitivity/resistance of different zones across the joint (PM1-HAZ1-UMZ1-WZ-UMZ2-HAZ2-PM2) as

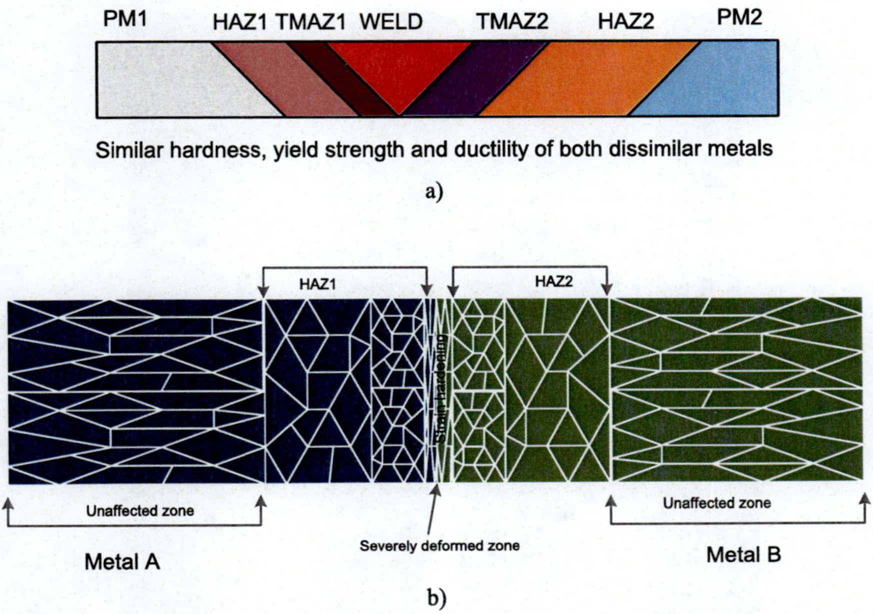

Fig. 1.15 Schematic showing different zones formed in dissimilar metal joints in terms of **a** regions formed and **b** grain structure

shown in Fig. 1.15. As per the technological requirement of the service, the mechanical performance of dissimilar metal joints is evaluated for tensile properties, impact resistance, fatigue and creep resistance.

The mechanical and corrosion performance of dissimilar metal joints developed using solid-state joining processes primarily depends on the soundness and metallurgical characteristics of the joint interface/weld. The interfacial characteristics of dissimilar metal joints developed using processes like diffusion bonding, ultrasonic and explosive welding allow the modification of metallurgical characteristics using suitable interlayer as per the need to improve the compatibility (Fig. 1.16). An application of interlayer in dissimilar metal joints, therefore, affects the microstructure at the joint interface and the mechanical performance of the joint.

The performance of dissimilar metal joints developed using solid–liquid joining processes like brazing and braze welding is mainly dictated by the soundness of the joint, stress raiser in the vicinity of the joint, IMCs developed at interfaces of brazing filler/weld metal and both the parent metals of dissimilar metal combination (Fig. 1.17). The parent metals of dissimilar metal combinations largely remain unaffected by these solid–liquid joining processes.

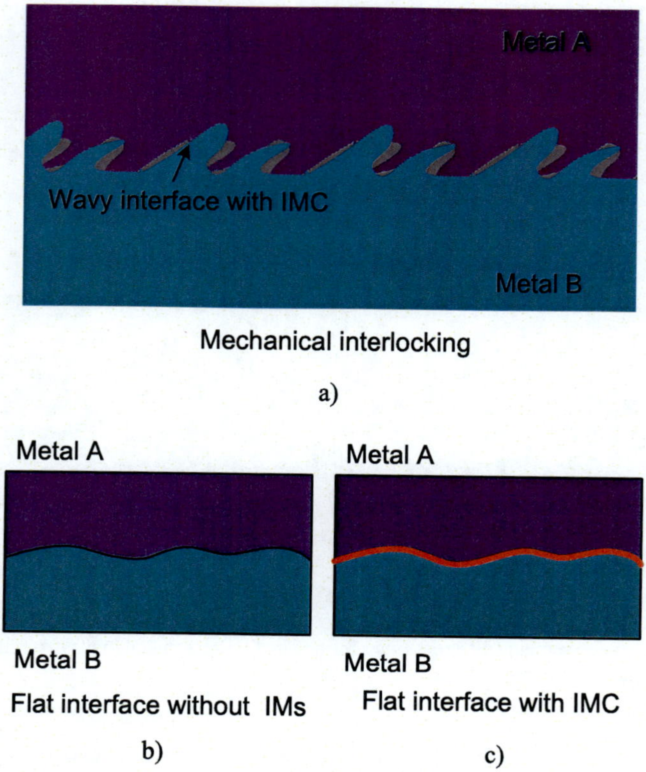

Fig. 1.16 Schematic showing interface morphologies with or without IMC in dissimilar metal joints developed by explosive/ultrasonic welding: **a** wavy interface with IMC, **b** flat interface without IMC and **c** flat interface with interlayer

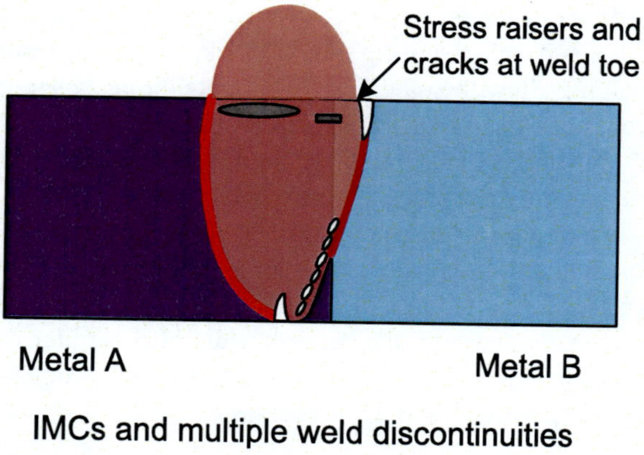

Fig. 1.17 Schematic showing features like poor soundness, IMC and stress raiser degrading the joint performance of dissimilar metal joints

References

Kaushik P, Dwivedi DK (2020) Selective induction heating in FSW of Al-steel combination. Presented in "28th international conference on processing and fabrication of advanced materials (PFAM 28)" held at VIT University, Chennai during 7th–9th December 2020

Kaushik P, Dwivedi DK (2021a) Effect of tool geometry in dissimilar Al-steel friction stir welding. J Manuf Process 68(Part B):198–208

Kaushik P, Dwivedi DK (2021b) Induction preheating in FSW of Al-Steel combination. Mater Today: Proc 46(1091–1095):2021

Kaushik P, Dwivedi DK (2022a) Influence of hook geometry in failure mechanism of Al-Steel dissimilar FSW lap joint. Archives of civil and mechanical engineering (accepted)

Kaushik P, Dwivedi DK (2022b) Al-steel dissimilar joining: challenges and opportunities. Mater Today: Proc, accepted

Kulkarni A, Dwivedi DK, Vasudevan M (2019) Dissimilar metal welding of P91 steel-AISI 316L SS with Incoloy 800 and Inconel 600 interlayers by using activated TIG welding process and its effect on the microstructure. J Mater Process Technol 274:116280

Kulkarni A, Dwivedi DK, Vasudevan M (2020) Microstructure and mechanical properties of A-TIG welded AISI 316L SS-Alloy 800 dissimilar metal joint. Mater Sci Eng: A 790(2020):139685

Kulkarni A, Dwivedi DK, Vasudevan M (2021) Novel functionally graded joint between P91 steel-AISI 316L SS. Weld J 100:269–280

Kulkarni A, Dwivedi DK, Vasudevan M (2018) Study of mechanism, microstructure and mechanical properties of activated flux TIG welded P91 Steel-P22 steel dissimilar metal joint. Mater Sci Eng A 731:309–323

Shankar R, Dwivedi DK (2017) Study of microstructure and mechanical property relationships of A-TIG welded P91-316L dissimilar steel joint. Mater Sci Eng A, 695(2017):249–257

Sharma G, Dwivedi DK (2017a) Microstructure and mechanical properties of dissimilar steel joints developed using friction stir welding. Int J Adv Manuf Technol 88(2017):1299–1307

Sharma P, Dwivedi DK (2019a) Comparative study of activated flux-GTAW and multipass-GTAW dissimilar P92 steel-304H ASS joints. Mater Manuf Process 34:1195–1204

Sharma P, Dwivedi DK (2019b) A-TIG welding of dissimilar P92 steel and 304H austenitic stainless steel: mechanisms, microstructure and mechanical properties. J Manuf Process 44:166–178

Sharma P, Dwivedi DK (2021a) Wire-feed assisted A-TIG welding of dissimilar steels. Arch Civ Mech Eng 21(2021a):1–20

Sharma P, Dwivedi DK (2021b) Flux assisted tungsten inert gas welding of bimetallic P92 martensitic steel-304H austenitic stainless steel using SiO_2–TiO_2 binary flux: welding arc/pool behaviour, microstructure and mechanical properties. Int J Pressure Vessel Piping 192:104423

Sharma P, Dwivedi DK (2021c) Improving the strength-ductility synergy and impact toughness of dissimilar martensitic-austenitic steel joints by A-TIG welding with wire feed. Mater Lett 285:129063

Sun Z, Ion JC (1995) Laser welding of dissimilar metal combinations. J Mater Sci 30:4205–4214

Chapter 2
Fundamentals of Dissimilar Metal Joining by Arc and Resistance Welding Processes

This chapter presents the fundamentals, causes and remedies related to the common issues (asymmetric weld, unmixed zone, intermetallic compound, cracking of weld and HAZ, residual stress and distortion) encountered in joining of dissimilar metals combinations by arc, resistance welding. Fusion welding processes, namely GTAW, Pulse, cold metal transfer and Narrow gap variants of GMAW, SMAW and SAW, have been described considering the dissimilar metal joining. Additionally, dissimilar metal joining using brazing and braze welding has also be elaborated.

2.1 Introduction

The joining of dissimilar metals using processes like arc and resistance welding involves the fusion of faying surfaces of the parent metals. The melting of parent metals during fusion welding imposes more issues and difficulties in developing technologically sound dissimilar metal joint than solid-state joining (friction-based joining, ultrasonic and explosive welding, forge welding) and solid–liquid joining processes (brazing and braze welding). The most of issues in dissimilar metal joining by fusion-based approaches occur due to the differences in thermo-physical and chemical properties of parent metals besides metallurgical incompatibilities. A part of heat applied (by arc or joule heating) is used for melting the faying surfaces of dissimilar parent metals while remaining heat is dissipated to the underlying parent metal as per the thermal conductivity of the respective parent metals. Therefore, parent metals experience the differential expansion and contraction as well (Vidyarthi and Dwivedi 2017; Kulkarni et al. 2018; Kulkarni et al. 2019). Thus, the quality and performance of dissimilar metal joints developed by fusion based joining processes are determined by two aspects (a) the relative contribution, intermixing and interaction of elements (in the molten weld pool) of two dissimilar metals in weld metal, and (b) the effects associated with the heat dissipated to the two parent metals.

© The Author(s), under exclusive license to Springer Nature Singapore Pte Ltd. 2023

D. K. Dwivedi, *Dissimilar Metal Joining*,
https://doi.org/10.1007/978-981-99-1897-3_2

The following section presents the common issues observed in fusion welding of dissimilar metal joining.

2.2 Asymmetric Weld

The development of asymmetric weld by arc and resistance welding of dissimilar metals is primarily attributed to difference in the following aspects (a) Thermo-physical properties are namely melting temperature and thermal conductivity affecting ease of fusion and heat dissipation to parent metals, (b) Chemical composition affecting pattern convection current of the molten metal in the weld pool and (c) net heat input applied for welding as per power density of the welding process affecting cooling rate and solidification time. The melting point of parent metal dictates the ease of melting and thermal conductivity determines how fast heat is dissipated to the underlying parent metal. Low melting point and low thermal conductivity of a parent metal in dissimilar combination result in faster melting to a wider distance than high melting point and high thermal conductivity metal, which in turn causes shifting the centre of weld towards low melting point and low thermal conductivity parent metal (Fig. 2.1).

The flow of molten metal in the weld pool affects the heat dissipation in fusion welding from the weld centre to the fusion boundary of the parent metal. The transfer of heat (delivered to the weld pool) during welding affects the flow of molten metal. Convection currents commonly observed in weld pool include (a) downwards, (b) outwards (centrifugal), (c) inwards (centripetal) and d) combination of these as per flow of the molten metal in the weld pool. The flow of molten metal primarily depends on the surface tension forces acting in weld pool besides other forces (buoyancy, electromagnetic force, etc.) acting in the arc zone. As per "Marangoni Convection", the flow of molten metal occurs from low surface tension (weld centre) to

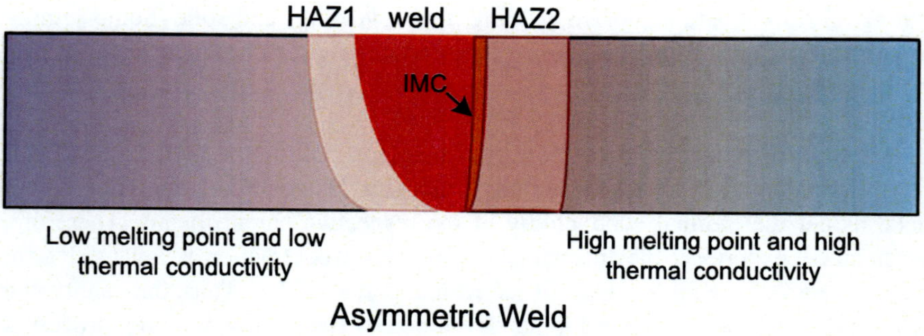

Fig. 2.1 Schematic of a dissimilar metal fusion joint showing asymmetric weld due to the difference in thermal conductivity and melting temperature of two parent metals

high-surface tension zone (fusion boundary). Chemical composition and temperature distribution in weld pool determine the viscosity and surface tension of the molten metal in the weld pool. In general, increase in temperature of the molten metal in weld pool reduces surface tension. Since, peak temperature occurs at the weld centre, and it decreases rapidly on moving away from weld centre towards the fusion boundary, therefore, in conventional fusion welding, molten metal flows from the weld centre towards the fusion boundary leading to a wider weld with shallow penetration (Fig. 2.2). Depending upon the type and amount of alloying elements in the molten coming from the two parent metals (as per dilution), the surface tension can increase/decrease, and accordingly, the flow of molten metal in the weld pool is affected. Therefore, dissimilar metals during arc welding (as per their ability) affect the surface tension of weld pool through alloying element (due to dilution), which in turn decides the flow pattern of the molten metal and heat so the penetration and weld pool geometry. The weld centre shifts towards the parent metal having higher surface tension in the weld pool due to dilution during fusion welding of dissimilar metals (Kulkarni et al. 2019; Sharma and Dwivedi 2019a).

Moreover, alloying elements and impurities like S, P and O_2 in Fe lower the surface tension while Cr, Ni, Co, Si increase the surface tension. Oxides of few elements like Si, Cr, Mo, Cu produce a reversal effect of Marangoni convection with rise of temperature, and this phenomenon is exploited in activated flux GTAW processes for realizing deeper penetration.

The net heat input during fusion welding affects the solidification time of the weld. Moreover, net heat input depends on the heat (as per welding current and voltage) delivered per unit time divided by the welding speed. The solidification time of molten metal in the weld pool determines the extent of heat transfer from the weld centre through convection current in weld pool, which in turn affects the weld geometry. Increase in heat input increases the solidification time, which in turn increases the effect of the surface tension, Marangoni convection and heat flow in the weld pool on weld bead geometry. Dissimilar metal joining by high power density welding processes uses very low heat input leading to high cooling rate and so very short solidification time limits the opportunity for effect on convection current on bead geometry (Fig. 2.3) (Sharma and Dwivedi 2019b; Kulkarni et al. 2020).

The dissimilar metal joints developed using arc and resistance welding (depending upon the parent metals) suffer from a wide range of issues like segregation of alloying element, formation of unmixed zone and intermetallic compound, hard and brittle zone, hardening/softening of heat-affected zone, unfavourable metallurgical transformation, etc. The fusion welding using processes like arc welding and resistance welding rely on melting the faying surfaces of parent metals being joined. Since the dissimilar metal joining by fusion welding involves the melting and then intermixing of alloying elements due to dilution from the parent metals, therefore, as per solubility and compatibility of parent metals, the weld metal may be sound and homogeneous, or contain defects and heterogeneity in the form of segregation and formation of unmixed zones.

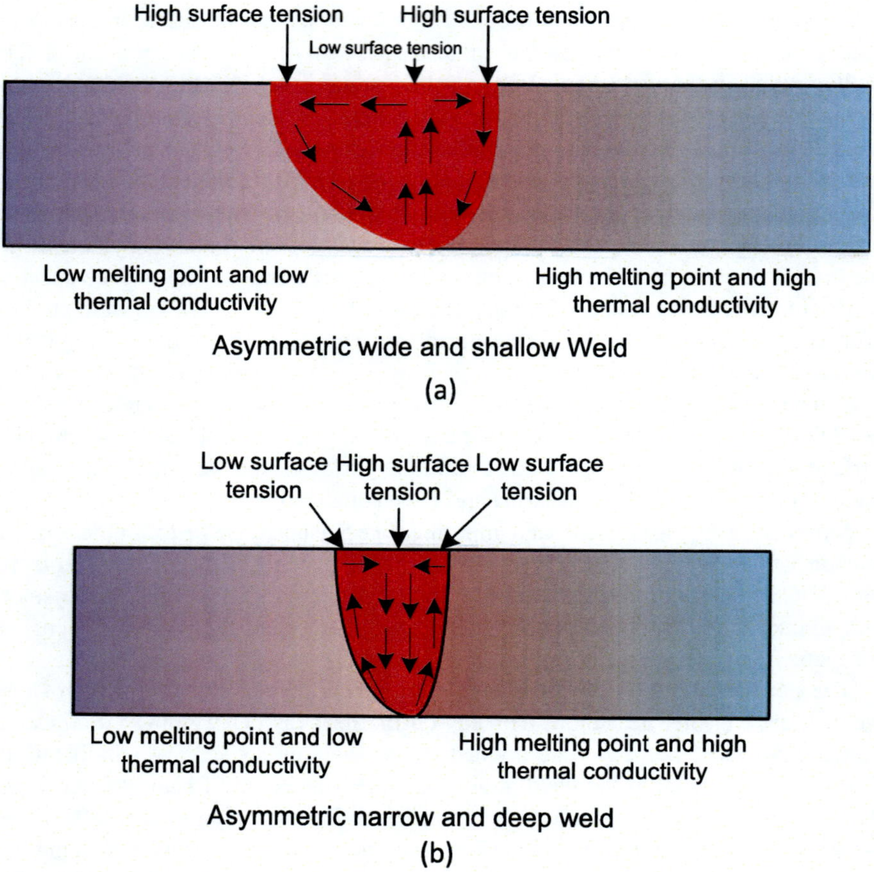

Fig. 2.2 Schematic showing the varying effect of convection currents in weld pool on bead geometry in dissimilar metal joining leading to asymmetric weld **a** wide and shallow weld and **b** narrow and deep weld

2.2.1 Segregation

The segregation is very common in dissimilar metal joining. It can be in the form of localized enrichment (normal segregation) or depletion (inverse segregation) of the certain element as per their solubility in liquid and solid states. The segregation increases the chemical heterogeneity, which in turn encourages the variations in microstructure, mechanical properties and corrosion resistance. Localized enrichment of low melting constituents at the weld centre, and along the grain boundary promotes the solidification cracking in weld zone (Fig. 2.4). Thermal transients during the welding, causing segregation, results in banded structure. The banded structure being a notch-sensitive lowers the fatigue and tensile strength of the weld joint (Sharma and Dwivedi 2021a, b; Kulkarni et al. 2021).

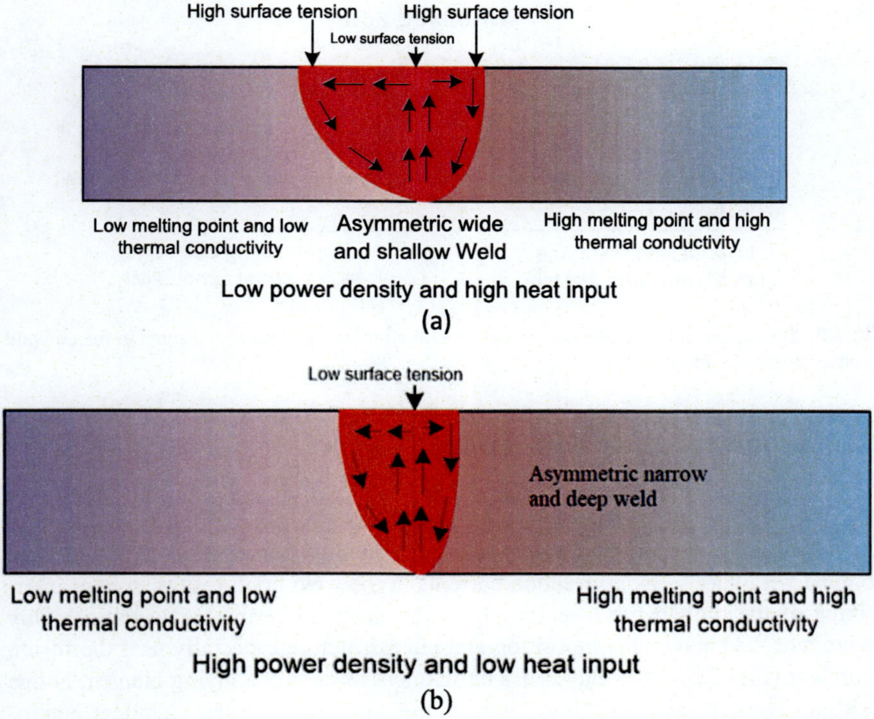

Fig. 2.3 Schematic showing the effect of the power density of fusion welding process on bead geometry in dissimilar metal joining **a** low power density and high heat input process and **b** high power density and low heat

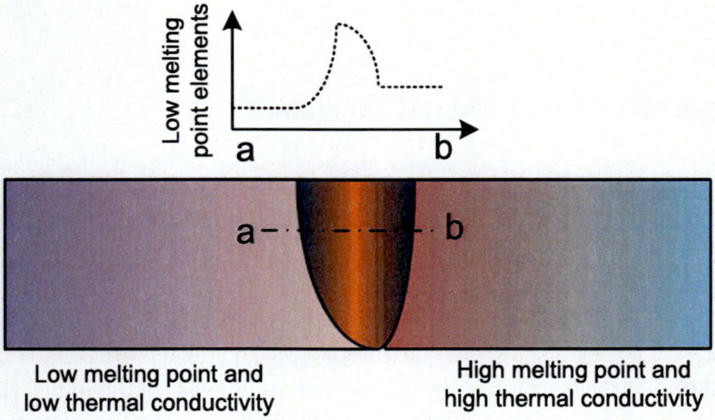

Fig. 2.4 Schematic showing segregation of low melting point element at the weld centre

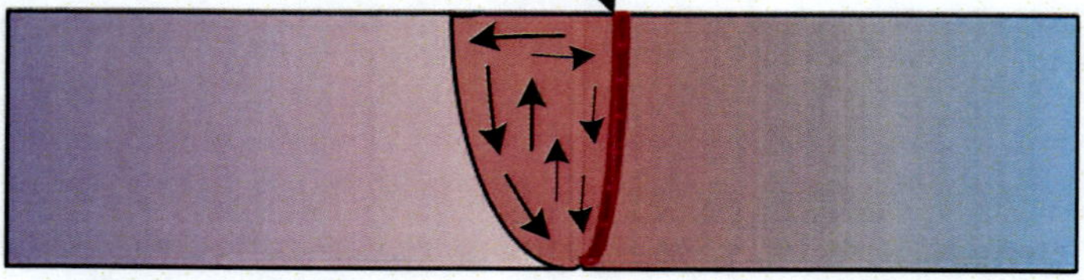

Fig. 2.5 Schematic showing the unmixed zone formation and convection currents in fusion weld joints of dissimilar metals

2.2.2 Unmixed Zone Formation

The dissimilar metal combination, having significantly different chemical composition, (alloying element) with very limited solubility (miscibility) in liquid state coupled with very weak convection currents in the weld pool, results in poor intermixing of the molten metal in the pool. The limited intermixing leads to highly heterogeneous chemical composition and microstructure especially near the fusion boundary (Fig. 2.5). This may cause enrichment of certain alloying element at one location while scarcity at other zones of the weld metal. Such chemical heterogeneity causes embrittlement/softening as per alloying elements and metal system in consideration. Choice of the suitable electrode/filler metal and increased convection currents in the weld pool can help to reduce the tendency of unmixed zone formation (Kulkarni et al. 2021).

2.2.3 Intermetallic Compound Formation

Dissimilar metal joining by arc and resistance welding processes frequently produces the intermetallic compounds (IMCs) especially when atomic sizes, crystal structures of two parent metals to be joined are different. Depending upon the parent metal combination of dissimilar metal joints a very wide range of IMCs is formed, e.g. Mg_2Al_3, $Mg_{17}Al_{12}$, TiFe, $TiFe_2$, Fe_2Al_5, $FeAl_3$, $TiAl_2$, $TiAl_3$, Ti_3Al, $CuAl_2$, CuAl and Cu_9Al_4, Cu_4Ti_3, $FeCu_4$, Cu_3Sn, Cu_6Sn_5, Ni_3Sn_4, etc. IMCs are usually hard and brittle; therefore, these deteriorate the ductility and toughness of the joints. However, the modification of IMCs using suitable controlled alloying can improve toughness and other mechanical properties for improved service performance. The dilution (%), heat input, electrode/filler composition, volume fraction, and morphology of IMCs all affect the mechanical performance of dissimilar metal joints. Fine, discrete and well-distributed globular shape IMCs are considered less harmful than coarse, localized, continuous and networked IMCs (Fig. 2.6). Therefore, considering dissimilar

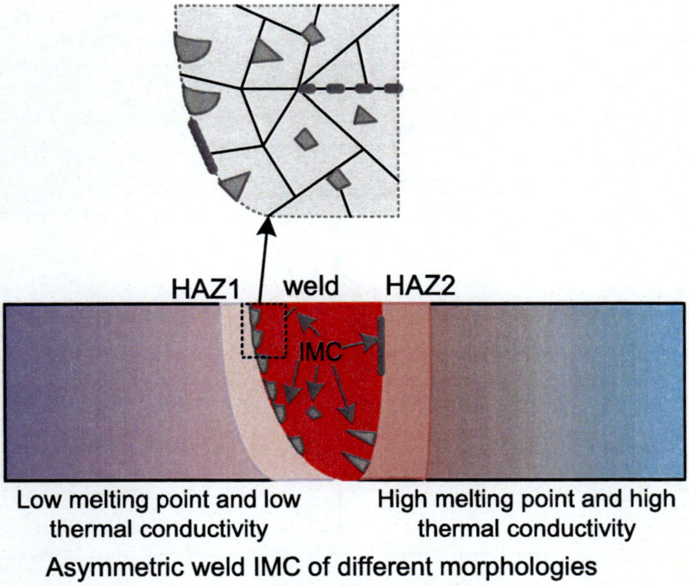

Fig. 2.6 Schematic showing the formation of IMCs of different morphologies, at different locations, namely within or at GBs, and fusion boundary

combination and welding parameter of a given welding process and expected dilution, suitable electrode/filler/interlayer is selected to realize the desired microstructure and mechanical properties of IMCs for improved performance of dissimilar metal joints. In general, increase in heat input increases the volume fraction and size of IMCs. The post weld heat treatment sometimes helps to modify the properties of IMCs favourably.

The mechanical properties of IMCs fall somewhere in between metals and ceramics. The IMCs formation can be desirable also because many IMCs offer very good resistance to hot corrosion, thermal stability and thermal softening. These characteristics considered good for high-temperature applications such as gas turbine and other thermal power plant components (Kulkarni et al. 2021; Sharma and Dwivedi 2021c).

2.2.4 Hardening of HAZ

The parent metals (like carbon steel, alloy steel, ferritic and martensitic stainless steel, Cu-W, Ti alloys) during arc/resistance welding exposed to unique weld thermal cycle (WTC) of high heating rate, peak temperature and subsequently high cooling rate. These conditions result in a typical metallurgical transformation by forming hard and brittle phase(s) like martensite in steels, which in turn causes embrittlement of weld

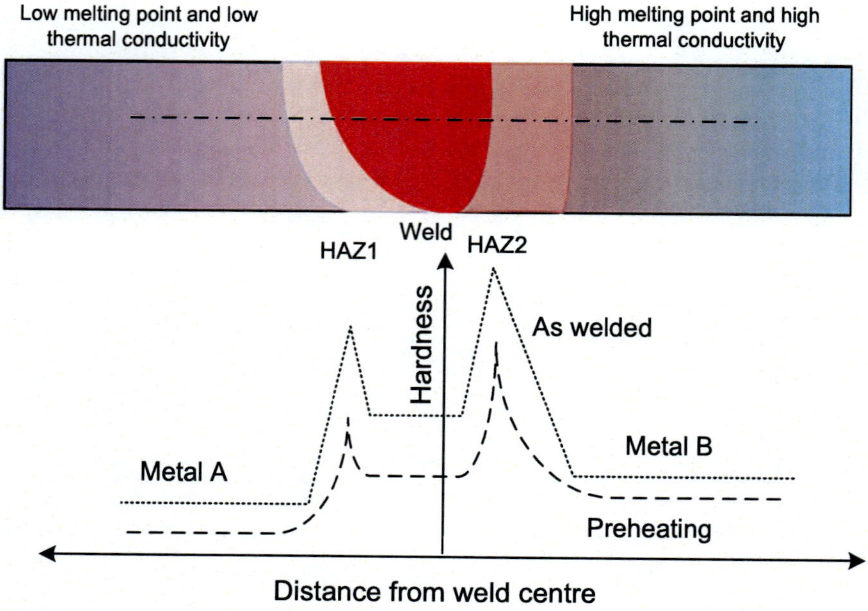

Fig. 2.7 Schematic showing hardening of HAZs and effect of preheat on hardness profile of a fusion weld joints of dissimilar metals

and heat-affected zone a region in vicinity of the fusion boundary. The hardening of HAZ of dissimilar metal joints is influenced by many aspects related to the joining, namely section thickness, groove geometry and joint design, preheat temperature, hardening mechanism of metals system, and net heat input according to weld thermal cycle, welding process and parameters (Fig. 2.7). The extent of hardening of HAZ is identified using micro-hardness profile of dissimilar metal joints, and accordingly, suitable post-weld treatment is applied to restore the properties of parent metals. A combination of high tensile residual stress and embrittlement of HAZ promotes cracking of dissimilar metal joints. The HAZ of parent metal (BM1) may show completely different mechanical behaviour than the weld zone, HAZ 2 of another parent metal (BM2).

2.2.5 *Softening and Weakening of HAZ*

The metals of dissimilar metal combination strengthened by work hardening, grain refinement and precipitation hardening (like Al, Mg, Cu, Q and T steel, PH-SS, Ti, Co alloys) during arc/resistance welding when exposed to unique weld thermal cycle (WTC) of peak temperature for long time followed by low cooling rate show softening of HAZ. These conditions result in grain coarsening, reversion, recovery

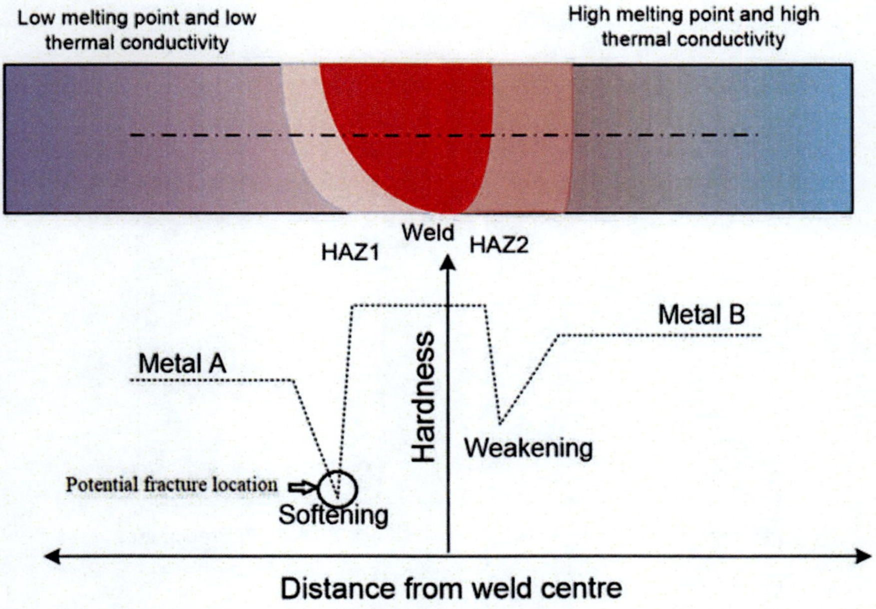

Fig. 2.8 Schematic showing softening of HAZs and effect of preheating on hardness profile of a fusion weld joints of dissimilar metals

and recrystallization, which in turn causes weakening of heat-affected zone in vicinity of the fusion boundary. The softening of HAZ of dissimilar metal joints (similar to the hardening) is influenced by many welding related factors affecting the WTC (heating and cooling) as per section thickness, groove geometry and joint design, metal strengthening mechanism, preheat temperature, net heat input, welding process and parameters (Fig. 2.8). High temperature retention for a longer time increases the softening of HAZ. The extent of softening of HAZ can be easily established using hardness profile of dissimilar metal joints and accordingly suitable post-weld treatment applied to restore the properties can be identified. Minimum hardness zone suggests potential fracture location in dissimilar metal joints.

2.2.6 *Unfavourable Metallurgical Transformation*

The weld metal and heat-affected zone of many dissimilar metal joints developed using arc/resistance welding are subjected to many undesirable metallurgical transformations. These transformations can be in the form of very coarse grains, high aspect ratio micro-constituents, banded structures, segregation and depletion of alloying element, precipitation of poor phases along the grain boundaries and joint

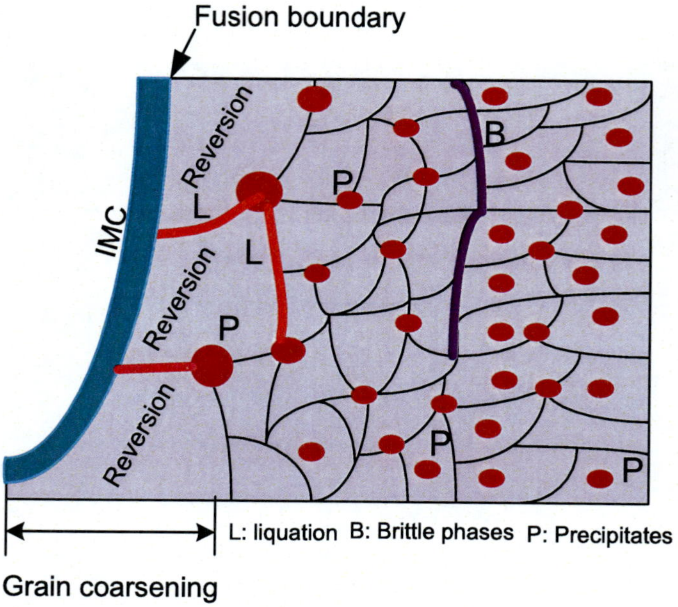

Fig. 2.9 Schematic showing the formation of different zones and micro-constituents adjacent to the fusion boundary in precipitation hardening metals of a dissimilar metal fusion weld joints

interfaces, dissolution of strengthening constituents, formation of brittle particle–matrix interfaces, etc. (Fig. 2.9). These microstructural features encourage crack/void initiation and their growth; which in turn reduces the ductility, toughness, tensile and fatigue strength of dissimilar metal joints. Some of these transformations are specific to certain parent metals and accordingly, their relative effect on deterioration of mechanical performance varies significantly.

2.3 Residual Stress and Distortion

The different thermal expansion behaviour of parent metals in dissimilar metal joining (after metallurgical incompatibility) is probably another major troublemaker especially in fusion welding (Fig. 2.10). Arc and resistance welding processes are high heat input joining process leading to the thermal expansion and contraction of parent metals up to greater distances from the fusion boundaries. The joining of dissimilar metals with large difference in thermal expansion coefficient causes differential expansion and contraction, which in turn results in high residual stress in one parent metal (of high thermal expansion coefficient) than the other (of low thermal expansion coefficient) as shown in Fig. 2.11. Section thickness, size of the weld, grove geometry, joint design, yield strength of filler metal and that of two parent metals, welding process and parameters affecting the net heat input, pre- and post weld heat treatment all affect residual and distortion in dissimilar metal joints (Table 2.1).

Increase in section thickness, weld size, yield strength of weld metal/filler/parent metal, heat input (as per welding process and parameters), in general, increases the residual stress and distortion while increase of preheat temperature and post weld heat treatment temperature may decrease the residual stress of dissimilar metal joints. Increase in differential residual stresses in two parent metals tends to the cause distortion due to imbalanced residual stress in addition to deterioration of mechanical performance like tensile strength and fatigue resistance (Fig. 2.12). The distortion

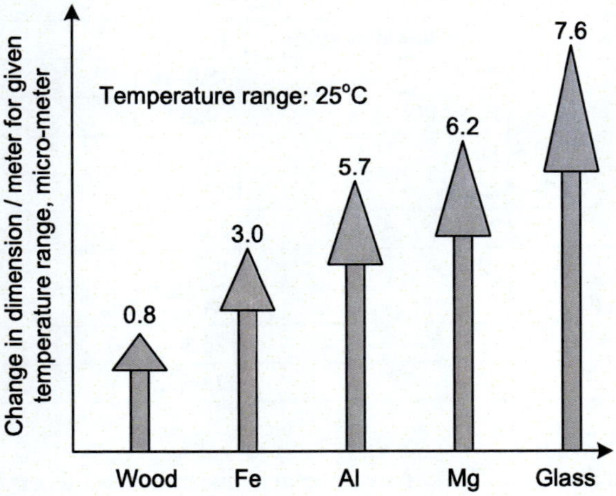

Fig. 2.10 Schematic showing the extent of linear-dimensional variation observed in different metal on heating

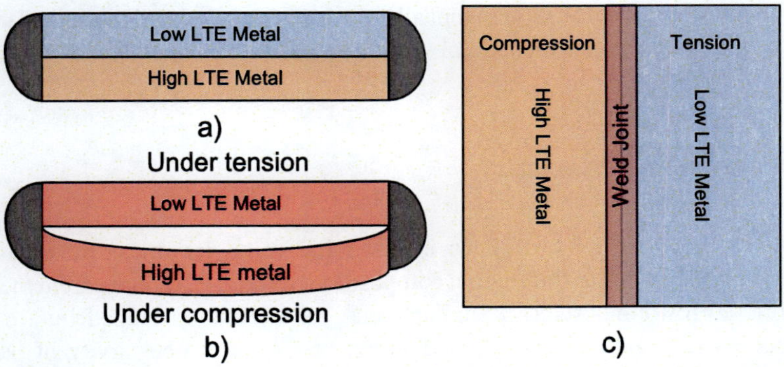

Fig. 2.11 Schematic showing **a** two different metal rigidly joined at the end, **b** differential expansion behaviour shown during heating leading differential stress conditions and **c** similar situation observed in a typical dissimilar metal joining

Table 2.1 Thermal expansion coefficient of common metals

S. No.	Metal system	Alloy	Linear thermal expansion coefficient, 10^{-6}/°C
1	Al	2xxx, 5xxx, 6xxx, 7xxx	22.9–23.8
2		AA356	21.8
3	Cu	ETP Cu	17
4		Cu alloys	16.2–20.5
5	CI	GCI	11.4
6		DI	10.6–11.2
7	Precious metal	Ag	19.7
8		Au	14.2
9	Fe–C	C and low alloy steel	11.3–12.3
10		ASS	15.9–18.5
11		FSS/MSS	10.2–11.0
12	Ti	Pure	8.6
13		Alloys	8.6–10.8
14	Mg	Alloys	26.0
15	Ni	Alloys	12.8–13.3
16	Pb–Sn (40/60)		24
17	Tin		23.8
18	Zn		23–32.5

mainly occurs in the low stiffness parent metal in dissimilar metal combination. High residual tensile stress increases tendency of cracks in the weld metal and HAZ. The residual stress development is further complicated if dissimilar metal combination has different section thickness as well. Reduction in heat input, application of low yield strength filler/electrode, small-narrow weld and heat-affected zone, preheat and stress relieving post weld heat treatment all reduce the issues related to residual stress and distortion in dissimilar metal joining.

2.4 Corrosion Behaviour

The loss of the corrosion resistance in dissimilar metal joining is inherent and inevitable primarily due to differential composition of weld, HAZ and parent metals. Therefore, corrosion-resistant dissimilar metal weld joints developed by arc and resistance welding need careful consideration of the electro-negativity of parent metals of dissimilar metal combination (Fig. 2.13). Increasing difference in electro-negativity of parent metals increases the corrosion sensitivity of dissimilar metal

Fig. 2.12 Schematic showing residual stress distribution across the dissimilar metal weld joint developed using **a** autogenous approach and **b** butter layers followed by welding using low strength filler

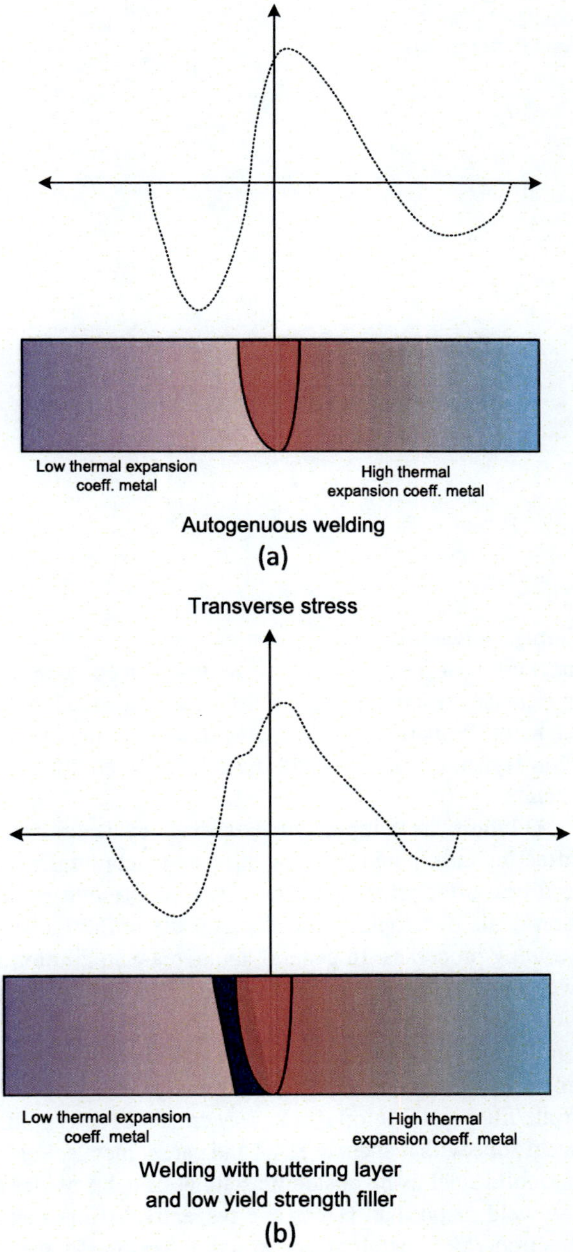

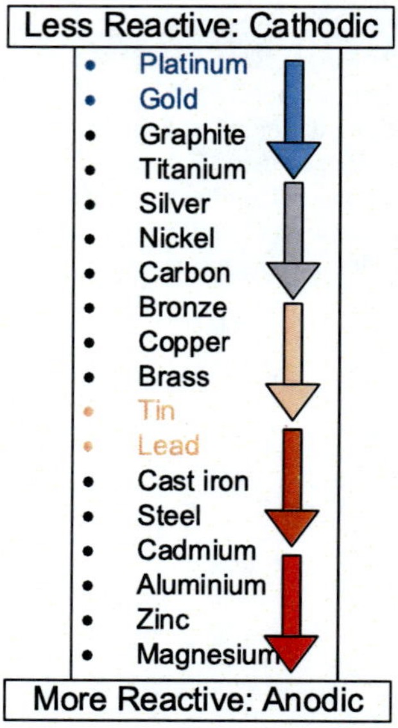

Fig. 2.13 Electro-negativity of different metals

joints. Galvanic corrosion commonly occurs in dissimilar metal joints as per electro-negativity of parent metals of dissimilar metal combination. The loss of corrosion resistance occurs not just in weld zone but also in HAZ due to segregation of alloying element, formation of IMCs, precipitates at grain boundaries. Relative loss of corrosion resistance of the weld and heat-affected zone is more with respect to parent metal.

Dilution, heat input, weld thermal cycle and electrode/filler metal/interlayer affecting the chemical composition and metallurgical transformations of weld metal and heat-affected zone are few very important aspects that need proper consideration while developing procedure for dissimilar metal joining by arc and resistance welding processes. In general, an increase in dilution and heat input in autogenous welding deteriorates the corrosion resistance of dissimilar metal joints. Effect of dilution (%) on corrosion performance of dissimilar metal weld joints largely depends on the composition of filler/electrode. The dilution may improve or deteriorate the corrosion resistance of dissimilar metal joints. Therefore, the choice of suitable electrode/filler metal/interlayer is very crucial in determining the corrosion resistance of weld zone with respect to HAZ and parent metals. For example, joining of alloy steel and mild steel using austenitic stainless steel provides better corrosion resistance of the weld metal than HAZs and respective parent metals. The electro-negativity of electrode/filler metal/interlayer acts as a bridge for the two parent metal having wider gap in this regard (electro-negativity).

Conversely, filler metal/electrode selection needs consideration of dilution as per welding process and heat input and welding position considering the fluidity of the molten metal. In fact, there is no perfect and closed answer regarding choice of filler for dissimilar metal joints. Filler for dissimilar metal joints for high-temperature application (thermal power plants) is chosen considering many other technological requirements such as matching the strength, oxidation resistance, carburization resistance, response to heat treatment, SCC, etc. with respect to the parent metals of dissimilar combination. Generally, it is preferred to select filler/electrode having chemical composition matching with one parent metal having better characteristics than other parent metals of dissimilar metal combination. The selection of filler/electrode becomes more complicated when the parent metals are brought together to satisfy entirely different or even contradicting service requirements such as yield strength, toughness, hardness, and resistance to oxidation, corrosion (acidic and basic environment).

2.5 Fusion Welding Processes for Dissimilar Metal Joining

The performance of fusion weld joints of dissimilar metals developed by arc and resistance welding processes depends on microstructure, weld metal composition, filler metal/electrode, dilution, net heat input, soundness, cleanliness and protection of weld pool. These aspects of dissimilar metal weld joint determined by the power density associated with welding process, welding process parameters, shield approach (gas, flow rate, welding speed and arc length, etc.), control over the weld pool and heat applied. Since different arc and resistance welding processes offer a wide range of power density with varying effectiveness of the weld pool protection, cleanliness, soundness and quality of the weld joint also vary significantly. Welding processes like pulse variants of GTAW and GMAW are preferred for dissimilar metal joining using (a) fusion welding with buttering layer, (b) braze welding. Resistance welding processes are usually autogenous welding process but can be applied suitable interlayers compatible to both the parent metals for dissimilar metal joining.

2.5.1 Gas Tungsten Arc Welding

The gas tungsten arc welding (GTAW) process produces the cleanest weld using minimum net heat input process among the arc welding processes primarily due to good arc stability even at low current (owing to W electrode), short arc length and use of inert shielding gases like (Ar, He). The low heat input results in small weld, and heat-affected zone weld is largely free from inclusion and spatters. Therefore, GTAW is commonly preferred for developing quality weld joints of dissimilar metals for critical applications. The weld joint design parameters like groove geometry, root gap, etc. affect the heat input requirement, number of passes, volume of the weld metal to be deposited, dilution and accordingly development of HAZ, microstructure, mechanical properties, residual stress and distortion are affected.

Depending upon the extent of difference in chemical composition, physical properties and metallurgical incompatibility between metals of dissimilar combination, suitable approach of GTAW applied for dissimilar metal joining (Fig. 2.14). GTAW performed by either direct fusion welding of parent metal with/without filler metal, or first applying buttering layer on the faying surface(s) one/both of the parent metals then developing weld joint using suitable filler. The former approach preferred when the extent of dissimilarity (in composition, thermo-physical properties, and metallurgical compatibility) between the parent metals of dissimilar metal combination is minor, else, the latter one is preferred.

The performance of dissimilar metal weld joint developed by GTAW is influenced by many more factors than that of similar metal weld joint because characteristics of weld metal are complicated by metallurgical incompatibility, segregation, varying % of dilution, differential thermal expansion and contraction besides normal expected metallurgical transformation in both the heat-affected zones. Development of weld metal in dissimilar metal joints, therefore, needs careful consideration in selection of filler metal, butter-layer and its thickness, if any; heat input and preheat affecting dilution from both parent metals. Choice of inappropriate butter layer or filler metal leads to (a) poor wettability and so porosity and limited bonding at fusion boundary, (b) formation of hard and brittle IMCs deteriorating the mechanical performance of the weld joint and (c) increased tendency of cracking of weld.

Welding procedure specification

Welding procedure specification including edge preparation and cleaning, welding parameters (current, voltage, travel speed), arc length, electrode tip angle and diameter, filler metal, butter layer, if any, and its thickness, shielding gas, flow rate of shielding gas all affect quality and so the performance of the dissimilar metal weld joint.

Edge preparation

Depending upon the section thickness of dissimilar metal combination suitable edge preparation is carried out which may be square, single/double V, U and bevel groove. The edge preparation determines whether GTAW will be autogenous or with filler and without butter layer according to the metallurgical compatibility of parent metals and expected dilution. All the aspects of edge preparation such as groove geometry, root gap affecting the weld metal deposition required to develop a weld joint in turn affect weld thermal cycle and so the properties of weld, HAZ, and so the mechanical and corrosion performance of the dissimilar metal weld joints. Further, chemical cleaning (using HCl/H_2SO_4) of faying surfaces of the parent metals especially transformation hardenable steel and cast irons increases the embrittlement and cold cracking of weld and HAZ.

Welding parameters

The welding parameters, namely welding current (50–100 A), arc voltage (10–15 V) and welding speed (50–200 mm/min), affect the net heat input which in usually less than 1.0 kJ/mm. Heat input using suitable control of the welding parameters must be kept to a minimum level while realizing desired fusion and penetration of the faying surfaces. Moreover, it also depends on whether it is root pass, welding with or

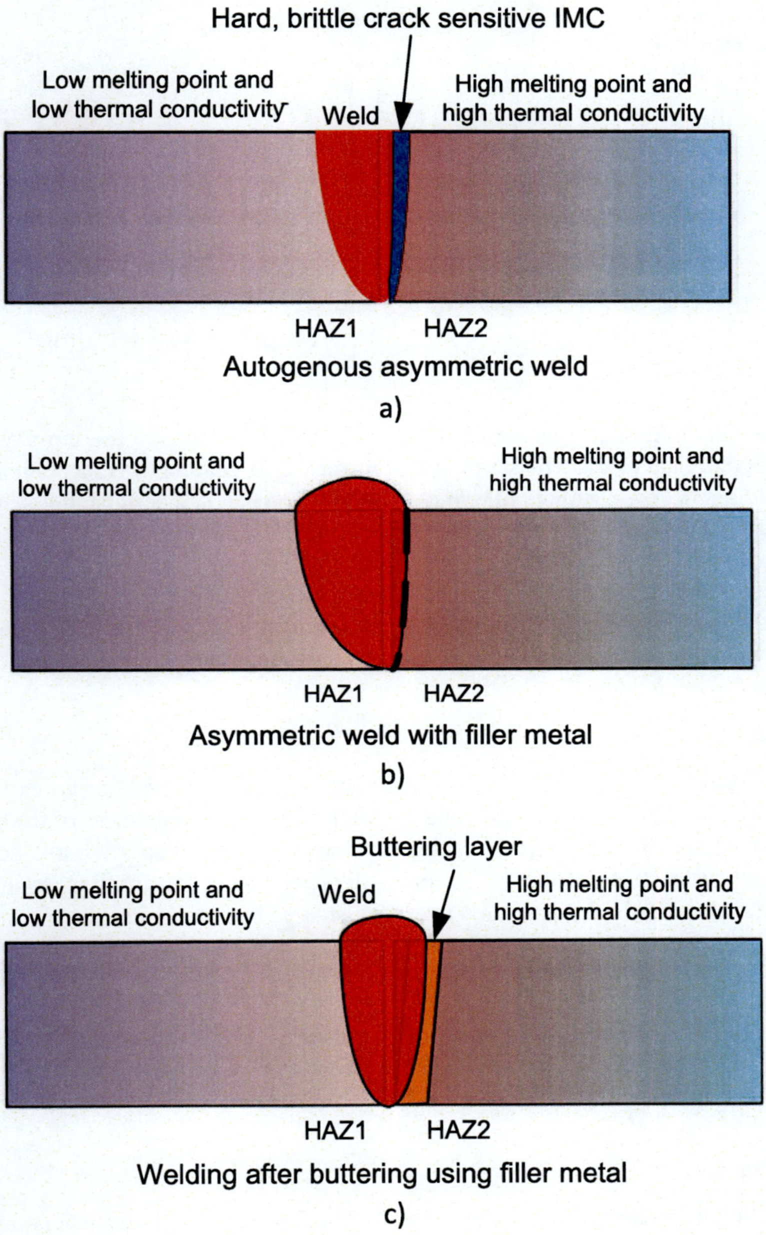

Fig. 2.14 Schematic of approaches for GTAW on dissimilar metals using **a** autogenous approach, **b** suitable filler metal and **c** buttering and filler metal by GTAW

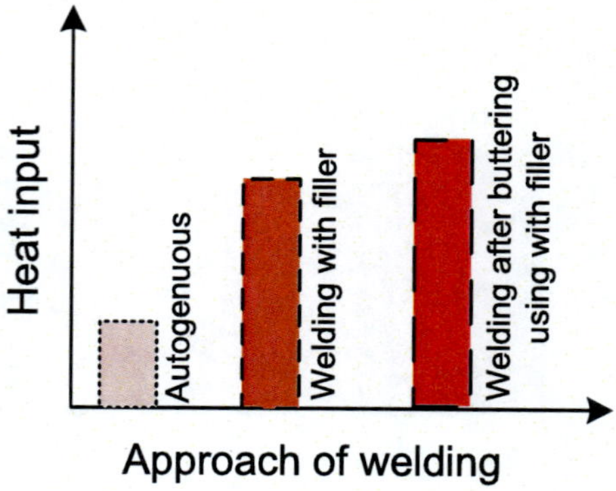

Fig. 2.15 Schematic showing trend of heat input as per approaches

without filler/buttering layer, degree of incompatibility in terms of thermo-physical and metallurgical characteristics of parent metals. Low heat input is recommended for welding root pass, without filler/buttering layer and parent metals having a significant difference in thermo-physical and metallurgical incompatibilities. Little high heat input can be used (to realize high welding speed and productivity using high deposition rate) for dissimilar metal GTAW using filler/buttering layer because these lower the risk of undesirable metallurgical transformation and mechanical properties of the weld metal (Fig. 2.15). Further, thermal damage of the parent metals in HAZ also kept in mind while developing welding procedures.

Arc length

Arc length directly determines the arc voltage and surface area of welding arc exposed to ambient air, which in turn affect the arc temperature and cleanliness of the weld metal. Increase of arc voltage increases the power of arc and so arc burns hotter. Increased surface area of the arc increases the heat dissipation from arc to air by convection besides heat distribution over a larger area by conduction.

Increase of arc length during GTAW of dissimilar metals having different electromagnetic behaviour increases the tendency of arc blow as arc tends to get deflected towards one parent metal leading to uncontrolled and skewed distribution of the heat and molten metal causing asymmetric weld coupled with spatter. Arc length therefore must be optimized, and even suitable arc offsetting is used to develop a symmetric weld joint (Fig. 2.16).

Shielding gas

The shielding gas determines the cleanliness of the weld metal and heat generation by welding arc, which in turn significantly affect the quality of the weld joint. Selection of the shielding gases like Ar, He and mixture of inert/inactive gases for GTAW during dissimilar metal welding depends on affinity of parent metals of dissimilar metal combination with atmospheric gases, and section thickness. Highly reactive metals like Al, Ti, etc. for high-quality weld joints should be welded by He, while for commercial grade, good-quality weld joints, Ar and Ar and He gas mixtures used.

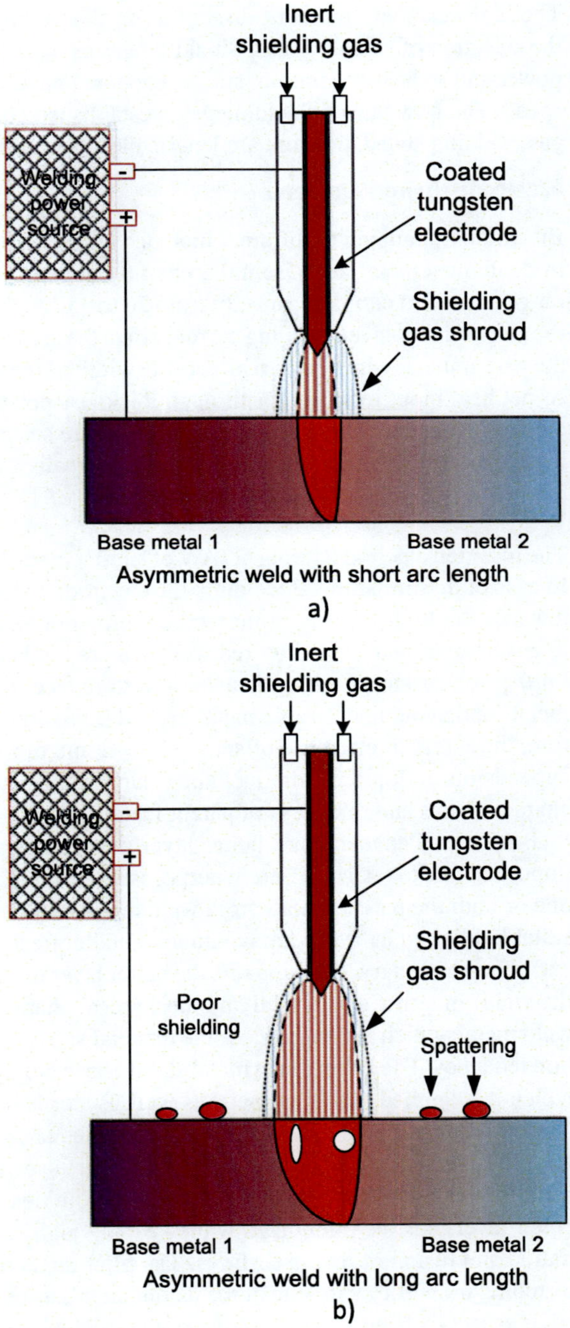

Fig. 2.16 Schematic showing asymmetric weld joint of dissimilar metals developed by GTAW using **a** autogenous approach and **b** with suitable filler

The high reactivity metal of dissimilar metal combination with air mainly dictates the selection of shielding gas. Shielding gas as per its ionization potential affect arc power and so heat generation, arc temperature and net heat input for a given welding speed. The flow rate of shielding gas is established considering the type of shielding gas, welding speed, positon, arc length, electrode size and nozzle design.

Electrode tip and diameter

Electrode tip tapered to different included angles like 30°, 45°, 60°, 90°, 120° electrode diameter can range from 0.25 to 4.8 mm, and accordingly, these offer a wide range of current carrying capacity from 20 to 350 A. The diameter and taper angle of the electrode for given welding current affect the area over which arc heat is spread on the parent metal which in turn affects the width of the weld, depth of penetration and so net heat input required for through thickness penetration (Fig. 2.17). Therefore, electrode diameter and electrode taper angle are selected judicially to minimize the heat input for dissimilar metal welding while realizing desired penetration.

Electrode offsetting

The tungsten electrode during GTAW of similar metal joining kept at the weld centre. In case of dissimilar metal welding, the electrode can be offset from the weld centre towards the high melting point temperature, and high thermal conductivity metal to ensure symmetric weld by redirecting more arc heat during welding (Fig. 2.18). Further, electrode offsetting is also used to reduce dilution from one of the parent metals and avoid undesirable metallurgical transformation in the weld zone. Therefore, the extent of electrode offsetting (0.5–2 mm or more) from the weld centreline depends on multiple factors including purpose/objective, mechanical and physical characteristics and thickness of parent metals.

Both the filler metal and butter layer metal in dissimilar metal joining play an important role to realize different purposes. The butter layer primarily helps to isolate one or both the parent metals from weld thermal cycle applied during GTAW using suitable filler so as avoid unfavourable metallurgical transformation in weld metal and fusion boundary. Additionally, the butter layer metal reduces the gradient (across the weld) in terms of physical and mechanical characteristics of parent metals and weld metal, which in turn reduces the residual stress, corrosion, cracking and distortion tendency. The butter layer metal should have compatibility with parent metal on which it is applied and filler metal as well. Butter layer metal, to be applied on two parent metals, as per metallurgical compatibilities can be different. In such a situation, the filler metal must be compatible with butter layer metals to avoid unfavourable metallurgical transformation and weld discontinuities. The heat input used to apply butter layer must be minimized while keeping high enough to achieve desire penetration and fusion of faying surfaces. The filler metal on the other hand should have matching properties with one of the parent metals as per service requirements. Many other aspects related to the selection of the filler metal have already been described.

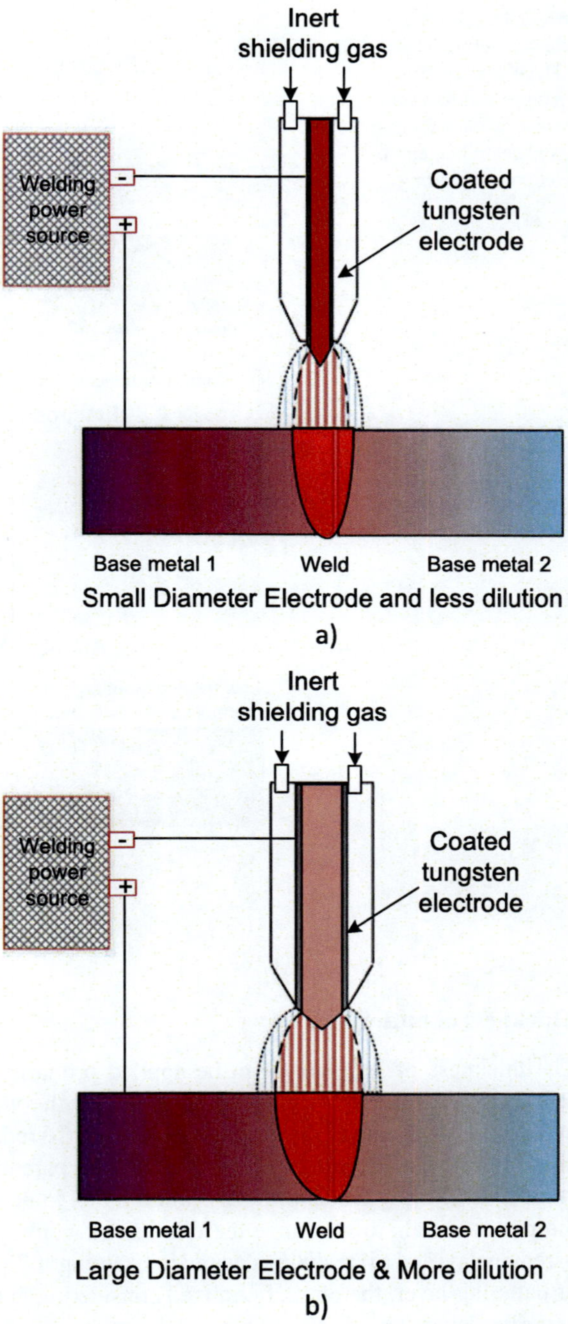

Fig. 2.17 Schematic showing effect of electrode diameter of GTAW on weld geometry of dissimilar metal joints developed using **a** small diameter and **b** large diameter

Fig. 2.18 Schematic showing effect of position of GTAW electrode on weld geometry of dissimilar metal joints developed using **a** at weld centreline and **b** electrode offsetting

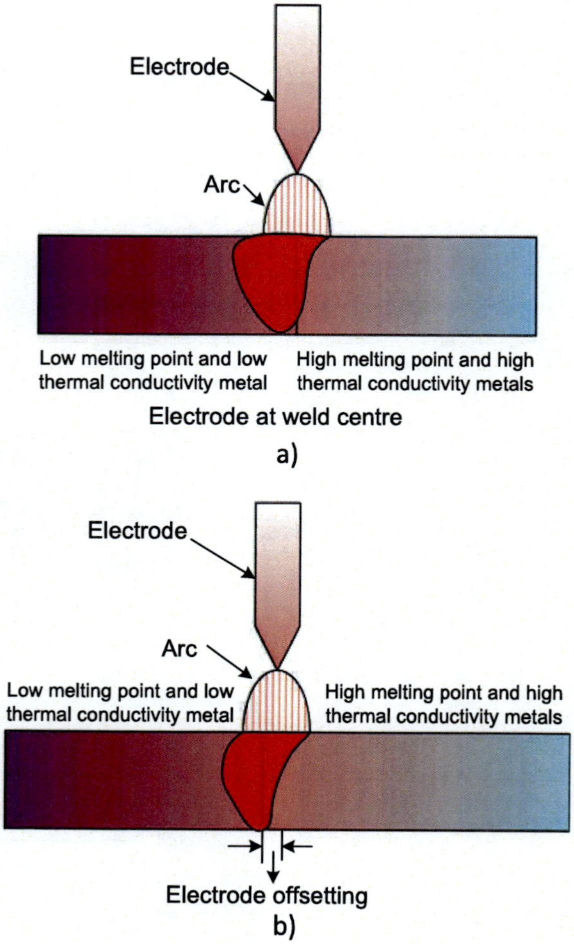

Thickness of buttering layer

The thickness of butter layer to be applied primarily depends on its purpose (a) reducing diffusion of alloying elements across the weld like carbon migration to avoid undesirable metallurgical transformation, (b) reducing gradient in mechanical and physical characteristics of weld metal and parent metal so as reduce residual stress and cracking tendency of dissimilar metal joints and (c) isolating one or both the parent metals to minimize the dilution of weld metal characteristics from the parent metals and (d) realizing desired joint strength (Figs. 2.19 and 2.20). Thickness of buttering layer, therefore, optimized considering application and purpose. Thick buttering layer (like 5, 10, 20 mm) reduces (a) the diffusion of alloying elements from one parent metal to another across the weld joint, (b) gradient in characteristics of joint properties and (c) dilution of weld composition from the parent metals. However, thick butter layer (metal of relatively low yield strength) compromises with the mechanical performance of the dissimilar metal weld irrespective of weld metal, both HAZs and parent metal characteristics. In such a situation, thickness of buttering layer should be optimized (say 1 or 2 mm).

In the following section, three variants of GTAW, namely pulse GTAW, hot wire GTAW and A-GTAW, used for dissimilar metal joining have been presented. These variants inherently offer the advantage of developing low heat input weld joints with or without use of filler metals. It is usually preferred to apply filler/butter layer to develop dissimilar metal weld joint of reasonably good characteristics and reduce deterioration in quality of dissimilar metal joints due to incompatibility of parent metals.

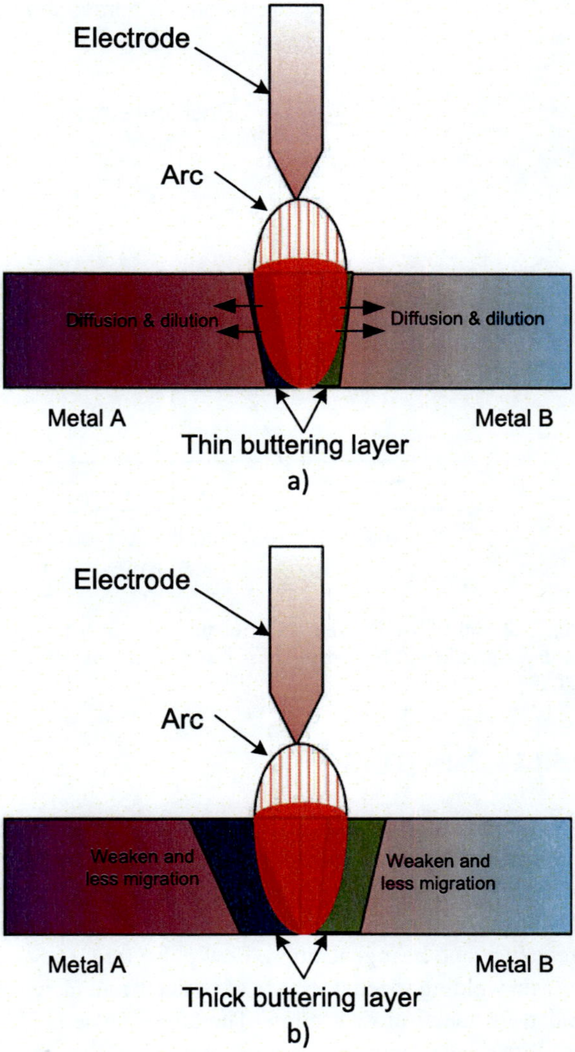

Fig. 2.19 Schematic showing effect of thickness of buttering layer on the parent metals on weld geometry of dissimilar metal joints developed by GTAW using **a** thin buttering layer and **b** thick buttering layer

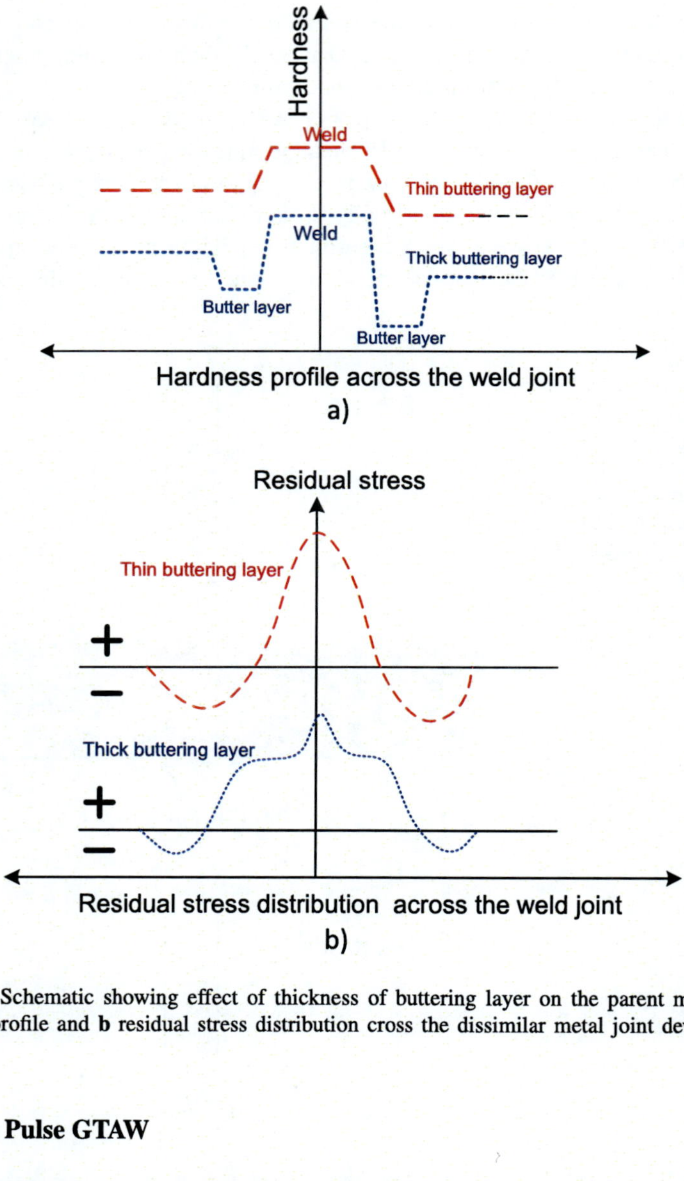

Fig. 2.20 Schematic showing effect of thickness of buttering layer on the parent metals on **a** hardness profile and **b** residual stress distribution cross the dissimilar metal joint developed by GTAW

2.5.1.1 Pulse GTAW

As compared to conventional GTAW, the pulse GTAW uses pulsation of welding current between two levels namely background current and peak current. The duration of these currents can be adjusted as per heat input requirement for dissimilar metal welding. The ratio of duration of peak and background current affects the pulse frequency and heat generation. Low pulse frequency results in lesser heat input for a given welding speed. Increase of peak current increases the heat input, penetration and weld metal cross section. Therefore, pulse GTAW provides opportunities to use lesser heat input (using suitable combination of peak current, pulse frequency and duty cycle) than conventional constant current GTAW (Fig. 2.21). The low heat input is considered very useful in dissimilar metal joining with regard to low dilution, narrow HAZ, fine grain structure and good mechanical properties. The pulse GTAW, however, offers low productivity primarily due to low heat input. Under identical net heat input conditions, the pulse GTAW offers deeper penetration and narrow weld than conventional GTAW.

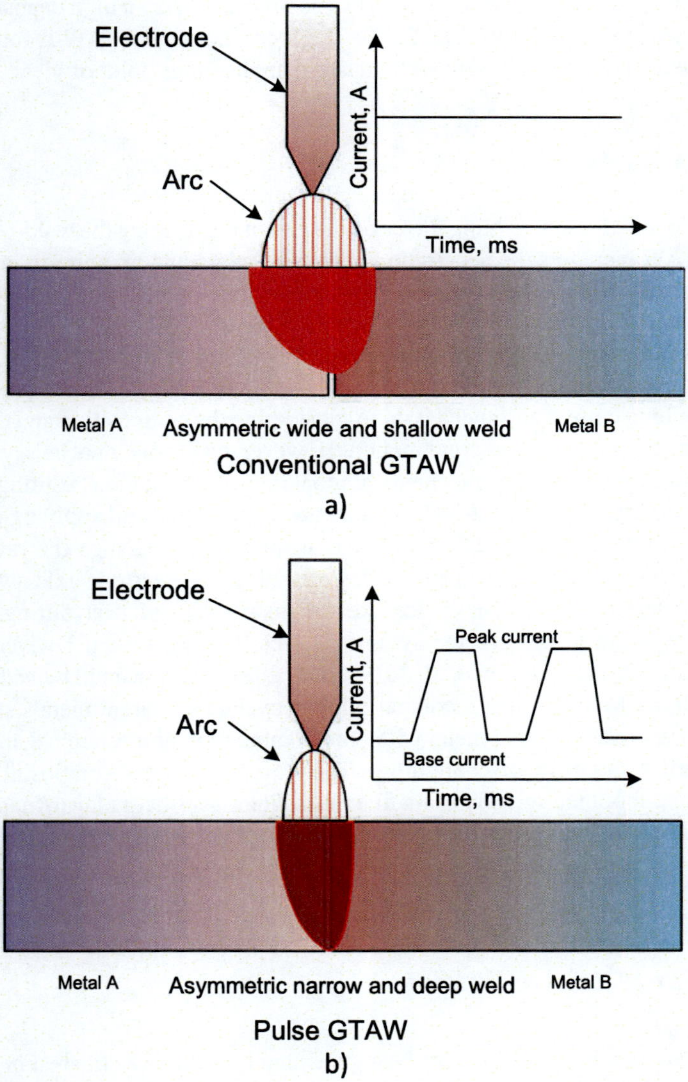

Fig. 2.21 Schematic showing effect of welding current on the weld geometry of the dissimilar metal joint developed by GTAW using **a** constant current and **b** current pulsation

2.5.1.2 Hot Wire GTAW

The hot wire GTAW is preferred for dissimilar metal joining of very thick sections. In this variant of GTAW, a preheated filler wire is fed into welding arc to increase the deposition rate. The preheating is achieved using a separate AC power supply while DC source is used for striking and maintaining welding arc. Low current, low heat input GTAW coupled with preheated filler wire helps in realizing dissimilar metal joints using very low arc heat while at the same time achieving a high deposition rate matching with GMAW/SAW (Fig. 2.22). Therefore, hot wire GTAW is found to be very attractive from productivity and industrial applications point of view.

2.5.1.3 A-GTAW

The application of few activating fluxes on the parent metal along the weld-centreline during GTAW increases the penetration many fold which in turn helps in increasing the productivity by facilitating welding of thick sections (10–15 mm) in single pass. Two mechanisms namely arc constriction and reversal of Marangoni convection in weld pool increase the depth of penetration (Fig. 2.23). The selection of suitable activated fluxes, flux application pattern and amount (mg/area), welding parameters, namely welding current and speed is done considering the composition and thickness of parent metals of dissimilar combination. However, weld zone may be asymmetric due to difference in physical and chemical properties of parent metals during dissimilar metal joining. Therefore, A-GTAW may need optimizing and applying different fluxes, in different coating patterns, thickness/amounts to balance the reversal in Marangoni convection as per need besides offsetting the electrode axis away from the weld centreline intentionally for skewed distribution of heat during dissimilar metal welding to achieve the symmetric weld (Vidyarthi and Dwivedi 2017; Kulkarni et al. 2018; Kulkarni et al. 2019, 2020, 2021; Sharma and Dwivedi 2019a, b, 2021a, b, c). Metallurgical incompatibility between two parent metals subjected A-GTAW may result in weld metal with poor combination of mechanical and corrosion properties and many discontinuities. Therefore, weld metal composition needs to be adjusted/modified suitably using filler metal/interlayer/butter layer/functionally graded weld metal to improve the mechanical performance and weld joint properties (Vidyarthi and Dwivedi 2019).

2.5.2 *Gas Metal Arc Welding*

Gas metal arc welding process uses heat generated by an arc established between a consumable electrode and parent metal for developing fusion weld joint between dissimilar metal combinations. The protection of weld pool and welding arc is achieved using inert or inactive shielding shroud/cover around the weld as per the affinity of the parent metals with atmospheric gases to form weld discontinuities

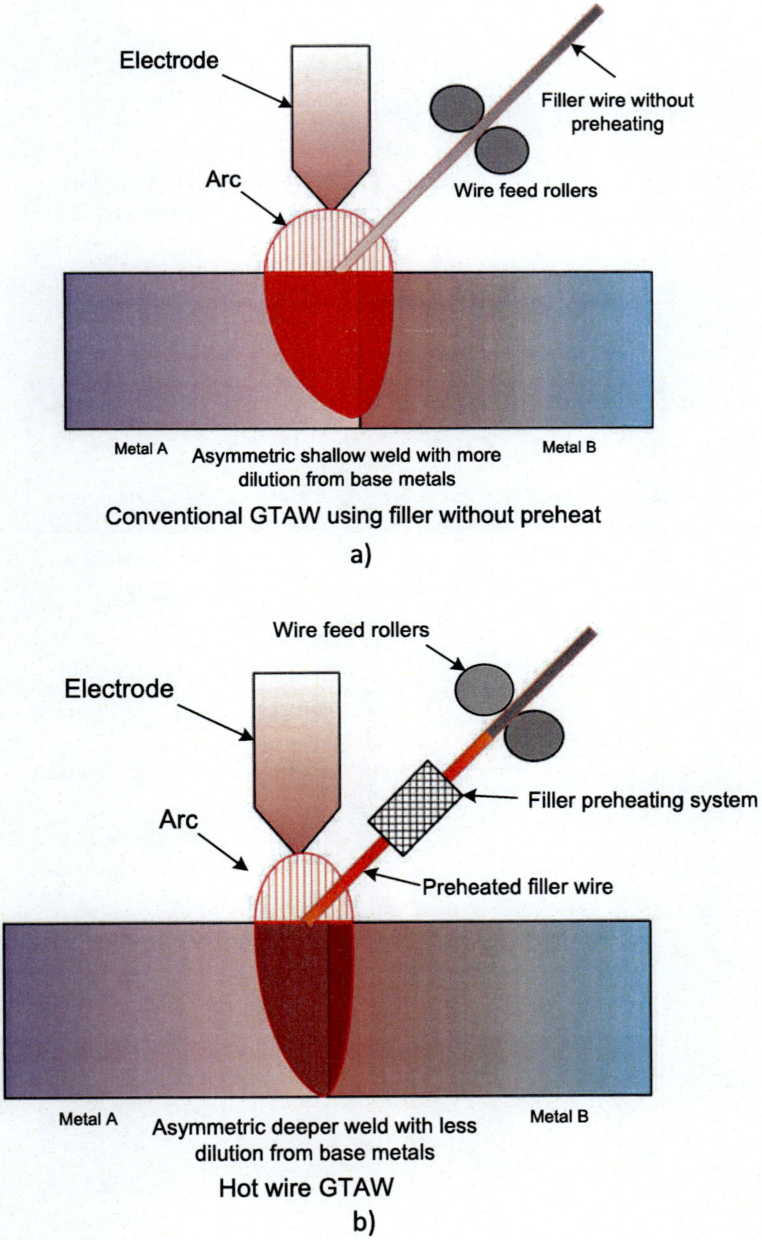

Fig. 2.22 Schematic showing effect of preheating filler wire on the weld geometry of the dissimilar metal joint developed by **a** conventional GTAW using filler and **b** hot wire GTAW

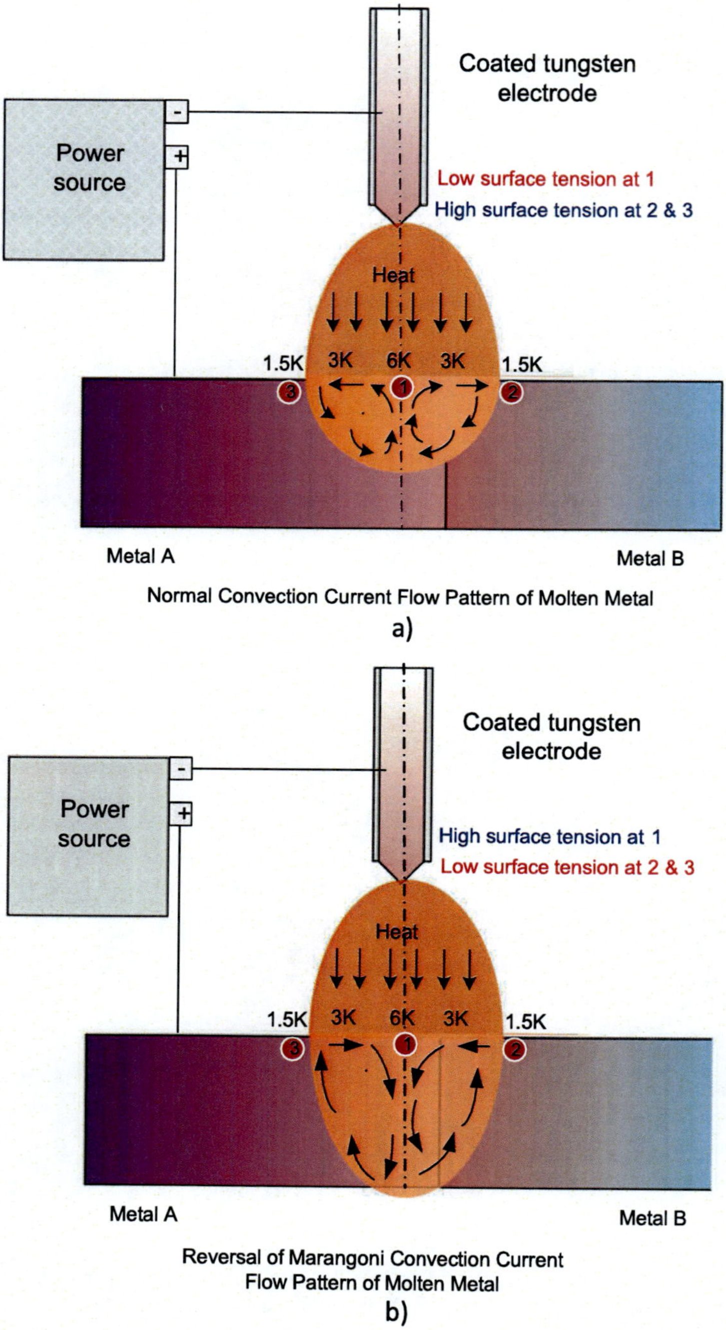

Fig. 2.23 Schematic showing patterns of convection currents and weld bead geometry in the dissimilar metal joint developed by **a** conventional GTAW and **b** A-GTAW

like pores and inclusions. The dissolved gases in the weld metal and net heat input applied during GMAW are higher than GTAW due to use of consumable electrode and somewhat less stable welding arc and fluctuating arc length in GMAW. Therefore, the quality of the GMAW joints is somewhat lower than GTAW joints. The dissimilar metal joining by GMAW developed in combination with GTAW for applying root pass buttering layering on the one or both the parent metals then groove is filled using GMAW. Inherent feature of using consumable electrode in GMAW allows not just higher deposition rates but also provides option to adjust the weld metal chemistry through controlled dilution, which is very crucial in dissimilar metal welding.

The GMAW, being a little higher heat input process than GTAW, many popular low heat input variants of GMAW, namely Pulse-GMAW, cold metal transfer welding (CMTW), narrow gap welding (NGW) can be extremely useful in dissimilar metal joining coupled with high productivity.

2.5.2.1 Pulse-GMAW

Like P-GTAW, welding current in pulse GMAW is pulsated between background and peak current level. The background current kept low enough just to maintain the welding arc at about 20–25% of peak current level. Peak current determines the penetration/melting of parent metal and electrode. The background and peak current and their duration determine the average heat input during Pulse-GMAW. In general, the net heat input applied during P-GMAW (for the given welding speed and penetration) becomes lesser than conventional GMAW. Thus, low heat input of P-GMAW helps in limiting the dilution from the parent metals and controlling weld metal composition in dissimilar metal joining favourably besides providing fine grain structure, improved mechanical properties while reducing adverse effects related to HAZs, residual stress and distortion.

2.5.2.2 Narrow Gap Welding

The narrow gap welding process is characterized by fusion welding of plates having placed with very narrow gap (8–20 mm) depending upon the section thickness using suitably designed V groove (< 8°) for joining of heavy section (20–200 mm) components and parts of dissimilar metals (Fig. 2.24). For example, groove width can be reduced from more than 15 mm to less than 8 mm. This typical feature of narrow gap (welding) for joining of thick section components reduces the volume of weld metal needed to complete the welding (producing typical I shape weld cross section); which in turn increases the productivity many folds, besides reducing adverse effects related to high heat input (Masatoshi et al. 2015).

The narrow gap welding is a unique variant of Pulse-GMAW, which uses an especially designed very narrow and small diameter welding torch/nozzle (6–8 mm) with power fed electrode feeder so that it can help to apply arc heat in the very narrow root gap using a consumable electrode. The consumable electrode extends

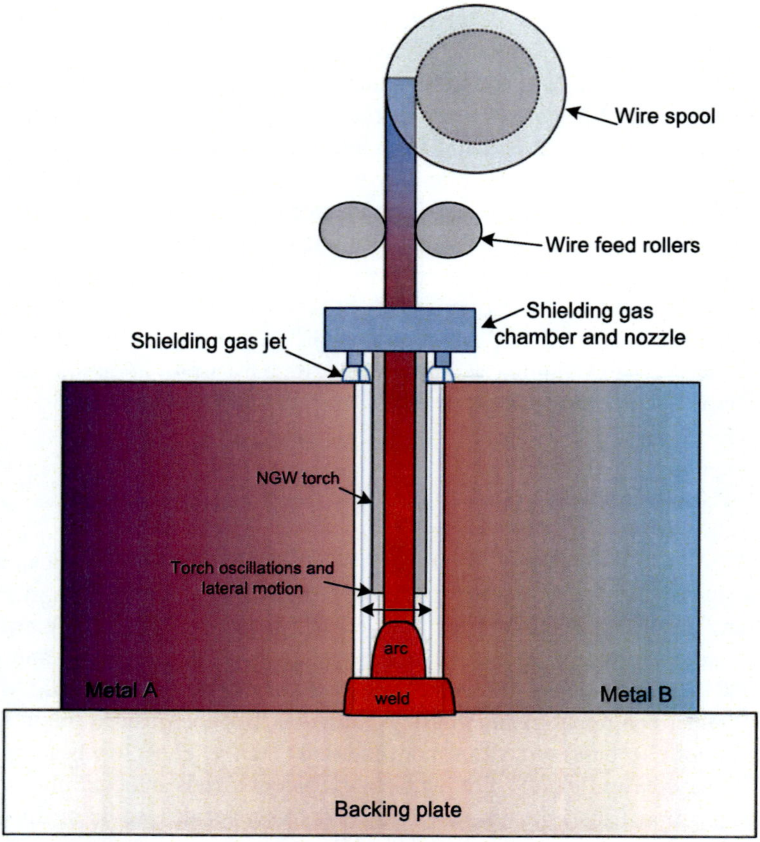

Fig. 2.24 Schematic showing set of narrow gap welding for the dissimilar metal joining

out of specially designed welding torch significantly (say by 50–60 mm), and welding torch is given circular or oscillatory weaving motion (amplitude of 3–5 mm) at low frequency (0.5–10 Hz). The narrow gap between components filled using multiple passes.

2.5.2.3 Cold Metal Transfer Welding

The cold metal transfer welding like the narrow gap welding is also a typical variant of Pulse-GMAW wherein unique approach for molten metal drop detachment from the electrode tip to the weld pool is attained. The droplet detachment is realized through stable and controlled short-circuit transfer using a combined action of electrode retraction (from the weld pool) and regulation of welding current/voltage (Fig. 2.25). The welding current is pulsated between the background and peak current. The duty cycle (ratio of peak and background current duration) is very low. The short peak current coupled long background current cycle avoids the formation of large molten metal drop. The fine drop as soon as touches the weld pool, electrode-wire retracts. Retraction of electrode helps in drop detachment from the electrode tip and facilitates the transfer of droplet to the weld pool. There are three main phases in cold metal transfer welding (CMTW), namely peak current phase, background current phase and short-circuiting phase. The melting of electrode wire and penetration of parent metal to the desired depth is realized during peak current phase, while the molten metal drop hangs on the tip until short-circuiting during background current phase. During the short-circuiting phase itself, electrode wire retracts to facilitate molten metal drop transfer to the weld pool. On completion of metal transfer, arc gap is established and then this cycle keeps on repeating.

The short circuiting current is very low, and voltage is almost nil; therefore, spatter during cold metal transfer welding (CMTW) despite of short-circuiting is

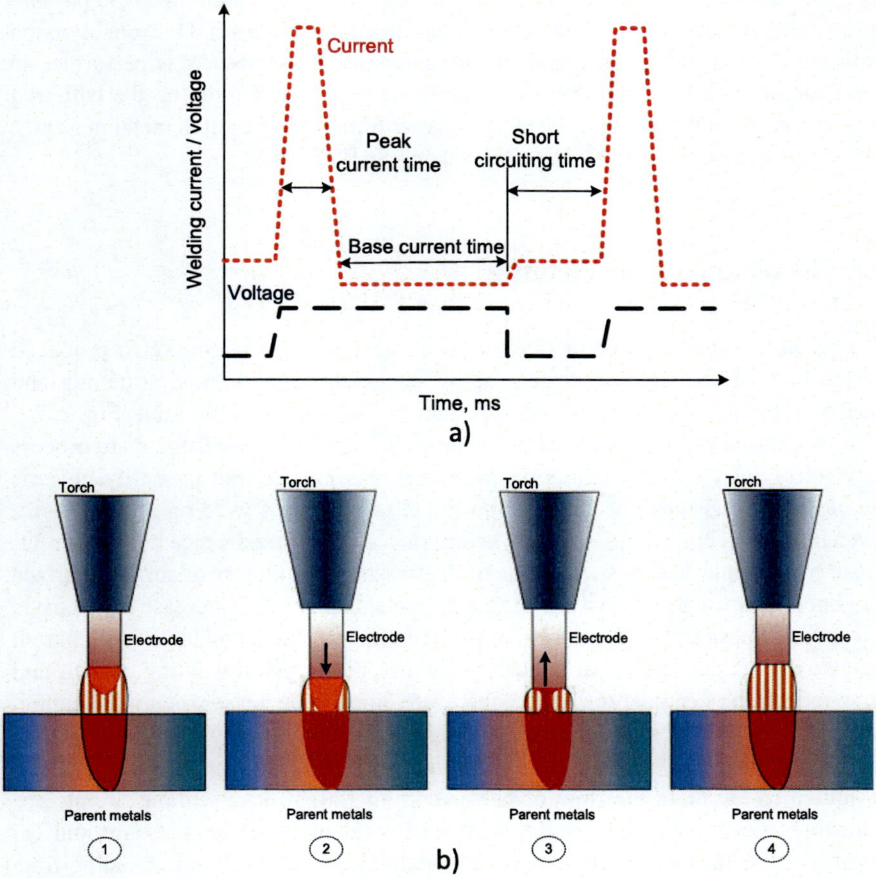

Fig. 2.25 Schematic showing **a** variation in welding current as a function of time during CMT welding and **b** stages of metal transfer during CMT welding

minimized. The welding current pulsation and low current metal transfer without spatter makes cold metal transfer welding(CMTW) attractive for dissimilar metal welding as it reduces the undesirable weld metal properties like IMC formation due to dilution from parent metals, narrow HAZ, reduced residual stress, distortion and related issues (Selvi et al. 2018).

2.5.3 Shielded Metal Arc and Submerged Arc Welding

The shielded metal arc welding (SMAW) and submerged arc welding (SAW) produce only commercial quality weld joints of similar and dissimilar metals primarily due to high gaseous content in the weld metal and high heat input. These technical aspects related to SMAW and SAW results in more weld discontinuities and wider heat-affected zone than the GMAW and GTAW (Fig. 2.26). Still both these are popular in the industry due to ease of operation, cost-effective welding and reasonably good productivity. The dissimilar metal joining using SAW and SMAW is performed in combination with GTAW. GTAW is used for root pass and applying the buttering layer on one or both the parent metals thereafter filling of the molten metal in groove between the dissimilar metals is realized using SAW/SMAW.

2.6 Resistance Spot Welding

The resistance welding process like spot welding uses joule heating (I^2Rt produced by the flow of the welding current across the sheets) for thermal softening and localized interfacial melting of sheets for the development of weld joint (Fig. 2.27). Welding current (I), electrode force (P) and welding time (t) are three main process parameters apart from electrode tip geometry. Welding current primarily governs the electrical resistance heating (proportional to square of welding current) for a given interfacial contact resistance. The interfacial contact resistance however influenced by multiple factors such as electrode pressure, cleanliness of surface, surface roughness, electrical resistivity of the parent metals. In general, increase of electrode force, cleanliness and surface finish of surfaces and electrical conductivity of parent metals reduces the interfacial contact resistance. For a given welding current and time, reduction in contact resistance, therefore, linearly decreases the joule heating. Interfacial contact pressure on the copper electrodes plays many roles in resistance spot welding such as (a) establishing firm metal to metal contact between sheets to be joined to facilitate the flow of heavy current (20–30 kA) without arcing, (b) achieving consolidation and solidification of weld metal under pressure and (c) forging action (at the interface) between thermally softened sheets. Welding time directly affects the heat generation by electrical resistance heating for a given electrode force, welding current. Joule heating should be optimized for maximum joint strength as it affects two aspects controlling the weld joint strength (a) weld nugget

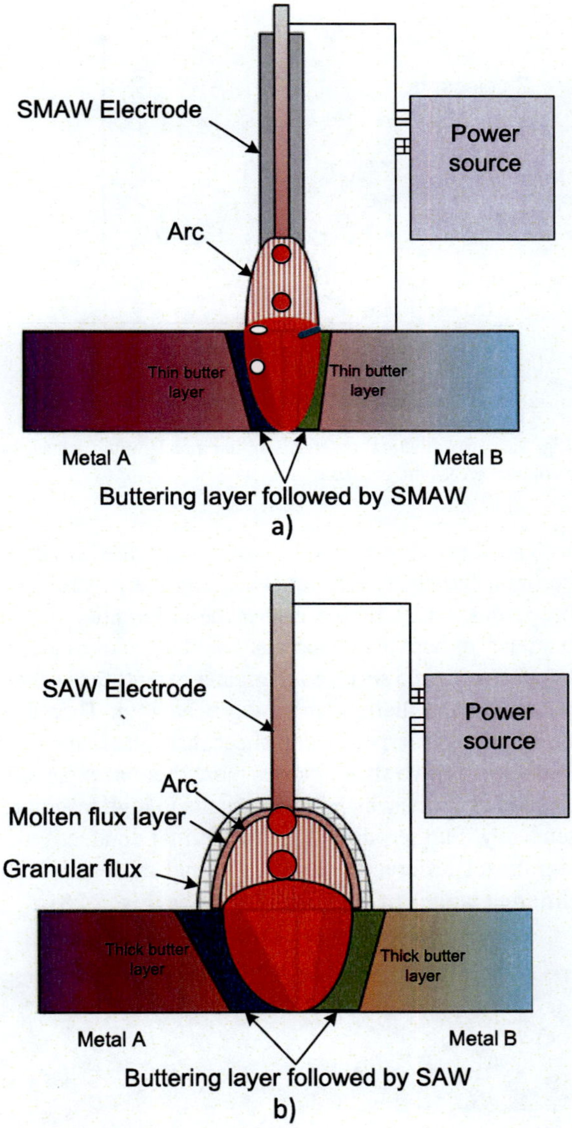

Fig. 2.26 Schematic showing dissimilar metal joining using butter layer developed by GTAW followed by **a** SMAW and **b** SAW

diameter and (b) expulsion. Increase of the weld nugget diameter increases the metallurgical bonding between sheets while expulsion increases the stress concentration at the joint periphery and weld defects. Therefore, two (weld nugget and expulsion) work in an opposite manner. Increase of heat generation continuously increases the strength up to a limit due to increase of weld nugget diameter thereafter reaching maxima, strength starts decreasing due to expulsion.

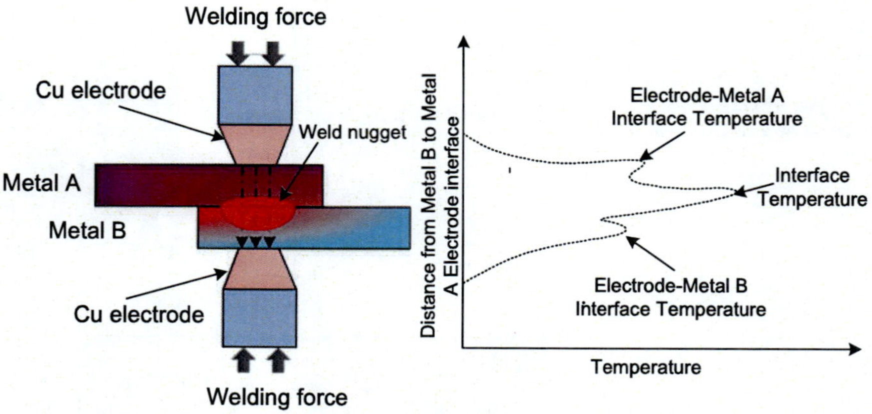

Fig. 2.27 Schematic showing dissimilar metal joining using spot welding **a** set up and **b** temperature variation across the interface

The dissimilar metal joining makes the situation in resistance welding little complicated due to differences in physical, chemical, mechanical and metallurgical properties, which in turn causes the differential joule heating and thermal softening tendency of both parent metals. The sheet of one parent metal may be heated to higher temperature (and softened to greater extent) than other due to difference in electrical resistivity and thermal softening behaviour. Therefore, weld nugget formed might be deeper and larger in one sheet than other sheet. Differential thermal expansion behaviour of two sheets causes more thermal expansion and contraction of one sheet than other sheet, which in turn results in high residual stress, cracking and distortion tendency. Further, difference in thermal conductivities of two parent metals offers significant difference in weld thermal cycle in vicinity of weld joints leading to different widths of heat-affected zones (Fig. 2.28).

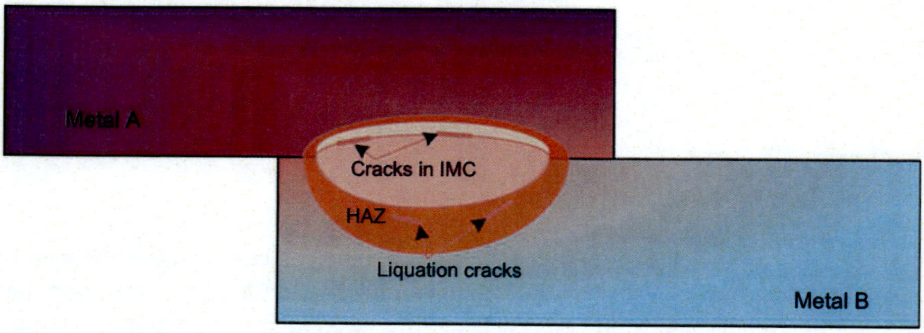

Fig. 2.28 Schematic showing various zones formed in a typical spot weld of a dissimilar metal joint

Additionally, metallurgical incompatibility of dissimilar metals during spot welding (being autogenous) frequently results in undesirable metallurgical transformation and formation of hard and brittle phases and intermetallic, which in presence of residual tensile stress causes cracking of the weld joint besides lowering strength, ductility and load carrying capacity of the joint. Therefore, a thin interlayer having metallurgical compatibility with both parent metals placed at the interface during the spot welding to encourage formation of favourable phases, micro-constituents and intermetallic. The selection of interlayer and its thickness is crucial for joint strength of dissimilar metal spot weld joints. These technological aspects are equally applicable for other resistance welding processes such as projection, seam welding.

2.7 Brazing and Braze Welding

The brazing and braze welding both are based on the same principle of applying low melting brazing metal between the dissimilar parent metals to develop a joint but two are different in terms of the placement of dissimilar metal component to be joined, i.e. type of joint to be developed. Both these processes theoretically do not involve any fusion of faying surfaces of parent metals but use comparatively very less heat input for brazing/braze welding than fusion welding processes, therefore, chemical composition, metallurgical incompatibility, mechanical and thermo-physical properties of dissimilar metal combination do not play a significant role in joint performance (Fig. 2.29). Further, strength and load carrying capacity of the brazed joint and brazed welds of dissimilar metal combination depends on (a) the soundness of the brazed zone, (b) metallurgical and mechanical properties of brazing metal itself and (c) the formation of interfacial intermetallic and metallurgical bond between brazed zone and two parent metals (Fig. 2.30). The development of braze weld/joint free from discontinuities and stress raisers results in good strength matching with that of brazing metal. The formation of equiaxed and fine grain structure results good joint strength and ductility. The most crucial aspect in brazing and braze welding determining the joint strength is interfacial zone between braze metal and two parent metals due to the formation of thin layer of intermetallic (Fig. 2.31). Additionally, the roughness of faying surfaces of two parent metals also affects the joint strength because molten brazing metal fills the peaks and valleys on the rough faying surfaces which subsequently on solidification offers mechanical interlocking of brazed metal and parent metals. Further, the low heat input results in a very narrow or nil heat-affected zone besides reduced issues related to differential thermal expansion and contraction causing residual stress and distortion.

The brazing process mainly uses a lap joint configuration with controlled clearance (0.025–0.25 mm) between parent metals to facilitate the capillary action for sucking and then uniform distribution of molten brazing metal at the joining interface. Depending upon the use of heat source available, brazing can be performed using different methods like furnace, molten bath, gas, arc, induction, laser, etc. Braze welding can be used to develop all type of joints butt, corner, T, and edge joint

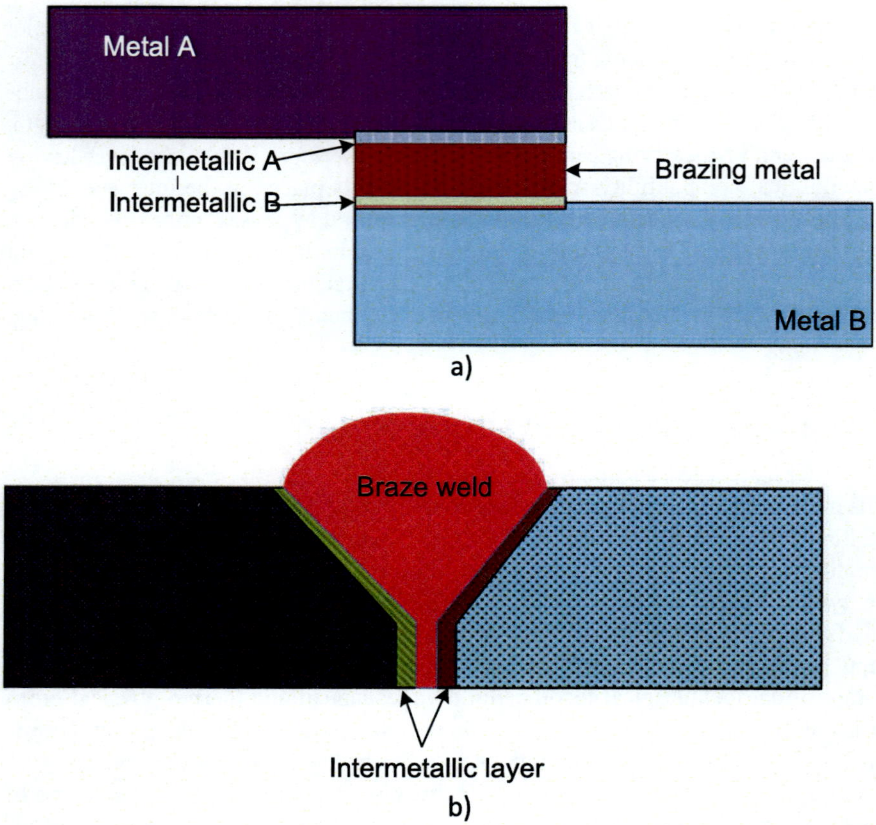

Fig. 2.29 Schematic showing formation of different zones in dissimilar metal joining using **a** brazing and **b** braze welding

using suitable edge preparation (groove geometry) followed by filling the molten brazing metal between the dissimilar metals to be joined.

Heat sources like oxy-fuel gas, induction, furnace and molten bath brazing do not cause melting of parent metals but high-energy and high-temperature heat sources like laser and arc cause little melting of faying surfaces of the parent metals. GTAW, CMTW and Pulse GMAW are common arc welding processes used for braze welding of dissimilar metals with due care. The melting of parent metals during brazing/braze welding should be minimized else, intermixing of parent metals with brazing metal may form undesirable micro-constituents including intermetallic which in turn can deteriorate the mechanical and corrosion performance of joint during the service. An increase of interfacial zone length between braze weld and parent metals increases the joint strength if fracture during tensile and tensile-shear loading takes place from the interface which suggests that brazed zone and both the parent metals and their HAZs are stronger than brazed joint/brazed weld (Fig. 2.32).

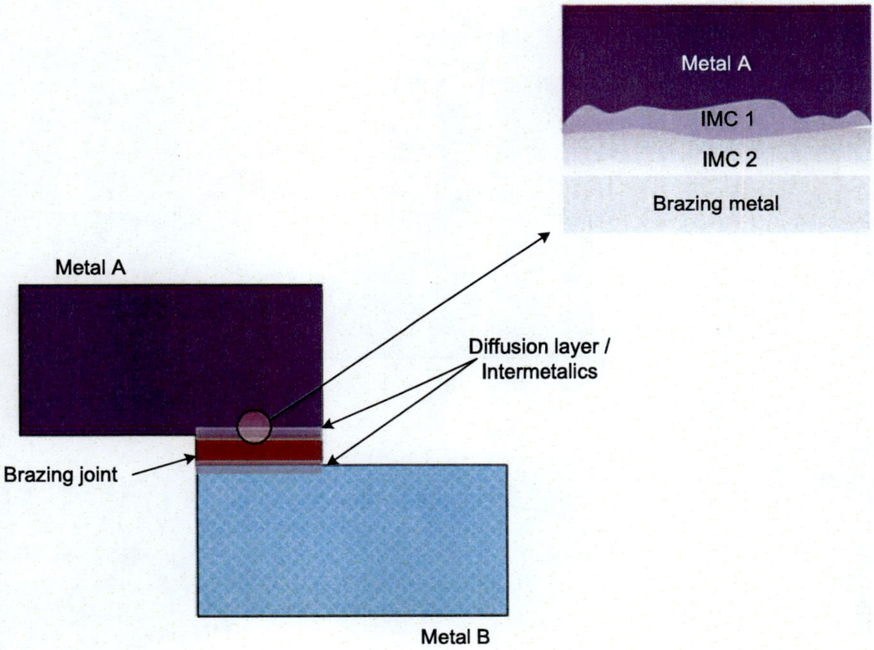

Fig. 2.30 Schematic showing microscopic feature of a dissimilar metal joint developed by brazing/braze welding

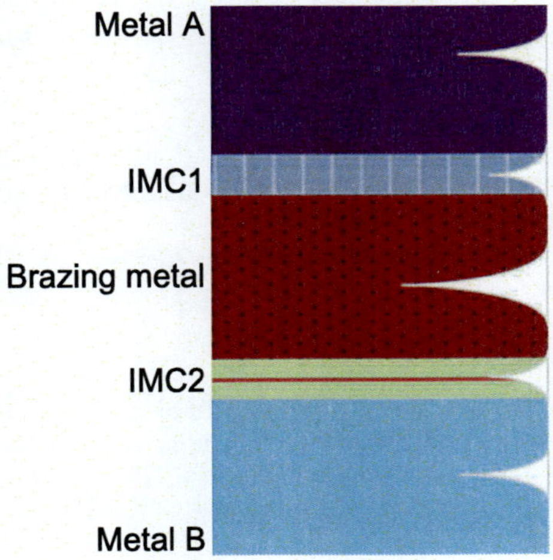

Fig. 2.31 Schematic showing various locations wherefrom fracture can imitate depending upon the weak zone dissimilar metal joint developed by brazing/braze welding

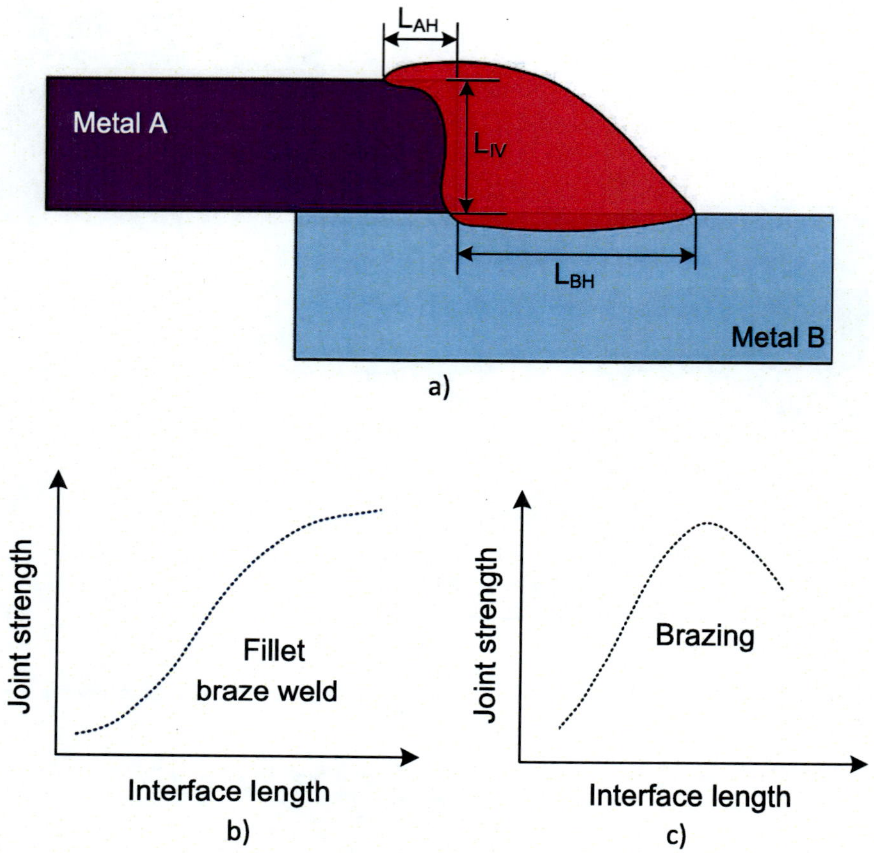

Fig. 2.32 Schematic showing **a** geometry of the brazed joint developed in arc-based process like CMT/GMAW/GTAW, effect of interface length on strength of dissimilar metal brazed joint developed by **b** fillet braze welding and **c** brazing

References

Kulkarni A, Dwivedi DK, Vasudevan M (2019) Dissimilar metal welding of P91 steel-AISI 316L SS with Incoloy 800 and Inconel 600 interlayers by using activated TIG welding process and its effect on the microstructure. J Mater Process Technol 274:116280

Kulkarni A, Dwivedi DK, Vasudevan M (2021) Novel functionally graded joint between P91 steel-AISI 316L SS. Weld J 100:269–280

Kulkarni A, Dwivedi DK, Vasudevan M (2020) Microstructure and mechanical properties of A-TIG welded AISI 316L SS-Alloy 800 dissimilar metal joint. Mater Sci Eng: A 790(2020):139685

Kulkarni A, Dwivedi DK, Vasudevan M (2018) Study of mechanism, microstructure and mechanical properties of activated flux TIG welded P91 Steel-P22 steel dissimilar metal joint. Materi Sci Eng A 731:309–323

Masatoshi M, Daisuke I, Kensuke OOE (2015) Narrow gap gas metal arc (GMA) welding technologies. Technical Report No. 20 (Mar. 2015)

Selvi S, Vishvaksenan A, Rajasekar E (2018) Cold metal transfer (CMT) technology—an overview. Defence Technol 14(2018):28–44

Sharma P, Dwivedi DK (2019a) Comparative study of activated flux-GTAW and multipass-GTAW dissimilar P92 steel-304H ASS joints. Mater Manuf Process 34:1195–1204

Sharma P, Dwivedi DK (2019b) A-TIG welding of dissimilar P92 steel and 304H austenitic stainless steel: mechanisms, microstructure and mechanical properties. J Manuf Process 44:166–178

Sharma P, Dwivedi DK (2021a) Wire-feed assisted A-TIG welding of dissimilar steels. Arch Civ Mech Eng 21(2021a):1–20

Sharma P, Dwivedi DK (2021b) Improving the strength-ductility synergy and impact toughness of dissimilar martensitic-austenitic steel joints by A-TIG welding with wire feed. Mater Lett 285:129063

Sharma P, Dwivedi DK (2021c) Flux assisted tungsten inert gas welding of bimetallic P92 martensitic steel-304H austenitic stainless steel using SiO_2-TiO_2 binary flux: welding arc/pool behaviour, microstructure and mechanical properties. Int J Pressure Vessel Piping 192:104423

Vidyarthi RS, Dwivedi DK (2017) Study of microstructure and mechanical property relationships of A-TIG welded P91-316L dissimilar steel joint. Mater Sci Eng A 695(2017):249–257

Vidyarthi RS, Dwivedi DK (2019) Effect of shielding gas composition and activating flux on the weld bead morphology of the P91 ferritic/martensitic steel. Mater Res Exp 6(8):0865f7

Chapter 3
Dissimilar Metal Joining by Laser Welding

This chapter presents the fundamentals of laser welding, commons mode of laser welding, issues encountered during laser welding of dissimilar metals. Factors affecting the ease of laser welding of dissimilar metals and methods to deal with common laser welding discontinuities have bene elaborated. Advances on laser welding of common dissimilar metal combinations line aluminium with steel, copper and copper with steel have been described.

3.1 Fundamental of Laser Welding

Laser welding process is characterized as high power density (10^2 to 10^8 W/cm^2) and high temperature (> 25,000 °C) joining processes wherein the radiation beam can be focused over very small area which in turn allows better control and application of heat at the desired location especially in dissimilar metal joining. These features help in performing welding of dissimilar metals using very less net heat input due to high power density, which in turn reduces (a) undesirable dilution of weld metal from the parent metals, (b) weld cross section and heat-affected zone and (c) residual stress and distortion. Laser welding is performed in ambient conditions; while electron beam welding usually is carried out in a vacuum chamber. The weld pool in laser welding is protected from atmospheric gases using a jet of inert gases (Ar, He). The laser welding uses heat produced by a laser beam directed on the parent metals along the weld centreline. Apart of the laser beam absorbed by parent metals while remaining is reflected in multiple directions as per reflectivity of metals. Laser absorbed by the metal is converted into heat, which is used for melting the faying surfaces of parent metals for metal joining. The high power laser beam allows focusing of energy over a small laser spot radius (100–2000 microns) as per need, which in turn facilitates adjustment of laser power density for welding from 10^6 to 10^8 W/cm^2 to achieve either melt in mode or key-hole mode of fusion welding (Fig. 3.1). Accordingly, the laser power density can vary significantly for dissimilar metal welding of thin sheet

© The Author(s), under exclusive license to Springer Nature Singapore Pte Ltd. 2023
D. K. Dwivedi, *Dissimilar Metal Joining*,
https://doi.org/10.1007/978-981-99-1897-3_3

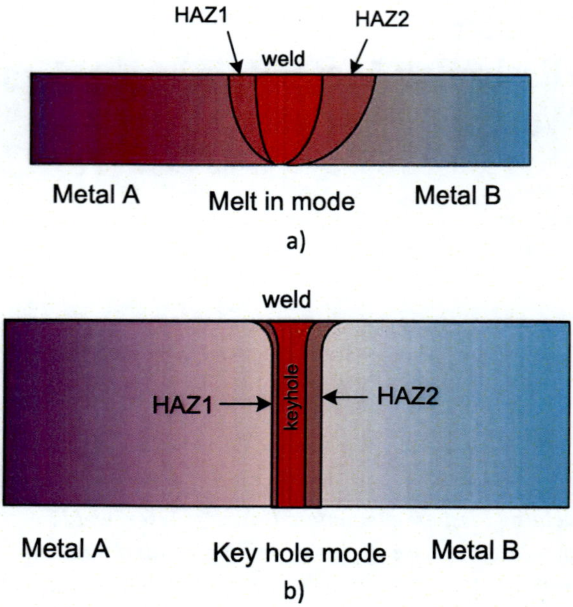

Fig. 3.1 Schematic of laser welding joint of dissimilar metals developed using **a** melt-in mode and **b** key-hole mode

to thick plates (up to 25 mm or more). The melt in mode of laser welding (< 10^6 W/cm^2) used for joining of thin sheets while keyhole technique applied for welding of thick plates (> 10^8 W/cm^2).

The evaporation of metal takes place at high laser power density (> 10^8 W/cm^2) during welding which in turn produces penetrating vapour cavity known as keyhole. A critical balance of evaporating metal vapour pressure, hydrostatic forces of thin molten wall surrounding metal vapours and surface tension forces need to be maintained for the keyhole in laser welding (Fig. 3.2). The movement of parent metal with respect to laser beam causes melting of parent metal at the leading edge of pool wherefrom the molten metal flows backwards while maintaining the keyhole. Heat dissipation from the molten metal by conduction to the underlying parent metals results in the rapid solidification of the weld metal at high cooling rate.

The laser power (kW), laser spot diameter (*d*, mm) and laser scanning speed during joining (*v*, mm/min) are few parameters determine the net heat applied during welding. The net heat input for laser welding is many fold lower (0.03 to 0.1 J/mm) than arc welding processes (0.5–5.0 kJ/mm) while realizing the same penetration. The low heat input to the parent metals of dissimilar combination is the most attractive feature of the laser welding processes. Low heat input during laser welding of dissimilar metal combinations offers many advantages (over other fusion welding processes) such as

(a) Reduced weld cross section,
(b) Narrow heat-affected zone,

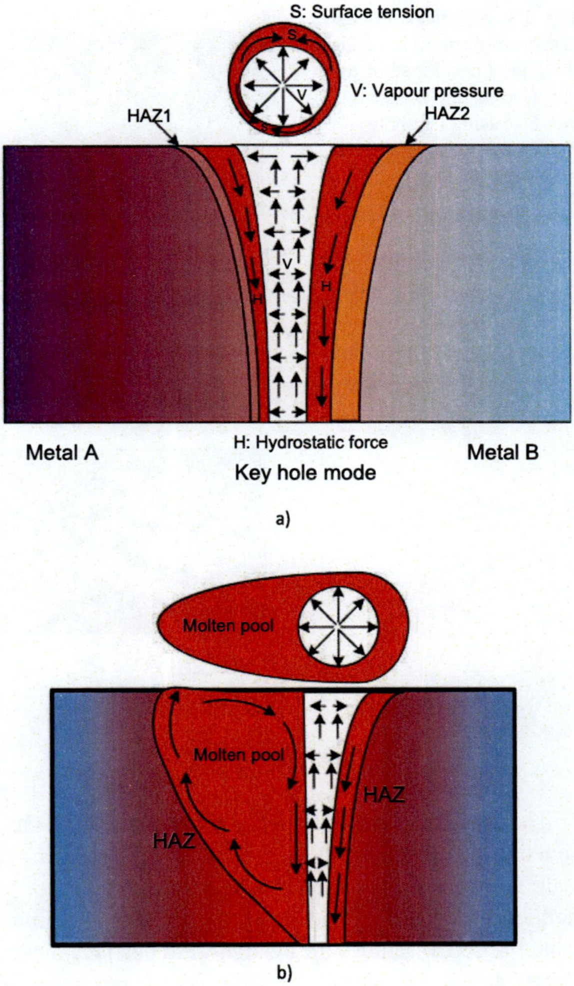

Fig. 3.2 Schematic showing structure of key-hole mode during laser welding of dissimilar metals: **a** front view in longitudinal direction and **b** side view transverse direction

(c) Reduced dilution of weld metal from the metallurgically incompatible parent metals,
(d) Reduced differential expansion and contraction of parent metals (and so reduced residual stress and distortion),
(e) High cooling rate and so short weld, poor solidification time leading fine grain structure, and
(f) Reduced segregation of alloying element decreases solidification cracking (Fig. 3.3).

Heat input for fusion welding and cooling rate of weld metal and HAZ are inversely related.

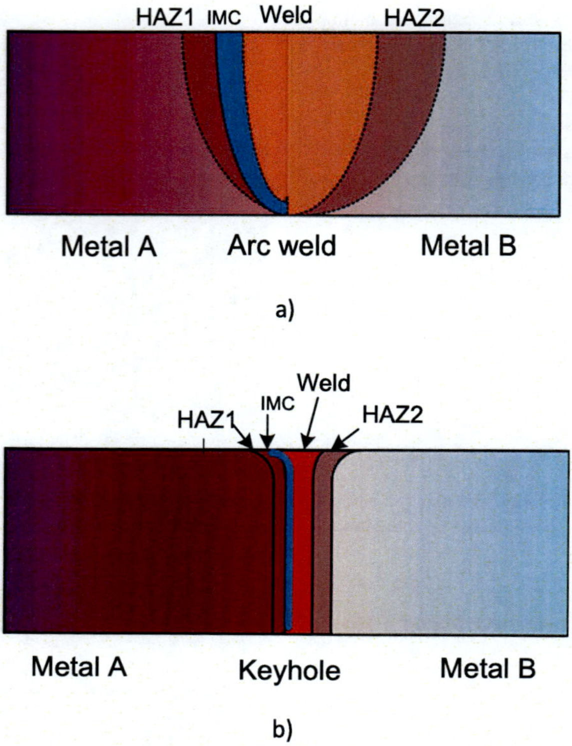

Fig. 3.3 Schematic of cross section of dissimilar metals weld joint developed using **a** arc welding and **b** laser welding using key-hole mode

The heat input in laser welding can be easily changed to achieve the desired penetration using parameters such as laser power (peak power, frequency, duty cycle), laser spot diameter and laser-scanning speed. Increase of laser power, pulse frequency, duty cycle and reduction in laser scanning speed increases the net heat input needed to achieve the desired weld bead geometry during welding. Laser beam allows deliver of heat towards the specific parent metal of dissimilar metal combination as per need with precision and accuracy, which in turn helps in achieving the desired weld metal chemistry with/without filler metal as per requirement. The filler metal can be preplaced or fed in the form wire, powder, interlayer and thin film during laser welding. The filler metal selection is critical in developing weld metal with a suitable gradient across the dissimilar metal weld joint in terms of physical, chemical and mechanical properties (Fig. 3.4).

The dissimilar metal components machined and smoothened for proper alignment to ensure a close gap between components for laser welding. This is critical in laser welding else, beam may pass through gap between parent metals without welding. Therefore, suitable fixture need to be designed and developed for easier maintenance of the close tolerance in fit-up, positioning and alignment of the components to be laser welded.

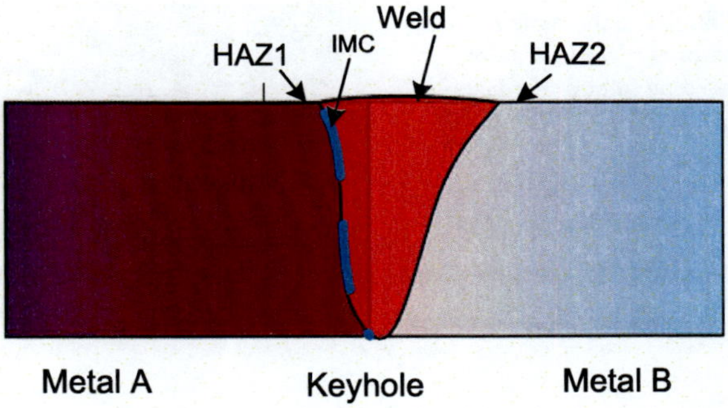

Fig. 3.4 Schematic of laser welding joint of dissimilar metals developed using suitable filler metal

3.2 Common Issues in Laser Welding

Common issues encountered during laser welding of similar and dissimilar metals include embrittlement, porosity formation, weld bead sagging, solidification cracking, misalignment and fit, lack of fusion and penetration. However, there are issues specific to laser welding of dissimilar metals only such as asymmetric weld, residual stresses due to differential expansion and contraction, undesirable metallurgical transformations such IMCs formation depending upon the metallurgical incompatibility between the parent metals of dissimilar combination (Sun and Ion 1995).

3.2.1 *Embrittlement*

The laser heat source of high power density results in unique weld thermal cycle of very high heating rate, peak temperature and followed by extremely high cooling rate primarily due to low net heat input for welding. However, limited heat input produces narrow weld and heat-affected zones. Hardenable metals like cast irons, medium and high carbon steel, alloy steel, martensitic stainless steel and many hardenable (CE > 0.3) ferrous metals during the laser welding experience a weld thermal cycle which causes austenitization, followed by cooling at a rate usually higher than the critical cooling rate (Fig. 3.5). Therefore, weld thermal cycle imposed by laser welding of hardenable ferrous metals invariably produces hard and brittle martensite. Embrittlement of weld metal and HAZ of laser weld joint reduces toughness and increases the cracking tendency in presence of residual tensile stress. Therefore, a suitable control of weld metal composition using filler and dilution from the parent metals, preheat and post weld heat treatment is used to reduce the embrittlement of the laser weld joint. Moreover, the high cooling rate experienced by the weld metal and HAZ restricts the grain growth, which in turn results in fine grain structure.

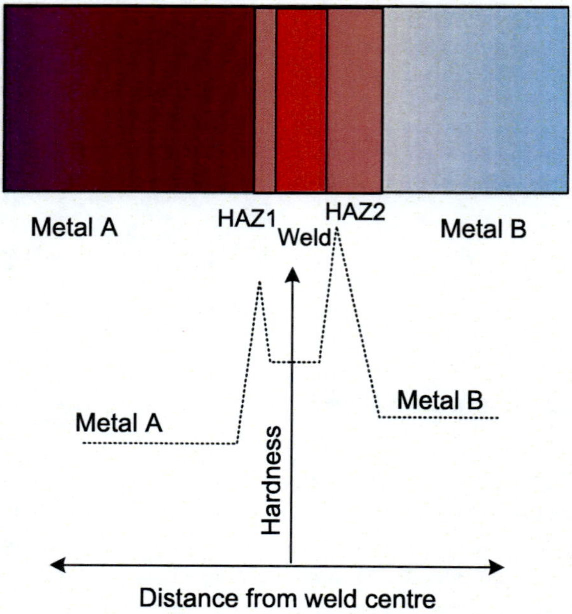

Fig. 3.5 Schematic showing hardness profile of laser welding joint of hardenable dissimilar metals

3.2.2 *Porosity*

The laser weld joints experience multiple types of pores. Porosity in form of micro/pin hole, air pockets and large cavity. All types of porosities reduce the actual load resisting cross-sectional area of the weld joint and act as stress raiser. The deterioration in mechanical performance and load carrying capacity due to spherical pores located internally in the weld metal is limited. The entrapment of gases, air pocket, metal vapour and instable key holes due to short time of weld metal solidification caused by high cooling rate are main causes of porosity in laser welding (Fig. 3.6). Gases like hydrogen, oxygen (> 100 pm) produced by thermal decomposition of moisture, chemicals and impurities at high temperature generated by the laser beam. The gases (if present) in parent metals in dissolved state may also contribute to porosity in weld metal. Metals like aluminium having large difference in liquid and solid-state solubility to hydrogen frequently show pinhole porosity. Low melting metals like Mg, Zn, Pb, etc. also evaporate at high temperature and contribute to metal vapour in weld pool. Further, poor control over laser welding parameters makes the keyhole instable. Collapsing of keyhole during laser welding followed by rapid solidification produces large unfilled cavity in weld metal like piping defect. Therefore, suitable control of welding parameters, proper cleaning, cleaned and baked welding consumables, and preheating of parent metals help to reduce the porosity.

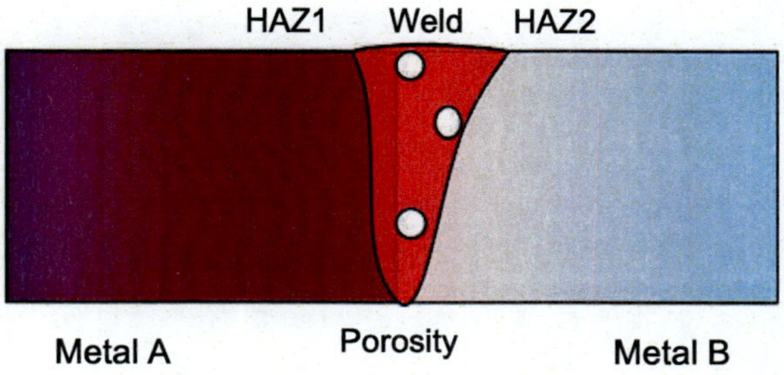

Fig. 3.6 Schematic of laser welding joint of dissimilar metals showing porosity

3.2.3 Solidification Cracking

The solidification cracking occurs at the terminal stage of solidification of the weld at centre (in presence of residual tensile stress) primarily due to segregation of alloying elements (Si, S, P, Pb, etc.) in the molten weld metal producing low melting point constituents (FeS, silicates, etc.). Presence of S, P and Pb in many steels causes solidification cracking. The typical weld thermal cycle of laser-welding causing high cooling rate leading to short weld metal solidification time, which reduces the segregation of alloying elements at the weld centre. Reduction in segregation in turn lowers the solidification-cracking tendency of laser weld joint as compared to other fusion welding processes for the same dissimilar metal combination. However, high depth-to-width ratio of the laser weld having typical I section geometry results in high angle abutting grains approaching the weld centre from fusion boundaries, which promotes segregation of low melting point constituents at the centre and so the solidification cracking (Fig. 3.7). The weld metal composition is adjusted suitably to control solidification cracking using suitable filler metal considering dilution from two parent metals, reducing residual stresses using preheating, and reduced restraint during welding. Addition of Mn in carbon and alloy steel welding and ensuring 3–5% delta ferrite in austenitic stainless steel welding using suitable filler metal selection reduces the solidification cracking.

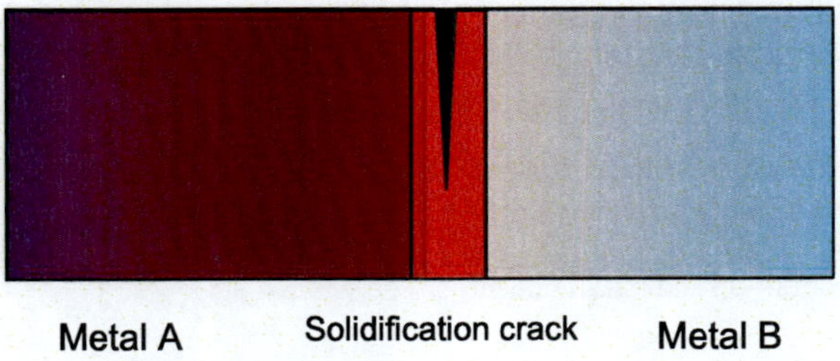

Fig. 3.7 Schematic of laser welding joint of dissimilar metals showing solidification crack

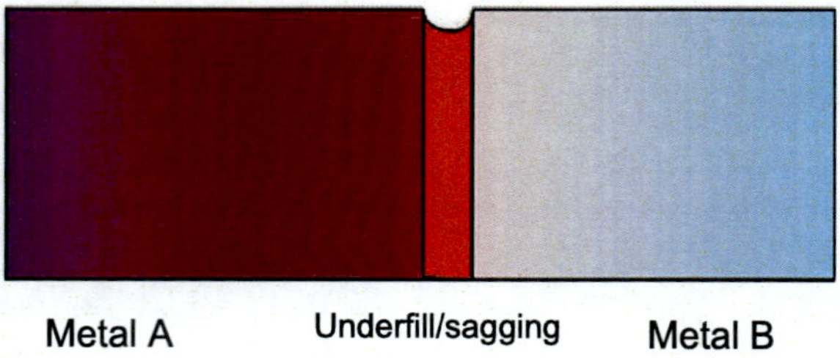

Fig. 3.8 Schematic of laser welding joint of dissimilar metals showing sagging

3.2.4 *Weld Bead Sagging*

The weld bead sagging in the form under-filling, crater, and concave bead mostly observed in autogenous welding especially when square edge preparation of parent metals (with wide root gap) is leading wide air gap along the weld line (Fig. 3.8). These gaps filled-in by the molten weld metal obtained through melting of edges of the parent metals during autogenous welding. Weld bead sagging reduces the throat thickness and acts as stress raiser. Both these factors reduce the load carrying capacity of weld joint. Proper edge preparation with straight edges and smooth surface coupled with application of filler metal can help to overcome the issues related to the weld bead sagging.

3.2.5 *Misalignment and Misfit*

The laser beam of very small diameters (from few micron to mm) used to apply heat for fusion of edges of the parent metals. Autogenous laser welding has very limited capability to accommodate the misalignment and misfit (Fig. 3.9). Laser beam may completely miss the faying surfaces of parent metals without or with very little melting. Therefore, very close tolerance, perfect edge preparation needed to maintain fit up and alignment of parent metals. Application of suitably designed fixture help significantly in maintaining fit up and alignment especially during welding of thin sheets of dissimilar metals.

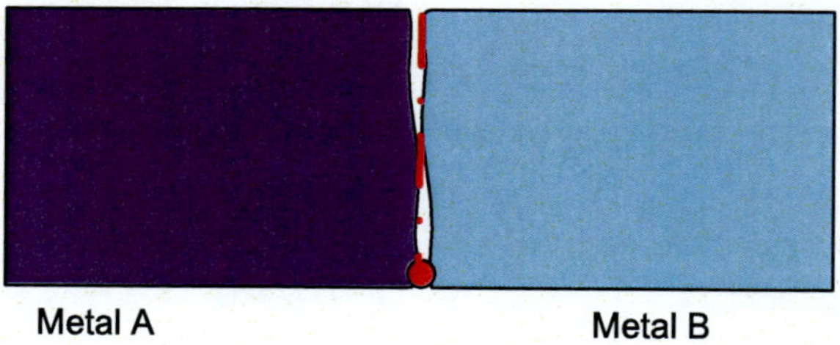

Fig. 3.9 Schematic showing issues in laser welding joint of dissimilar metals due to poor fit-up and misalignment

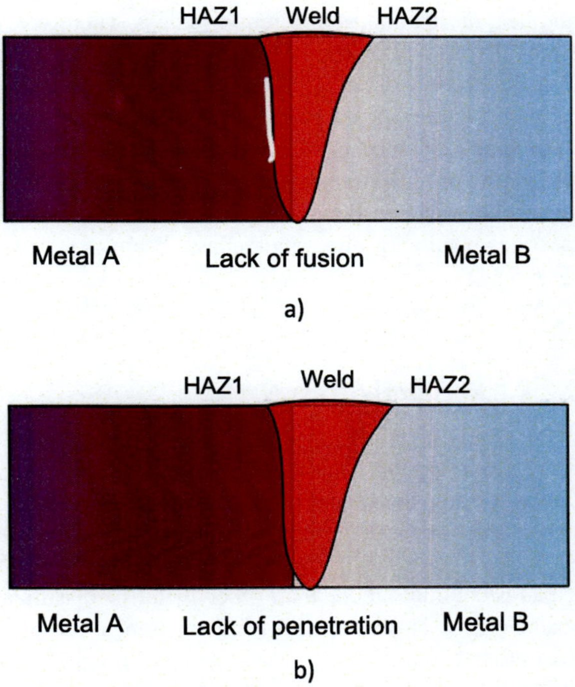

Fig. 3.10 Schematic showing issues in laser weld joint of dissimilar metals **a** lack of fusion and **b** lack of penetration

3.2.6 Lack of Penetration and Fusion

Welding of few highly laser reflective and high thermal conductivity metals (Al, Cu) does not only cause problems of lack of penetration and fusion but also low thermal efficiency of laser welding process itself. Laser beam applied on the parent metal needs to get absorbed to provide heat enough for melting the parent metal. The reflection of high proportion of laser beam results in limited heat generation and high thermal conductivity dissipates the heat rapidly to underlying parent metal. Therefore, high thermal conductivity coupled with reflection in laser welding makes fusion and penetration up to desired depth difficult (Fig. 3.10). The laser welding of thick sections of high thermal conductivity/reflectivity metals makes it further difficult to initiate and maintain keyhole. Differential thermal conductivity and laser reflectivity in dissimilar metal joining by laser welding therefore pose many challenges.

3.2.7 Asymmetric Weld

The differences in thermo-physical properties, namely melting temperature, thermal conductivity and laser absorptivity of two parent metals of dissimilar combination, lead to not just differential heat generation in two metals but also different melting and heat dissipation to different distances. Therefore, a large variation in fusion of edges of the two dissimilar metal components takes place, which in turn results in asymmetric weld (Fig. 3.11a). In case of excessive asymmetry, weld may miss through thickness penetration at the centreline of the weld. Few suggested approaches to develop a symmetric laser weld joint in dissimilar metal joining includes: selective preheating of a parent metal, applying suitable laser absorbing/reflective coating on surface of a parent metal, roughening of surface of a parent metal, off-setting the laser beam suitably, directing the laser beam as per need at the desired location, etc. (Fig. 3.11).

3.2.8 Residual Stresses

The expansion and contraction of the metals due to weld thermal cycle imposed during welding depends on the thermal expansion coefficient of the metals of dissimilar combination. Metals exhibit different thermal expansion coefficient ranging from 8 to 25 μm/m/°C (8–25 $\times$ 10^{-6}/°C). Metal of high thermal expansion coefficient expands/contracts more than other metal of dissimilar metal combination during laser welding. However, laser welding, being a comparatively low heat input process than other fusion welding processes (like gas, arc, plasma welding), results lesser differential expansion and contraction-related issues.

The residual stress (tensile/compressive) developed in a dissimilar metal laser weld joint may be good or bad from the mechanical performance point depending upon the type of externally applied stress during the service (Fig. 3.12). Presence of tensile residual stress in weld joint deteriorates the tensile strength and fatigue strength under zero-tension tension–tension, compression-tension loading condition while compressive residual stress improve the tensile and fatigue strength under the said loading conditions. Still weld zone and near fusion boundary region of HAZ experience residual tensile stress making weld joint more prone to cracking and distortion as per stiffness and metallurgical transformation experienced by the parent metals in weld and HAZ. High heating and cooling rate experienced by the weld and HAZ of hardenable steel may exhibit cracking due to embrittlement and hydrogen induced cracking. Similarly, dissimilar metal combination forming the hard and brittle IMCs can also cause cracking in weld zone.

Further, post weld heat treatment of dissimilar metal weld joint to improve the structure and properties may result in a new set of residual stress due to differential expansion and contraction experienced by two parent metal and weld metal during heat treatment as per heat treatment cycle applied. These stresses mostly tend to localize at the parent metal A—Weld metal—Parent metal B intersection. Therefore, mechanical methods of residual stress relieving namely shot peening, ultrasonic vibrations are considered to be more useful in case of dissimilar metal weld joints. Application of soft filler (if acceptable) having yield strength lower than parent metals reduces residual stress in the weld joint and related issues.

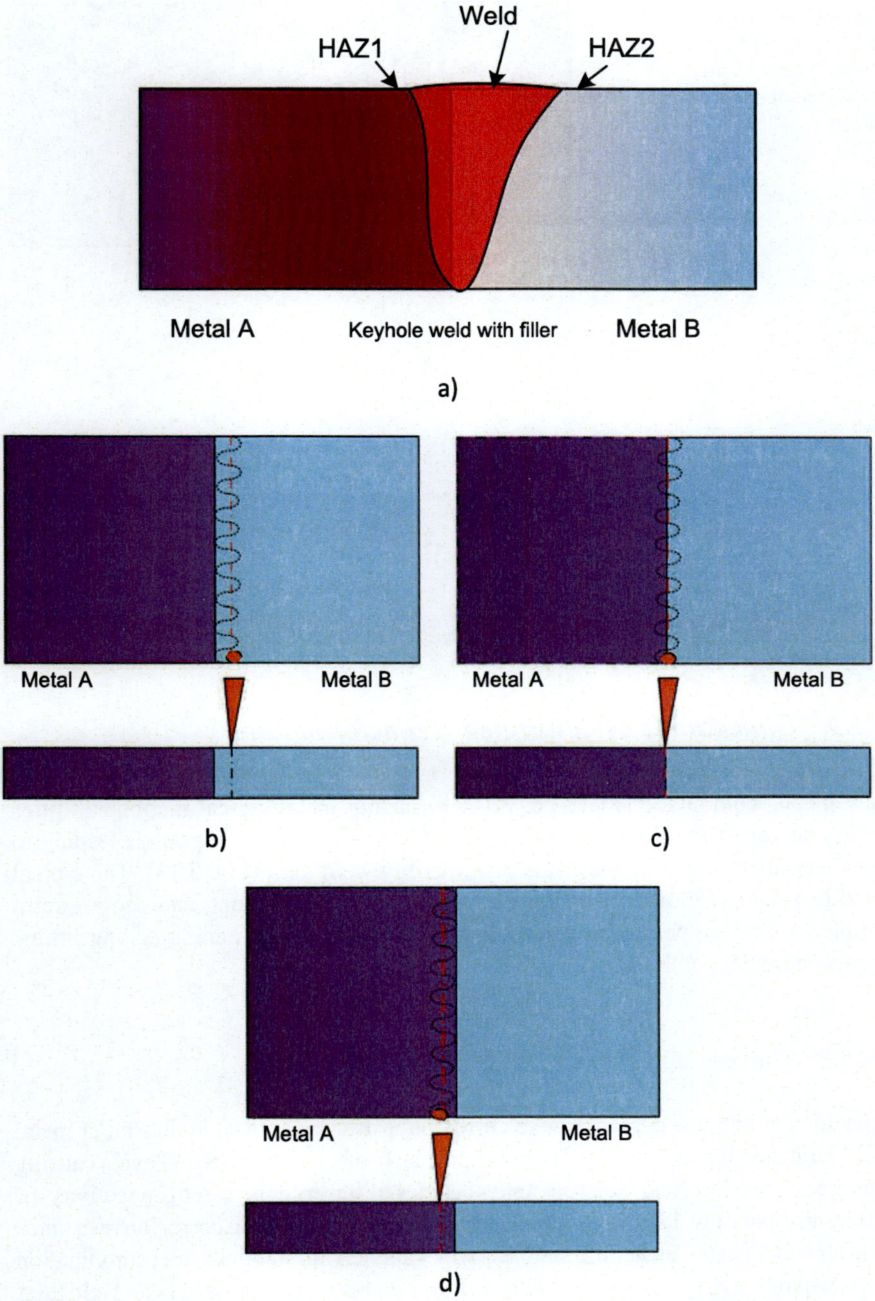

Fig. 3.11 Schematic showing laser-weld joint of dissimilar metals **a** asymmetric weld and **b–d** various possible locations (with respect to weld centreline) where laser beam can be focused to realize the symmetric weld as per thermo-physical properties of the parent metals

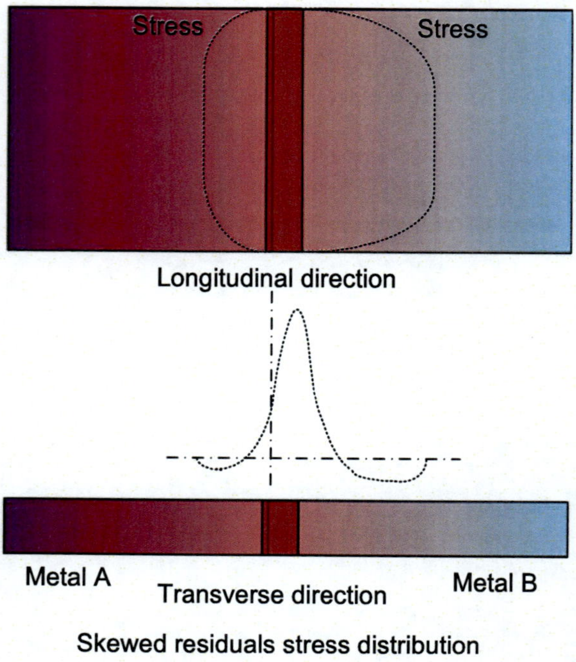

Fig. 3.12 Schematic showing skewed residual stress distribution in laser weld joint of dissimilar metals

3.2.9 Undesirable Metallurgical Transformations

Few metals combinations (Al–Fe, Pb–Mg) having metallurgical incompatibilities on interactions form complex, hard and brittle intermetallic compounds leading to crack-sensitive, low toughness and low ductility weld joint (Fig. 3.13). The typical weld thermal cycle of laser-welding involving very high heating and cooling rates coupled with high peak temperature can cause embrittlement and cracking in few dissimilar metal combination.

3.2.9.1 Corrosion Resistance

Like arc welding, the degradation in corrosion resistance noticed in dissimilar metal weld joint developed by laser welding is primarily due to weld thermal cycle causing chemical, metallurgical heterogeneity besides difference in electro-negativity of different zones of weld joint. A typical example of degradation in corrosion resistance of laser weld joint of austenitic stainless streel and ferritic stainless steel combination is shown in Fig. 3.14. However, the loss of the corrosion resistance of laser weld joint is somewhat lesser than arc weld joint because of limited thermal damage in case of laser welding.

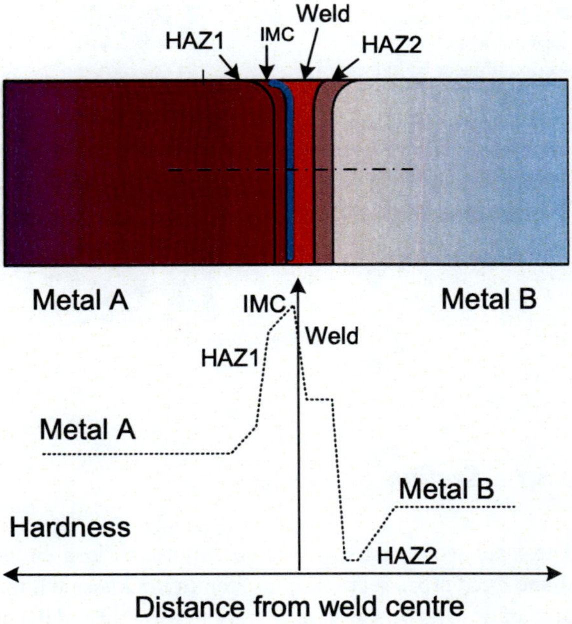

Fig. 3.13 Schematic showing hardness distribution in along with different zones across the laser weld joint of dissimilar metals

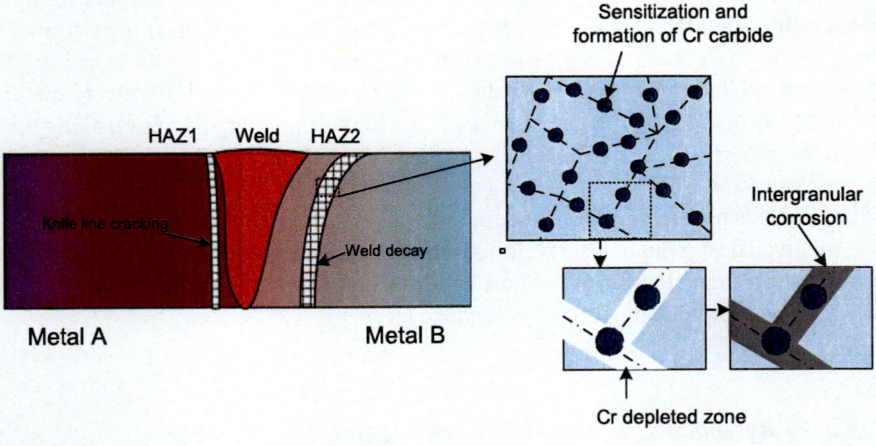

Fig. 3.14 Schematic showing knife line cracking and weld decay issues of HAZs of laser weld joint of dissimilar stainless steels due to chromium carbide formation

3.3 Weldability by Laser Welding

The weldability of dissimilar metals by laser welding depends on many factors, namely plate section, joint design, physical and chemical properties of parent metals of dissimilar combination and process parameters of laser welding.

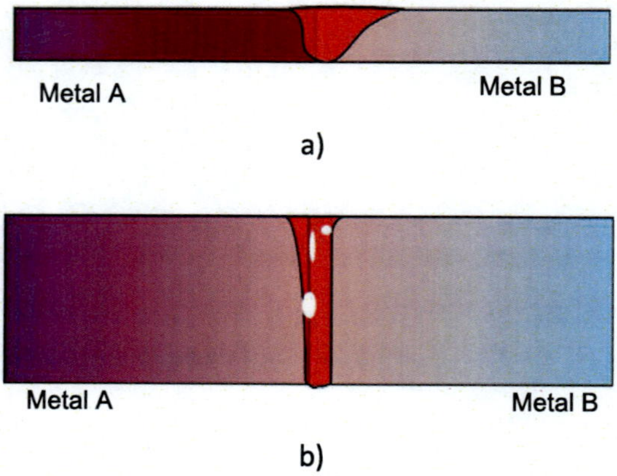

Fig. 3.15 Schematic showing weld bead cross section of laser weld joint of dissimilar metals of different section thicknesses **a** thin sheet welded using melt in mode and **b** thick plates welded using key-hole mode

3.3.1 *Section Thickness*

The laser welding using keyhole approach in a single pass can produce almost I shape weld cross section up to 25 mm thickness without any need of V groove edge preparation; however, square, smooth, straight and flat edge preparation are important for laser welding. The presence of gap along the weld centreline during welding of thick section results in multiple discontinuities such as misalignment, under filling, concave bead, undercut. Laser welding of very thin sheet (< 0.5 mm) is performed using melt-in (conduction) mode needs proper fixtures to maintain the position of sheets, especially in autogenous welding, as there is no additional filler/weld metal available to compensate the gaps/issues related to misalignment and fit up along the weld centreline (Fig. 3.15). The maximum thickness joined by laser welding depends on type of metal. The ferrous metals can be welded up to about twice the thickness of non-ferrous metals. Further, increase in thickness of section increases the cooling rate for given heat input of laser welding, which in turn decreases (already very short) solidification time significantly. Short solidification time in turn increases the gas, inclusion and air pocket entrapment tendency.

3.3.2 *Physical Properties of Parent Metal*

Thermo-physical properties, namely melting point, boiling point, solidification temperature range, linear thermal expansion coefficient, thermal conductivity and laser beam absorptivity all affect the weldability of the dissimilar metal joining by laser welding. Differential melting temperature, thermal conductivity, absorptivity and specific heat of parent metals encourage asymmetric weld while boiling temperature difference promotes the porosities in the weld metal. Difference in thermal conductivity and thermal expansion coefficient causes residual stress, distortion and cracking tendency.

3.3.3 *Metallurgical and Chemical Properties of Parent Metals*

The issues related to the fusion welding of dissimilar metal combination primarily governed by solubility of two parent metals in liquid and solid states, which in turn depends on their crystal structure, valency, size of atom and chemical affinity. In general, increase of solid solubility between two metal increases the ease of welding. Metals (Ni, Cu) having perfect solubility in both liquid and solid state are easy to weld even by fusion welding processes. Metal combinations (Pb–Sn, Si–Cu, Al–Si) having almost nil solubility in solid state during laser welding form fine eutectic structure in weld zone, and these offer reasonably "fair ease" of welding. Metals having partial solid solubility can offer good to fair ease of welding while taking proper precaution for welding like preheating, filler metal, controlled heat input at the desired location. Dissimilar metal combinations like Al–Fe, Fe–Ti, Fe–Cu, Al–Cu forming complex, hard and brittle intermetallic compounds lower the weldability significantly.

However, the development of hard and brittle undesirable IMCs and micro-constituents in weld metal can be minimised by adjusting the weld metal composition using suitable filler metal considering the estimated dilution during the welding. The factors need to be considered for selection of filler metal for laser welding are similar to that of dissimilar metal joining by arc welding process as described in Chap. 2. Filler metal can be used to isolate the one or both the parent metals of dissimilar combination by buttering layer (Fig. 3.16). Over-matching filler metal (with respect to both the parent metals) should be avoided to control the residual stress and related issues. Moreover, the filler metal having the minimum acceptable strength (considering application/service requirement) lower than both the parent metals can also be used. For example, the filler metal C having minimum acceptable strength of 420 MPa can be used to laser weld two metal A and B say having strength of 650 MPa and 520 MPa, respectively.

3.4 Few Approaches for Joining of Specific Dissimilar Metal Combinations

The dissimilar metal joining helps to exploit the specific properties related to each of parent metals for improved efficiency and performance of the engineering system of automotive, thermal power, aerospace and electronics industry. The joining aluminium with alloy steel, stainless steel and copper is frequently needed. Similarly, dissimilar metal joining of copper with steel, stainless steel and aluminium is also needed.

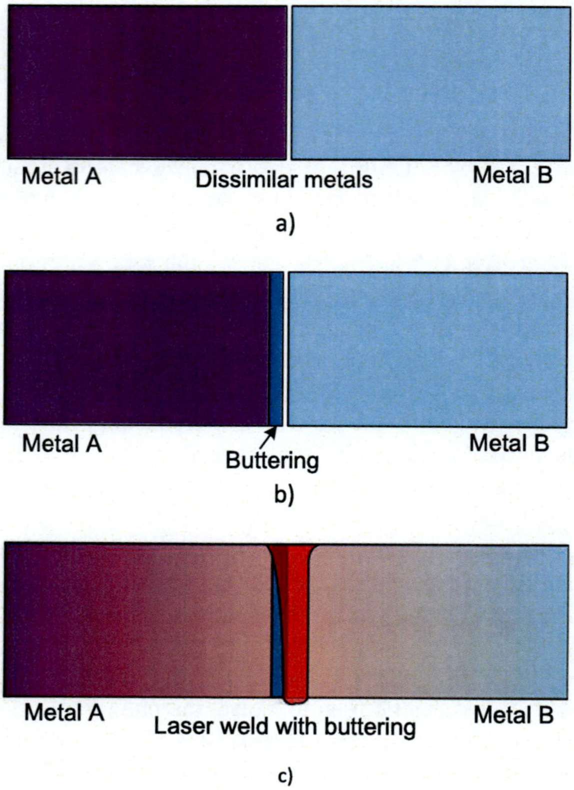

Fig. 3.16 Schematic showing the steps for reducing issues related to metallurgical incompatibilities of dissimilar metals to be laser-welded **a** plates to be welded, **b** applying butter layer in one of the parent metal to be isolated from weld metal and **c** developing laser weld

3.4.1 *Joining of Al with Other Metals*

Dissimilar metal joints of aluminium and steel can be developed by fusion welding using buttering layer of zinc on steel to deal with issues caused by difference in thermo-physical properties (melting point, thermal conductivity and thermal expansion coefficient) of two metals. Zn shows compatibility with aluminium. Dissimilar metal joining of aluminium and stainless steel is achieved using buttering layer of pure aluminium on the stainless steel. Joining of aluminium and copper combination is achieved using interlayer/filler of Al–Cu alloy (Fig. 3.17).

3.4.2 *Joining of Cu with Other Metals*

Dissimilar metal joining of copper with mild and alloy steel, stainless steel is achieved using two approaches depending upon the section thickness. Thin section dissimilar metal joint of Cu and steels is developed using high copper filler with minimum melting of steel side to avoid the formation of hard and brittle IMC. While thick

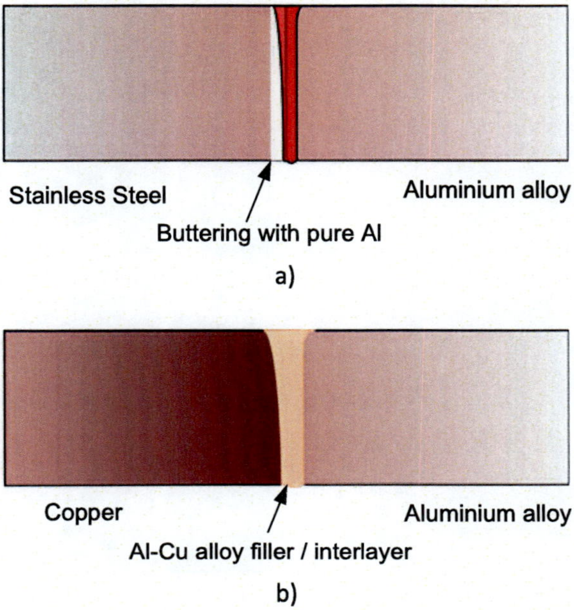

Fig. 3.17 Schematic showing laser welding of aluminium with steel and copper using different butter layer/interlayer

sections need isolation of the one of the parent metals, either copper side by buttering using high nickel alloy followed by welding or steel side by buttering using high copper alloy followed by welding (Fig. 3.18).

3.5 Few Advances on Laser Welding and Brazing

3.5.1 Laser Welding of Alloy Steel and Stainless Steel

Thermal power plant components operating at temperature above 550–600 °C are designed using heat-resistant austenitic stainless steel while those subjected at lower temperature are made of ferritic alloy steels. Therefore, it is frequently needed to weld ferritic steel and austenitic stainless steel (AISI 304, 316) instead of making entire system of austenitic stainless steel due to economic reasons. However, dissimilar nature of these two metals in respect of mechanical and chemical, thermo-physical properties (melting temperature, thermal conductivity and thermal expansion coefficient) poses many issues during the welding and subsequently during service, especially at high temperature. The common problems encountered during laser welding include asymmetric weld, solidification cracking, embrittlement and cold cracking of weld and HAZ due to martensitic transformation, sensitization, unmixed zone formation and segregation of alloying elements. While prolong high temperature exposure of such dissimilar metal joints during service results in SCC, corrosion,

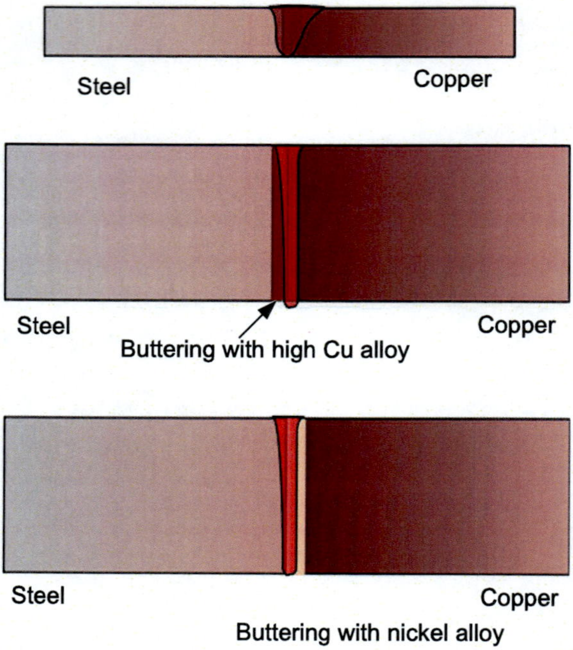

Fig. 3.18 Schematic showing laser welding of steel with copper using different butter layer/interlayer

cracking of HAZ of austenitic stainless steel (ASS) due to carbon migration, weakening of ferritic steel due to grain coarsening and depletion of alloying elements, and thermal fatigue (Krishnaja et al. 2018; Dwivedi 2022).

Due to low heat input associated with high power density of laser welding process, very high cooling rate is imposed in weld and HAZ. The high cooling rate reduces the width of zone of both austenite and ferrite formation for given Cr and Ni equivalent, and increases tendency of unmixed zone formation and segregation of alloying elements in weld metal (Fig. 3.19). Appropriate application of focussed laser beam on austenitic stainless steel side reduces the dilution of weld from ferritic steel side and width of HAZ as well. Further, net heat input by laser welding can be 10–20 lesser than other common arc welding processes such as GTAW and PAW leading to reduction in width of HAZ and shrinkage by 4–6 times besides narrowing zone of weld joint experiencing the reduced residual stresses. The narrowing of austenite–ferritic zones in Schaeffler Diagram due to high cooling during laser welding causes higher solidification cracking than arc welding primarily due to reduced ferrite formation (Fig. 3.20). Presence of 3–5% ferrite in austenitic weld metal avoids solidification crackling of the weld metal.

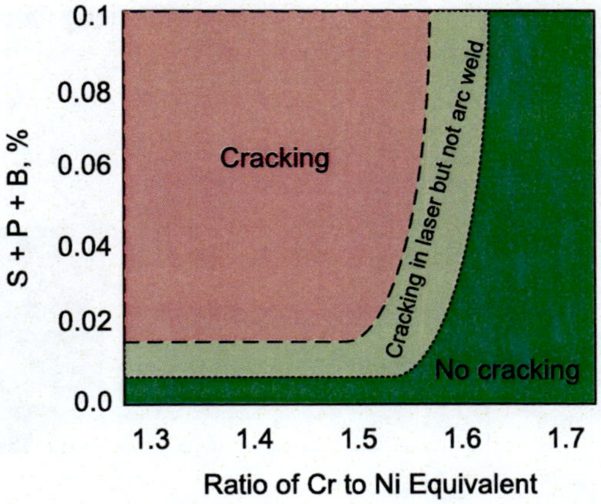

Fig. 3.19 Plot showing the effect of S, P and B impurities and Cr/Ni equivalent ratio on solidification cracking tendency of austenitic stainless steel laser weld joints

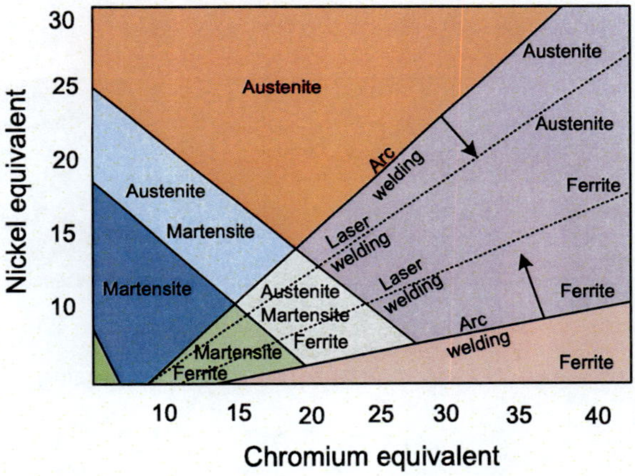

Fig. 3.20 Schaeffer diagram showing effect of high cooling rate during laser welding on shrinking of austenite-ferrite zone

3.5.2 *Laser Welding of Stainless Steel and Copper*

The copper and stainless steel combination is used in marine, mining, and electrical industries to exploit high corrosion resistance, high thermal and electrical conductivities of copper and excellent corrosion resistance and stiffness of stainless steel. Two issues are commonly encountered during fusion welding of steel–copper dissimilar metal combination, namely cracking of steel and IMC formation ($FeCu_4$) in the weld metal. Cracking in steel–copper dissimilar metal joining takes places due to diffusion of copper into the steel grain boundary during fusion welding which in turn makes it crack sensitive in presence of tensile residual stresses (Fig. 3.21). Further, the usage of nickel and silver brazing alloy as filler metal reduces the IMC-related issues due to good mutual solid solubility and metallurgical compatibility.

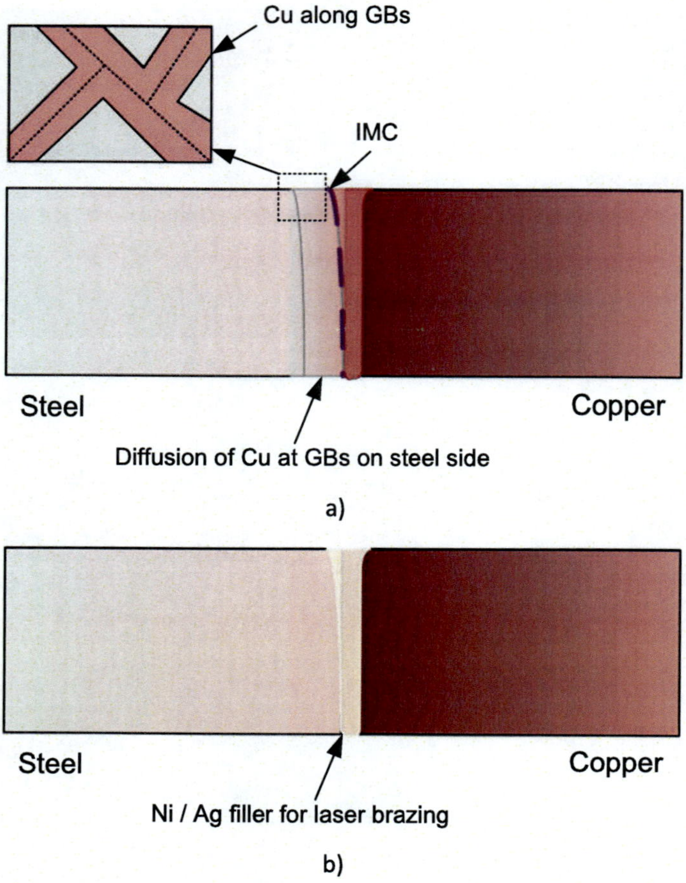

Fig. 3.21 Schematic showing laser weld joints of steel with copper **a** cracking and IMC formation and **b** reducing issues using suitable filler

3.5.3 Laser Welding of Alloy Steel/Stainless Steel and Aluminium

Joining of stainless and aluminium is frequently needed to exploit the attractive properties of both metals such as lightweight, thermal conductivity, stiffness, corrosion resistance, retention of impact resistance even in cryogenic conditions for spacecraft applications. Automotive industry also using alloy steel and aluminium combination to take advantage of stiffness and lightweight. Evaporation of Al and formation of Al-rich Al–Fe IMC are two common issues encountered during laser welding of Al and steels. Appropriate application of focussed laser beam on aluminium side reduces the dilution from steel and so IMC formation is minimized (Meng et al.

2020). Further, the application of filler metal/interlayer of Ag, Cu, Ni helps to form more favourable IMCs.

The thickness of IMC layer in Al-Steel welding must be optimum and uniform across the section thickness for maximum joint efficiency. The thickness of IMCs is determined by the reaction kinetics, which in turn depends on temperature of weld metal. The variation in temperature of the weld pool in vertical (thickness) and horizontal (radially) direction is inherent feature of any fusion welding process. Maximum temperature at top surface while minimum found at the root. Accordingly, thicker IMC layer is formed (say 10 μm) at the top zone of steel-weld metal interface than the root zone (say 1.9 μm). A large variation in IMC layer thickness triggers the fracture from minimum IMC thickness zone. An oscillation of the laser beam during welding reduces variation in IMC layer thickness across the section thickness at the steel-weld metal interface, which in turn increases the load carrying capacity of the weld joint significantly (Fig. 3.22).

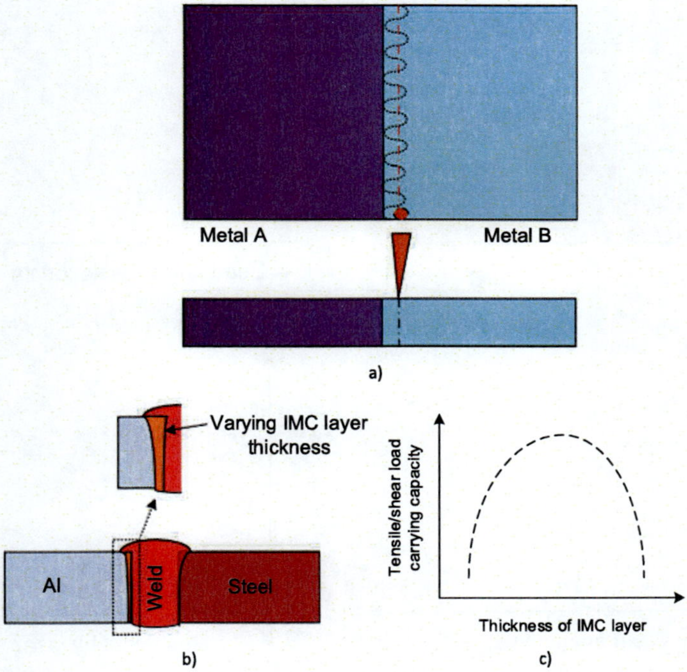

Fig. 3.22 Schematic showing laser welding of steel with aluminium **a** controlled movement of laser weld with suitable offsetting, **b** cross section of Al-steel weld joints and **c** effect of thickness of IMC layer on joint strength (Meng et al. 2020)

3.5.4 Laser Brazing of Aluminium and Steel

During laser brazing of Al-steel using suitable Al and Zn filler metal, an increase of laser scanning reduces the heat input, which in turn decreases the weld length and IMCs thickness layer of lap joint (Fig. 3.23a). Zn filler metal increases the fluidity of molten metal that results in longer weld length and higher load carrying capacity than Al filler. Increase of heat input increases the fluidity and solidification time and so reaction time at filler metal and steel interface. Therefore, increasing heat input of laser brazing increases the weld length and IMCs thickness and decreases the bead angle (Fig. 3.23b, c). A combination of low bead angle and longer brazed length offers high tensile/shear load carrying capacity of laser-brazed joints of dissimilar metals. It is important to develop an optimum IMCs thickness layer to achieve maximum joint strength for a given dissimilar metal combination (Fig. 3.23d).

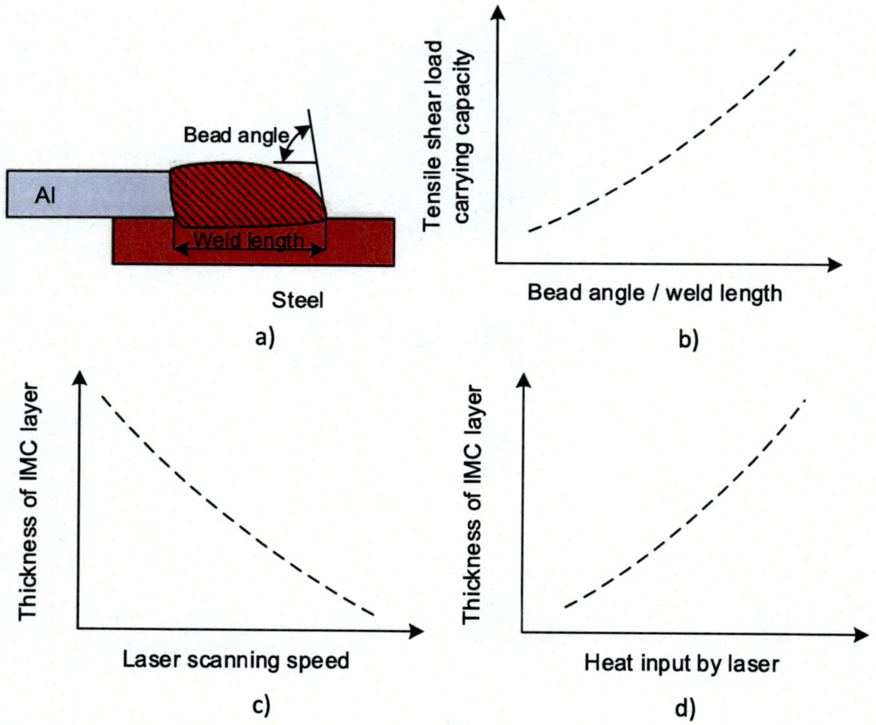

Fig. 3.23 Schematic showing laser brazing of steel with aluminium **a** lap joint, **b** effect of bead length and bead angle on joint strength, **c** effect of laser scanning speed and **d** heat input on thickness of IMC layer

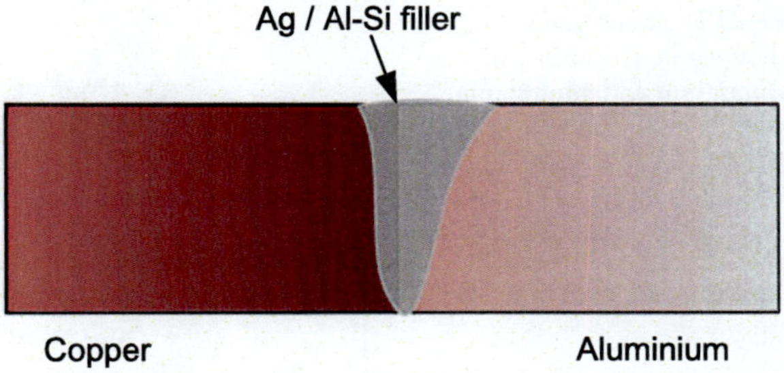

Fig. 3.24 Schematic showing laser welding of copper with aluminium using Ag/Al–Si filler

3.5.5 *Laser Welding of Copper and Aluminium*

The manufacturing of heat exchangers, electrical machines commonly use joining of copper with aluminium to take advantage of high thermal and electrical conductivities. The difference in thermo-physical properties of two metals results in asymmetric weld while metallurgical incompatibilities cause hard and brittle IMC formation in fusion weld joint. An application of suitable filler metal/interlayer of solders Ag, Al-Si filler metal reduces cracking tendency of Al and Cu laser weld joints (Fig. 3.24).

3.5.6 *Laser Welding of Aluminium and Aluminium Matric Composite*

The laser welding of aluminium 6061 alloy with AMC (reinforced with TiB_2) results in the formation of Al_2Ti, Fe_2Si and $Al_{0.5}Fe_3Si_{0.5}$ TiB_2 particles are decomposed followed by reaction with molten Al. TiB_2 particles observed at the grain boundaries in weld metal region (Fig. 3.25). The corrosion resistance of Al-AMC is therefore compromised in order of weld metal > 6061 Al > AMC (Dai et al. 2019). Ease of joining of various dissimilar metal combinations by laser welding is shown in Table 3.1.

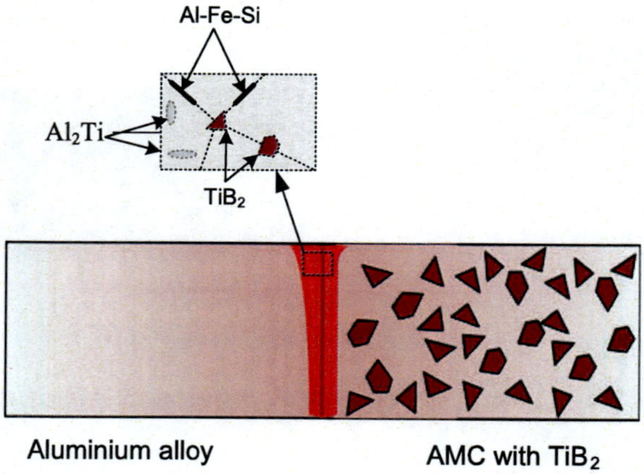

Fig. 3.25 Schematic showing laser welding of aluminium with aluminium matric composite (Dai et al. 2019)

Table 3.1 Laser weldability of binary metal combinations

	W	Ta	Mo	Cr	Co	Ti	Be	Fe	Pt	Ni	Pd	Cu	Au	Ag	Mg	Al	Zn	Cd	Pb
Ta	E																		
Mo	E	E																	
Cr	E	P	E																
Co	F	P	F	G															
Ti	F	E	E	G	F														
Be	P	P	P	P	F	P													
Fe	F	F	G	E	E	F	F												
Pt	G	F	G	G	E	F	P	G											
Ni	F	G	F	G	E	F	F	G	E										
Pd	F	G	G	G	E	F	F	G	E	E									
Cu	P	P	P	P	F	F	F	F	E	E	E								
Au	*	*	P	F	P	F	F	F	E	E	E	E							
Ag	P	P	P	P	P	F	P	P	F	P	E	F	E						
Mg	P	*	P	P	P	P	P	P	P	P	P	F	F	F					
Al	P	P	P	P	F	F	P	F	P	F	P	F	F	F	F				
Zn	P	*	P	P	F	P	P	F	P	F	F	G	F	G	P	F			
Cd	*	*	*	P	P	P	*	P	F	F	F	P	F	G	E	P	P		
Pb	P	*	P	P	P	P	*	P	P	P	P	P	P	P	P	P	P	P	
Sn	P	P	P	P	P	P	P	P	F	P	F	P	F	F	P	P	P	P	F

E = excellent, *G* = good, *F* = fair, *P* = poor, * = no data available

References

Dai J, Yu B, Jiang W, Htun HM, Liu Z (2019) Laser welding of dissimilar metal joint of 6061 Al alloy and Al matrix composite. Adv Mater Sci Eng 2019:6, Article ID 2359841
Dwivedi DK (2022) Fundamentals of metal joining. Springer
Krishnaja DR, Cheepu M, Venkateswarlu D (2018) A review of research progress on dissimilar laser weld-brazing of automotive applications. IOP Conf Series: Mater Sci Eng 330(2018):012073
Meng Y, Gong M, Zhang S, Zhang Y, Gao M (2020) Effects of oscillating laser offset on microstructure and properties of dissimilar Al/steel butt-joint. Opt Lasers Eng 128:106037
Sun Z, Ion JC (1995) Laser welding of dissimilar metal combinations. J Mater Sci 30:4205–4214

Chapter 4
Dissimilar Metal Joining by Solid-State Joining Technologies

This chapter presents fundamentals of solid-state joining processes such as friction welding, friction stir welding, impact welding and diffusion bonding considering joining of dissimilar metals. Further, mechanism involved in solid-state joining, process parameters, common features of joint interfaces, microstructural transformation and discontinuities, formation of heat-affected zone have been elaborated. The different variants of friction stir welding, impact welding and diffusion bonding are described.

4.1 Introduction

The solid-state joining processes primarily use compressive force/pressure involving macro/micro-scale deformation with or without the application of external heat to develop a metallic joint between similar and dissimilar metal combination. The unique features such as deformation/forging action at the joining interfaces, almost absence of melting of parent metals coupled with low heat input, make the solid-state joining processes very attractive for dissimilar metal joining. Most of the solid-state joining processes are autogenous; however, some of these allow the application of filler in the form of film, interlayer, buttering layer, etc. that makes these processes furthermore suitable for dissimilar metal joining. Application of interlayer/filler/film between the metallurgically incompatible parent metals reduces the issues related to differences in physical, mechanical, chemical and metallurgical properties of parent metals in dissimilar metal joining which not only improves the joint performance but also ease of dissimilar metal joining. Low heat input and absence of fusion of faying surfaces of parent metals significantly reduce the (a) width of heat-affected zone, (b) extent of differential thermal expansion and contraction of parent metals, (c) residual stress and distortion tendency, (d) the dilution and intermixing of incompatible parent metals and (e) formation of hard and brittle intermetallic compound. Further, inherent feature of macro-/micro-plastic deformation and forging action of solid-state joining

© The Author(s), under exclusive license to Springer Nature Singapore Pte Ltd. 2023
D. K. Dwivedi, *Dissimilar Metal Joining*,
https://doi.org/10.1007/978-981-99-1897-3_4

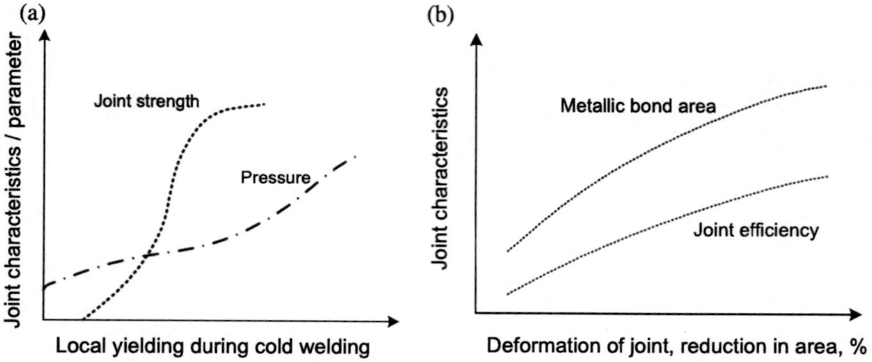

Fig. 4.1 Schematics showing effect of interfacial deformation/local interfacial yielding as a function of bonding pressure on **a** joint strength, **b** bonding area and joint efficiency

processes results in one or more of the following (a) cleaning by fracturing brittle surface oxides, (b) work hardening, (c) grain refinement, (d) mechanical interlocking and (e) elimination of defects. For example, in cold welding, interfacial yielding depends on joining pressure applied that facilitates metallic continuity and joining which in turn directly affects the joint strength. The extent of the bonding (bond area) determined the load carrying capacity and joint efficiency. Some minimum pressure must be applied to ensure the joint formation during cold welding (Fig. 4.1).

4.2 Mechanism(s) of Solid-State Joining Processes

The solid-state joining processes based on the approach can be grouped as metallurgical, mechanical, adhesive, interlocking and hybrid joining. Metallurgical joining processes include fusion joining, solid-state joining (involving friction, impact, diffusion) and brazing/soldering. Fusion and solid-state joining processes sometimes offer joint strength even higher than the parent metal without appreciably raising the weight of assemblies. However, fusion joining processes suffer from various issues like solidification related defects, wider HAZ, hard and brittle IMC formation, high residual stress and distortion tendency. Therefore, solid-state joining processes based on friction (rotary/linear, friction stir and ultrasonic), impact (explosive, electromagnetic force) and diffusion (roll and diffusion bonding) suit better for dissimilar metal joining. Because either these processes do not involve any fusion or if it occurs, it is very negligible at the interface especially when joining is performed beyond the optimized window of process parameters. Therefore, it is important to understand the mechanism(s) leading to the metallurgical bond formation during the solid-state joining of dissimilar metals. Depending upon the location of the weak zone, the fracture of adhesive joint developed in solid state may be triggered from the either of the

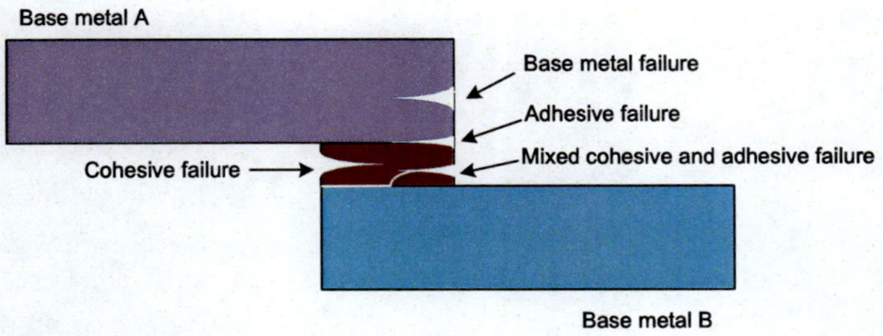

Fig. 4.2 Schematic of dissimilar metal joints developed using adhesive showing different ways by which failure can occur

parent metal, adhesive itself, joint-parent metal interfaces, etc. (Fig. 4.2). Following mechanisms singly or in combination are expected to cause metallurgical joining.

4.2.1 Metallic Bonding

The metallic bonding between two parent metals by solid-state joining, under favourable conditions (surface finish, cleanliness, pressure and temperature), takes place only if mating surfaces come in very atomic level close contact (within a few angstroms) with each other (Fig. 4.3). Mechanical, chemical and metallurgical properties of parent metals in dissimilar metal joining affect the metallic bonding.

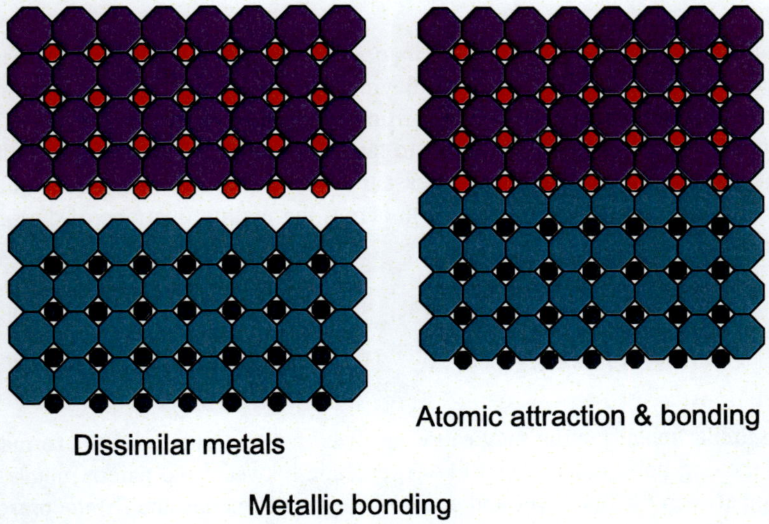

Fig. 4.3 Schematic of metallic bonding in dissimilar metal joining in solid state

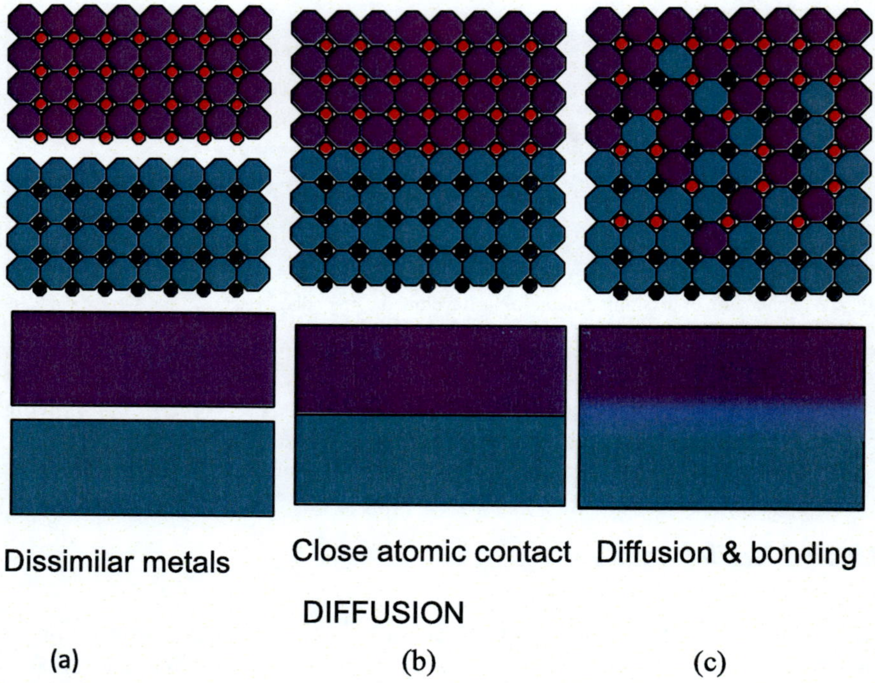

Fig. 4.4 Schematic showing stages of dissimilar metal joining in solid-state through diffusion

4.2.2 Diffusion

The compositional gradient across the mating interface of parent metals in dissimilar metal joining leads to diffusion of alloying elements when direct metal-to-metal contact established at the interface is free from oxides and other impurities at a reasonably high temperature for long enough to facilitate diffusion (Fig. 4.4). This diffusion results in the formation of grain, phase transformation, new micro-constituents, IMCs, recrystallization and grain growth across the mating interface of dissimilar metal joint.

4.2.3 Localized Melting

The dissimilar metal joining may cause localized melting due to (a) the formation of low melting point constituents like (i) eutectic between two parent metals, (ii) eutectic between the interlayer and one/both the parent metals and (b) the presence of one of the parent metal having very low melting temperature than other parent metal. This kind of localized melting is usually unexpected but if occurs, then it can degrade the joint performance (Fig. 4.5).

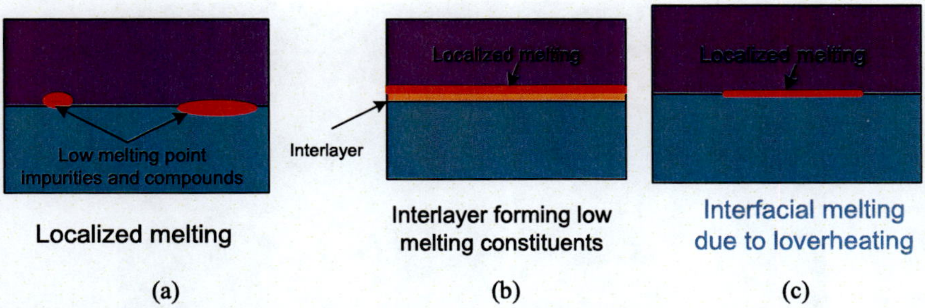

Fig. 4.5 Schematic showing various possibilities in dissimilar metal joining involves localized interfacial melting

4.2.4 Dynamic Recrystallization

The solid-state joining processes involving severe plastic deformation of surface layers of parent metals (which would be different in dissimilar metal joining) coupled with high enough friction and deformational heat generation cause dynamic recrystallization during the joining process itself. The localized micro-/macro-scale deformation of surface layers of parent metals facilitates the recrystallization at the interface producing new fine grains (Fig. 4.6).

Fig. 4.6 Schematic showing stages of dissimilar metal solid-state joining involving recrystallization

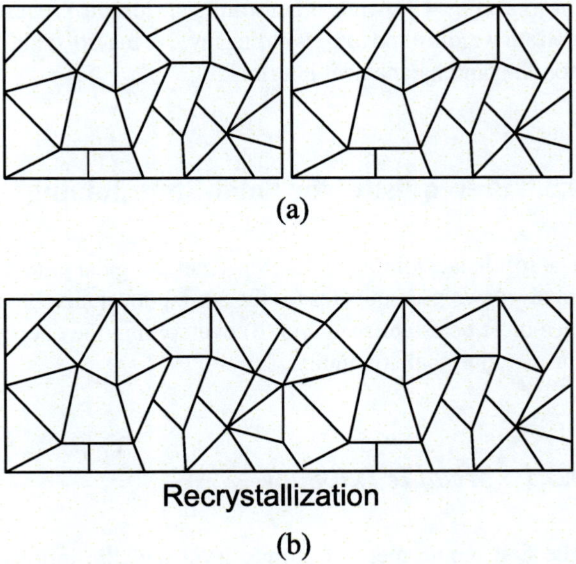

Mechanical interlocking

Fig. 4.7 Schematic dissimilar metal solid-state joining involving interfacial mechanical interlocking

4.2.5 Mechanical Interlocking

The joining processes like explosive, electromagnetic and ultrasonic welding cause localized micro-scale interfacial deformation by forming a wavy interface. The clean mating surface in combination with very intimate interlocking between the parent metals results in metallurgical bonding and improves the mechanical performance of the joint (Fig. 4.7), while other solid-state joining processes like friction welding, friction stir welding and forge welding involve macro-scale deformation for metallurgical bonding and mechanical interlocking.

4.3 Prerequisites for Solid-State Joining

The solid-state joining to develop a metallurgical joint between two dissimilar metals involves two prerequisites (a) the atomic level metallic intimacy at the mating interface must be established and (b) overcoming the energy barrier across the interface for metallurgical continuity.

4.3.1 Metallic Intimacy

The first requirement of atomic level metallic intimacy between the two dissimilar metals is influenced by the presence of thin surface oxides (20–100 angstrom), surface-contaminant films (moisture, greases, paint, oil, etc.), surface roughness, yield strength and ductility of parent metals (Fig. 4.8). Metallic intimacy is mandatory to develop a metallurgical bonding through atomic diffusion across the interface.

Fig. 4.8 Schematic showing various surface impurities cleaned before solid-state joining of dissimilar metals

However, the cleanliness of mating surfaces in respect of surface oxides during solid-state joining in turn depends on (a) the extent of differential micro-level surface layer plastic deformation of two parent metals, (b) relative hardness of oxides and that of two parent metals and (c) mechanical properties of oxides. An increase of (a) differential surface deformation at mating the surfaces of two dissimilar metals as per their stacking fault energies and work hardening tendencies, (b) differential hardness of oxides and their parent metals and (c) brittleness of oxides increases fragmentation of surface oxides which in turn encourages the metal to metal between two parent metals. The degree of fragmentation of surface oxides during solid-state joining, therefore, depends on the extent of interfacial movement (shear deformation) of the two mating surfaces of dissimilar metals under pressure apart from the characteristics of surface oxides and both the parent metals. Rough surfaces reduce metallic intimacy at the mating interface due to localized contacts between parent metals here and there. Shear deformation of peaks and valleys present on rough surfaces of soft and ductile parent metal during solid-state joining increases the metallic intimacy.

Further, surface contaminants like moisture and organic compounds must be removed using combined chemical and mechanical cleaning processes like wire brushing and acid pickling. Therefore, all factors related to parent metals, joining process, surface cleaning, heat generation, etc. affect the metallic intimacy, which in turn determine the ease of dissimilar metal welding by solid-state joining processes.

4.3.2 Overcoming the Energy Barrier

There are multiple theories including energy requirement for diffusion, recrystallization and misorientation of atoms at mating interfaces between two dissimilar metals that have been proposed to explain the precise mechanism responsible for metallurgical continuity during solid-state joining. According to the first two theories, solid-state joining must overcome the energy barrier for diffusion and recrystallization by developing or imposing favourable weld thermal cycles (high temperature for long enough duration) to establish a metallurgical bond between two dissimilar metals. However, those two are not fully accepted by welding technologists because researchers have also developed solid-state joints of aluminium, copper, etc. by putting them under pressure with close metallic contact at liquid nitrogen temperature. Because diffusion and recrystallization under such subzero conditions of solid-state joining are not expected to play any major role in metallurgical bonding. Therefore, it was suggested that the energy barrier comes from the misorientation of the crystal of two dissimilar metals at the contact interface.

Summary of mechanism

Technically, the direct metal-to-metal contact without any interfacial impurities at mating surfaces of the dissimilar metals to be joined under pressure for a long time enough is the only requirement for developing a metallurgical joint. Moreover, other aspects like micro- or macro-scale surface layer deformation, impact, heat generation due to friction and/or deformation and conversion of kinetic energy into thermal energy that are process-specific features affect the metallic intimacy. Because these aspects affect the removal of surface impurities (interfacial cleanliness), metallic intimacy, diffusion, metallurgical transformations, recrystallization and grain growth, reversion, IMC formation, softening and hardening of HAZ, localized melting, etc. as per weld thermal cycle experienced by parent metals during solid-state joining. Localized surface layer deformation results in work hardening, grain refinement due to fracture of micro-constituents and dynamic recrystallization, strain-induced metallurgical transformation and easy recrystallization as per metal system. Therefore, all the factors related to parent metals of dissimilar combination like ductility, yield strength, stacking fault energy, thermal softening and work hardening behaviour and process-related features like impact/shear/compressive force, the relative speed of the mating interfaces/tool, etc., external heating/cooling, if any, affecting the deformation and strain at mating surfaces. These interfacial changes in turn govern the characteristics and the performance of dissimilar metal joints developed by solid-state joining. Heat generation as per weld thermal cycle (peak temperature, high-temperature retention period and cooling rate) imposed during the solid-state joining affects the metallurgical characteristics of dissimilar metal joints in the form of grain and phase structure, metallurgical transformation, reversion, recrystallization and recovery. These metallurgical changes may cause hardening/softening of joint interface and heat-affected zone, which in turn determine the mechanical performance of dissimilar metal joints. The weld thermal cycle of solid-state joining depends on many parent metal characteristics (thermal conductivity, specific heat, section thickness, external heating and cooling, if any) and joining process-related parameters (impact, friction, deformation rate of heat generation, rotational and translational speed, impact velocity, amplitude and frequency, etc.). These aspects as per joining process and dissimilar metal combination are considered while analysing dissimilar metal joint performance (Mohamed and Washburn 1975).

4.4 Solid-State Joining Processes

The solid-state joining processes, namely friction welding, friction stir welding, ultrasonic welding, explosive and electromagnetic welding, diffusion bonding and roll bonding have been presented within the following scope, mechanism, process parameters and dissimilar metal joint characteristics.

4.4.1 Friction Welding

The friction welding process uses a combination of friction and localized deformation of mating surfaces for metallic intimacy, cleaning and thermal softening followed by forging action. The process involves rubbing through controlled rotational or linear relative motion between the mating surfaces under a normal force for dissimilar metal joining. There are two main stages in friction welding: (a) rubbing and (b) forging. Rubbing of mating surfaces under normal force (a) cleans the surfaces by fracturing and removing surface oxides and other impurities, (b) deforms the surface layers to achieve atomic level closeness between the mating surfaces and (c) generates the frictional/deformation heat (Sharma and Dwivedi 2017; Kaushik and Dwivedi 2021a). Once enough cleaning and thermal softening of mating surfaces are realized, the relative movement between mating surfaces is stopped and additional forging force is applied and maintained till the cooling off and development of a joint (Fig. 4.9). Normal force during rubbing and forging, relative speed (rotational/linear), surface condition (roughness, cleanliness), thermo-physical properties, friction pressure and time, forging time and pressure, axial shortening and shortening rate affect the frictional and deformation heat generation and subsequently forging action which in turn dictates the mechanical and metallurgical characteristics of the friction weld joint (Fig. 4.9d).

4.4.1.1 Thermo-mechanical Aspect

The interface of friction weld joint is subjected to maximum temperature and strain due to friction and deformation. The heat is dissipated to underlying parent metals on both sides as per their thermal conductivities. A combination of heat and deformation during friction welding at interfaces of joints results in four zones, namely weld zone, thermo-mechanically affected zone, heat-affected zone and parent metal (Fig. 4.10). Temperature and strain experienced by weld zone are much higher than thermo-mechanically affected zone, while the heat-affected zone experiences changes in

Fig. 4.9 Schematic from **a–c** various stages of rotary friction welding and **d** effect of welding time on axial shortening (reduction in length)

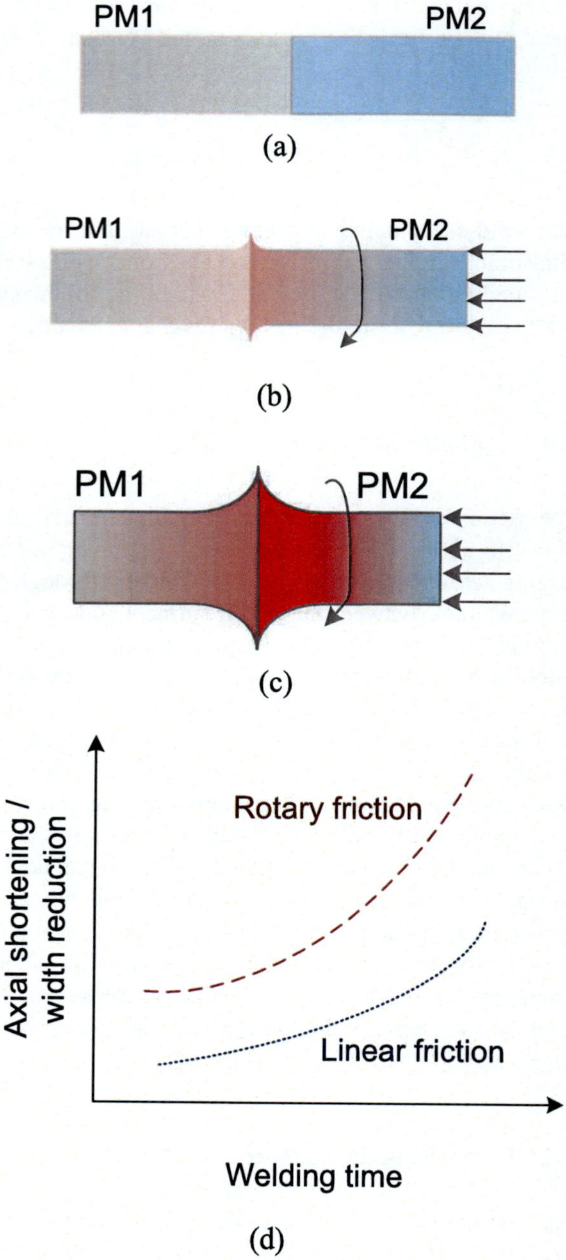

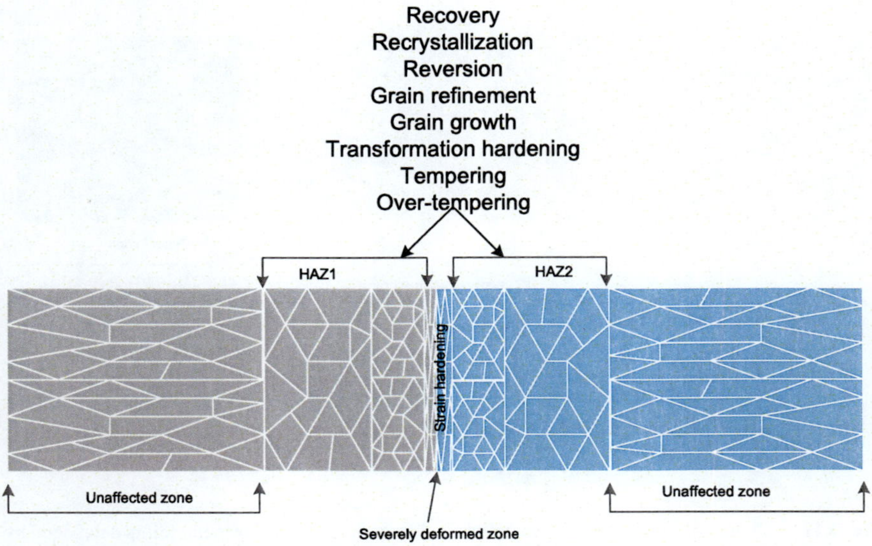

Fig. 4.10 Schematic showing thermos-mechanical aspects of dissimilar metal joints developed by friction welding

mechanical, corrosion and metallurgical characteristics only due to heat generated at the joint interface (Kaushik and Dwivedi 2021a, 2022).

4.4.1.2 Metallurgical Aspect

A weld zone of friction weld joint is produced due to severe plastic deformation coupled with forging action at a high temperature, which in turn results in fine recrystallized grains, and even formation of a thin layer of the intermetallic compound at the interface as per the dissimilar metal combination. The HAZ as per the metal system may show grain growth, recovery and recrystallization, reversion and metallurgical transformation leading to thermal softening/hardening. An increase of friction pressure, time and relative speed increases the heat generation, which in turn results in wider weld and heat-affected zones. However, as per metal system width and properties of thermo-mechanically affected zone, heat-affected zone may differ significantly. The metallurgical changes in the weld zone, TMAZ and HAZ in totality determine the joint performance (Fig. 4.11).

4.4.1.3 Performance of Friction Weld Joints

The weak link can be anywhere from one side parent metal, HAZ1, TMAZ1, WZ, TMAZ2, HAZ2, or other side parent metal as per the response of the metal system to

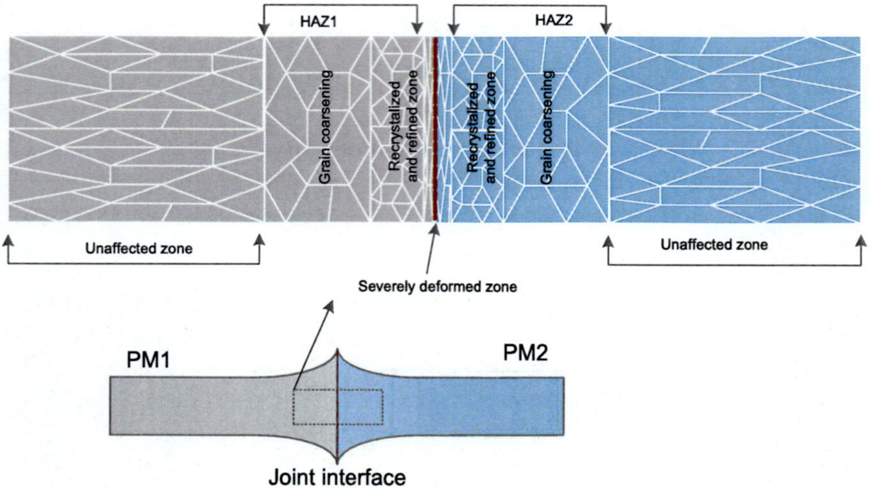

Fig. 4.11 Schematic showing various metallurgical aspects of dissimilar metal joints developed by friction welding

thermo-mechanical stresses imposed during friction welding. Simple micro-hardness distribution across the weld joint interface can suggest a weak location, which would fail under tensile loading (Fig. 4.12). The formation of hard and brittle IMC in the weld zone causes weld failure from the interface itself, while softening of HAZ encourages failure from heat-affected zone.

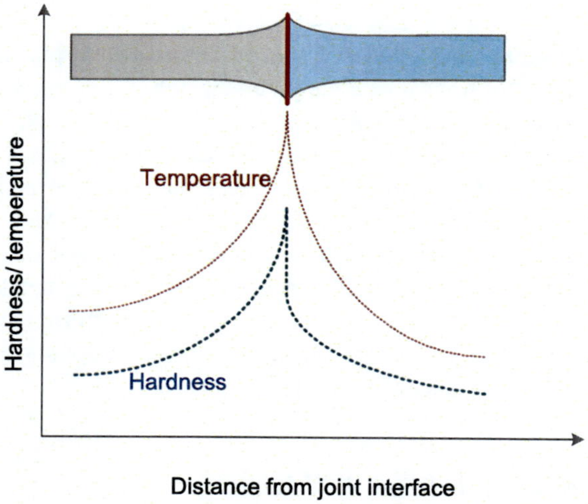

Fig. 4.12 Schematic showing hardness profile of a typical dissimilar metal joint developed by friction welding

The metallurgical transformations in dissimilar metal friction weld joint at weld zone and HAZ including IMCs formation lower the resistance to all galvanic and pitting corrosion and stress corrosion cracking of the friction weld joint of dissimilar metals. Further, the formation of the IMC layer at the weld interface in friction weld joint of dissimilar metals (Al, Cu) increases the resistance to the flow of electric current. Typical IMCs observed in friction weld joint of the various dissimilar metal combinations include Al-steel (FeAl, Fe_3Al, Fe_2Al_3), Al–Cu ($CuAl_2$, Cu_2Al), Al–Mg (Al_3Mg_2, $Al_{12}Mg_{17}$) and Ti-steel (FeTi).

4.4.2 Friction Stir Welding

The friction stir welding is a comparatively new solid-state joining process, which is based on the principle of thermal softening of parent metals joined using friction and deformation heat followed by the controlled material flow and forging action to achieve a metallurgical joint. The frictional heat and controlled plastic flow of metal during friction stir welding are achieved using a non-consumable tool and a suitable combination of process parameters, namely tool rotational speed, plunging rate, plunging time, tool plunge depth, normal load, traverse speed, tool shoulder and pin diameters, tool tilt angle and external cooling conditions, if any (Mohamed and Washburn 1975; Sharma and Dwivedi 2017; Kaushik and Dwivedi 2021a, b, 2022).

Stages of friction stir welding include (a) plunging of rotating tool up to the desired depth, (b) plunge (dwell) time to generate enough friction and deformational heat to facilitate the thermal softening of two parent metals to be joined, (c) traversing the rotating tool to achieve the metallurgical continuity along the weld line and (d) retracting the tool after completion of the welding. Friction stir butt and lap welding for joining dissimilar metals is completed in the above four stages, while the third step of tool traversing is not needed in the case of friction stir spot welding (Sharma and Dwivedi 2017; Kaushik and Dwivedi 2021a, 2022).

Thermo-physical properties (thermal conductivity, specific heat), flow stress and ductility, stacking fault energy, work hardening and thermal softening behaviour of both parent metals affect the weld thermal cycle (heating rate, peak temperature, high-temperature retention period, cooling rate) and metal flow. Increasing the difference in weld thermal cycle and metal flow behaviour increases the tendency of weld defect formation and asymmetric weld formation during friction stir welding (Kaushik and Dwivedi 2022). Therefore, friction stir welding of dissimilar metals needs extra care with regard to (a) placement of parent metal with respect to the axis of tool rotation and welding direction during welding and (b) location of the tool with respect to weld centreline, i.e. offset if any.

Friction stir butt/lap welding

The parent metal of high yield strength, low ductility, resistant to thermal softening, high hardness and high thermal conductivity (steel, copper) is usually placed on the advancing side, while low yield strength, high ductility, low resistance to thermal softening, low hardness and low thermal conductivity metal (Al, Mg) is placed on retreating side. This is done to facilitate the thermal softening of hard metal through excessive heat generation so that a sound FSW joint can be developed. Putting hard metal like steel on advanced side during welding with aluminium removes the oxides layer and activates the surface, which in turn helps to form IMCs needed for metallurgical joining. Advancing and retreating sides are determined by the direction of tool rotation and that of welding. During the FSW in butt welding configuration, the side having the same direction of welding and tool rotation is called the advancing side, while the other one is called the retreating side (Fig. 4.13). The FSW tool in case of dissimilar metal joining is not set at the weld centreline, but it is shifted toward the softer metal side at a certain distance called tool offset to deal with issues related to differential weld thermal cycles and resistance to thermal softening of parent metals. The tool offset for dissimilar metal joining can be negative, zero and positive as per the location of the tool axis with respect to the weld centreline (Fig. 4.14). Positive tool offset causes rubbing of tool pin with hard metal leading to the increased tool wear and inclusions in weld nugget due to fragmentation of wear out particles from the hard metal.

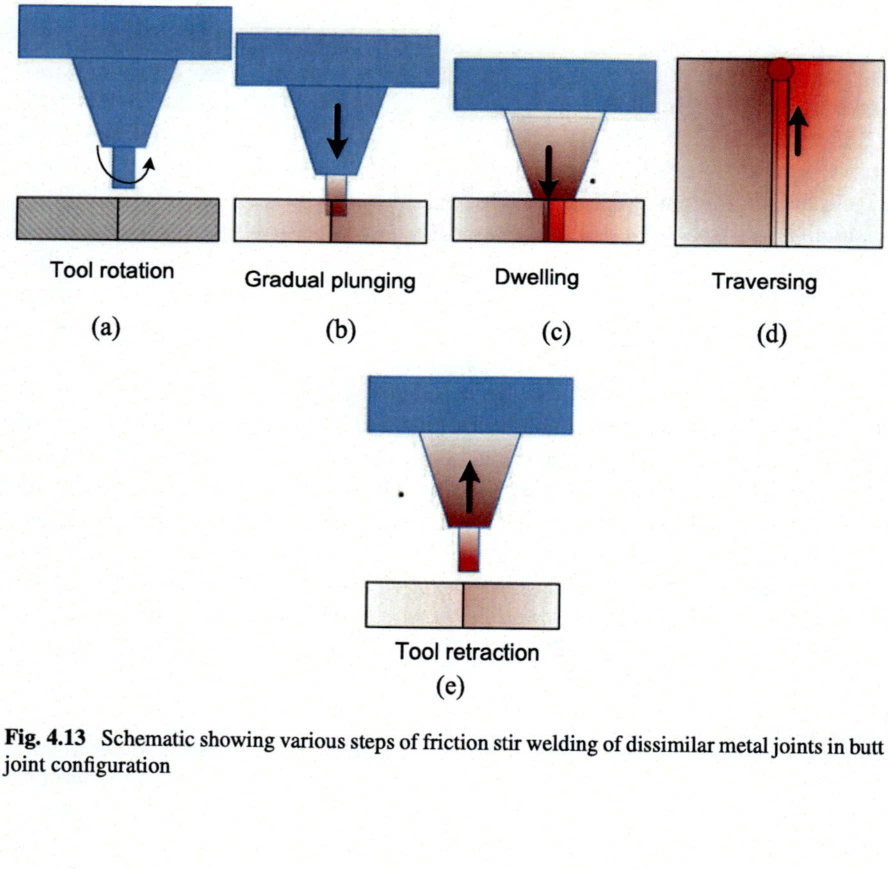

Fig. 4.13 Schematic showing various steps of friction stir welding of dissimilar metal joints in butt joint configuration

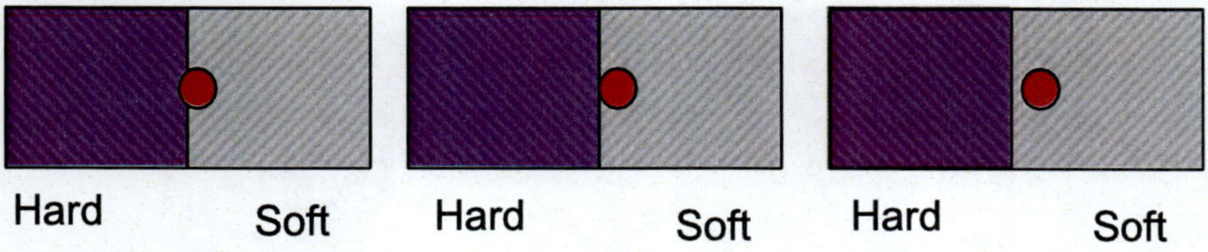

Fig. 4.14 Schematic showing increasing offset of FSW tool from the weld centreline in dissimilar metal joining

Dissimilar metal joining using friction stir lap and spot welding is performed by placing softer metal at the top, and harder one is kept at the bottom if the difference in yield strength and ductility of two parent metals is large. The tool is allowed to penetrate the bottom sheet (through the top sheet) up to desired depth to ensure the formation of optimum hook length with the desired orientation (Figs. 4.15 and 4.16). A hook of too short or long size lowers the load carrying capacity of friction stir spot/lap weld joints (Kaushik and Dwivedi 2021b; Kaushik and Dwivedi 2020). Many metallurgical transformations in the weld zone as per metal systems, intermetallic compound formation and hook geometry determine the joint performance (Figs. 4.17 and 4.18).

4.4.2.1 FSW Process Parameter

The FSW process parameters primarily affect two aspects (a) heat generation and (b) metal flow during welding which in turn affects the soundness, microstructure and mechanical properties of the FSW joint. Further, the difference in chemical, metallurgical, mechanical and thermo-physical properties of parent metals in dissimilar metal joining further complicates the heat generation and metal flow behaviour. Therefore, FSW of dissimilar metals needs careful consideration of various process parameters with regard to their effect on metal flow and heat generation due to mechanical, metallurgical properties and weld thermal cycle imposed.

Tool Rotational Speed

Tool rotational speed (rpm) directly affects the rate of heat generation (kJ/min) and the rate of deformation (strain rate) of metal during welding. An increase in rotational speed increases the heat generation rate, heat localization and net heat input for a given welding speed. Increased heat input affects the weld thermal cycle of weld nugget, thermo-mechanically affected zone and heat-affected zone (Fig. 4.19). Increase in rotational speed in general increases the thermal softening and flowability of metal in the weld zone and the width of heat-affected zone and reduces the width of TMAZ.

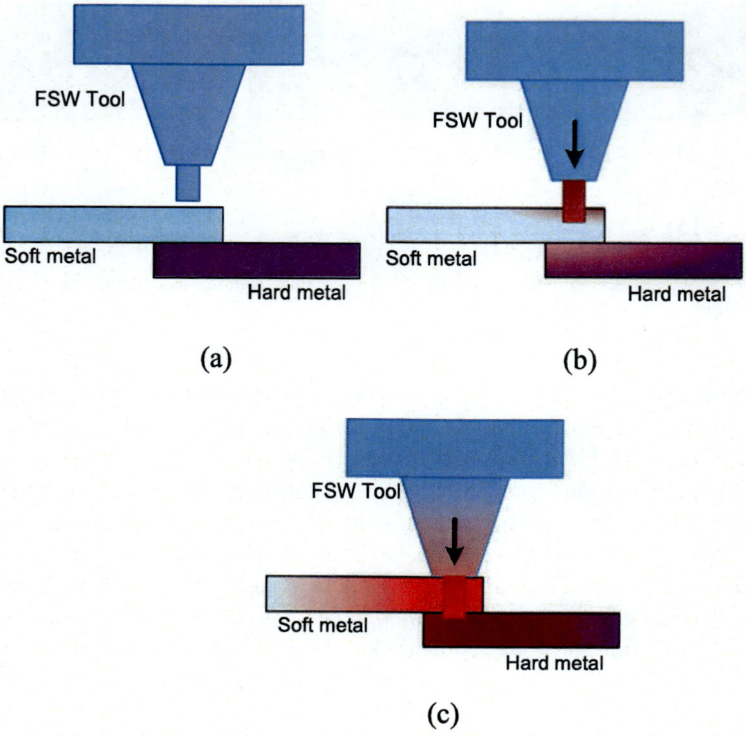

Fig. 4.15 Schematic showing various steps of friction stir welding of dissimilar metal in lap joint configuration

Further, an increase in rotational speed increases the strain rate experienced by the metal during FSW making metal in the weld zone flow like a viscous fluid. Tool rotational must be optimized considering the tool traverse speed because too high or low rotational speeds compromise the metal flow, the soundness and mechanical and corrosion performance of FSW joints.

Traverse Speed

The tool transverse speed (mm/min) corresponds to welding speed. Traverse speed mainly determines the net heat input to parent metals during welding which in turn affects the metal flow behaviour, metallurgical transformation and width of weld zone, TMAZ and HAZ. An increase of traverse speed (for given tool rotational speed) reduces the net heat input, and accordingly weld thermal cycle results in a narrow width of HAZ but increased tendency of weld defect formation due to reduced flowability of metal during welding (Fig. 4.20). While the low traverse speed increases, the heat input and so increased thermal damage to weld, HAZ and

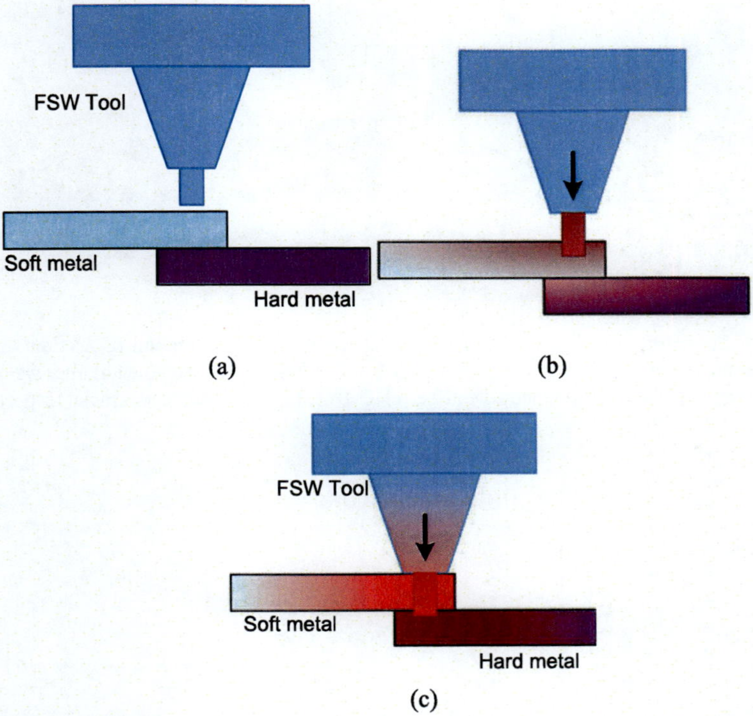

Fig. 4.16 Schematic showing various steps of friction stir spot welding of dissimilar metal joints

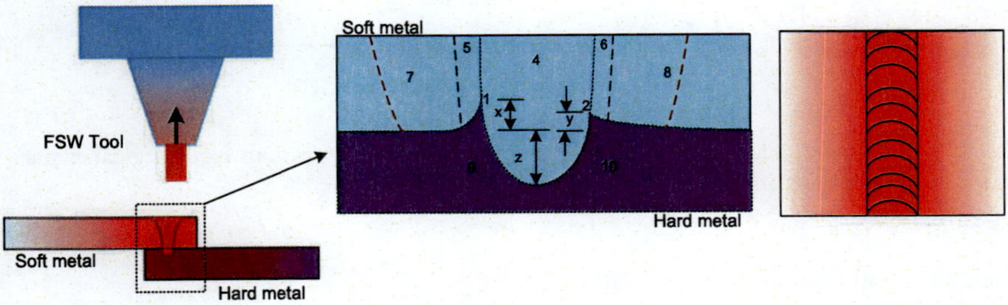

Fig. 4.17 Schematic showing various geometrical features of cross section of friction stir dissimilar metal joints developed in lap joint configuration (where in 1 & 2 shows hook, 4 weld nugget, 5 & 6 TMAZ, 7 & 8 HAZ of soft metal, 9 & 10 HAZ of hard metal respectively)

parent metals in the form of undesirable metallurgical transformation and residual stress development. Therefore, an optimum ratio of tool rotational speed and tool traverse speed is used for developing the sound weld joints as per dissimilar metal combination.

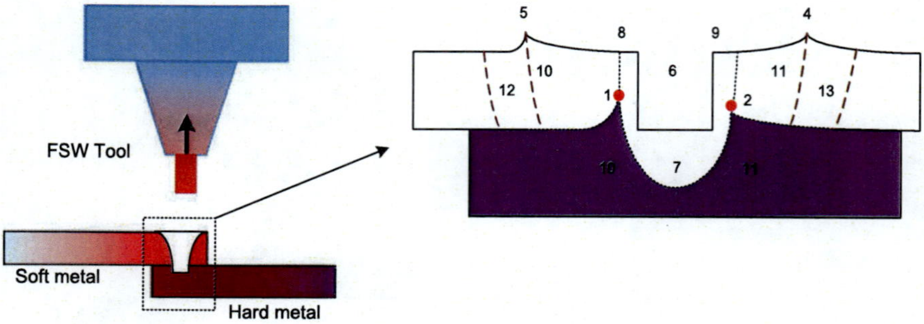

Fig. 4.18 Schematic showing various geometrical features of cross section of friction stir spot dissimilar metal joints (where in 1 & 2 shows hook, 4 & 5 flash, 6 unfilled hole left after welding, 7 weld nugget, 8 & 9 tool pin affected zone, 10 & 11 TMAZ, 12 & 13 HAZ of soft metal respectively)

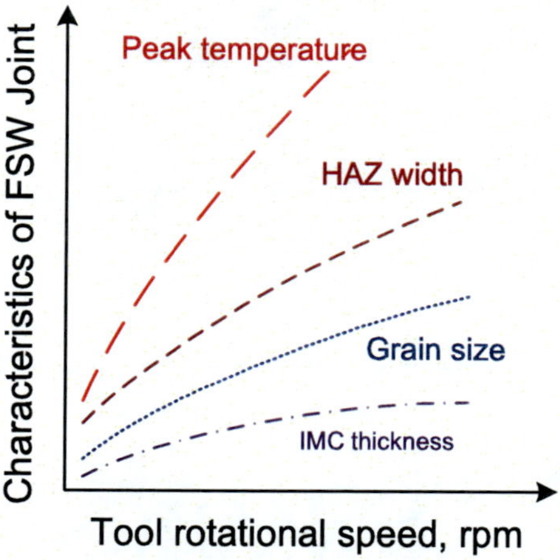

Fig. 4.19 Schematic showing effect of FSW tool traverse speed on various metallurgical characteristics of a typical dissimilar metal joint developed by FSW

Geometry and Size of Tool Pin and Shoulder

The design of the tool pin and shoulder affects the heat generation and flow of metal both during welding for developing a metallurgical joint (Fig. 4.21). Tool pin size (diameter), cross section (cylindrical, taper, threaded, square, triangular, etc.) and its length must be optimized considering the requirement of heat generation and metal flow for dissimilar metal joining of given section thickness. Tool pin length kept about 0.2–0.5 mm less than the thickness of the plates is to be joined. Tool pin controls the metal softening by heat generation followed by flow of metal through the thickness of weld nugget. Poor metal softening and limited metal flow during welding result in weld discontinuities leading to reduced performance of the weld joint.

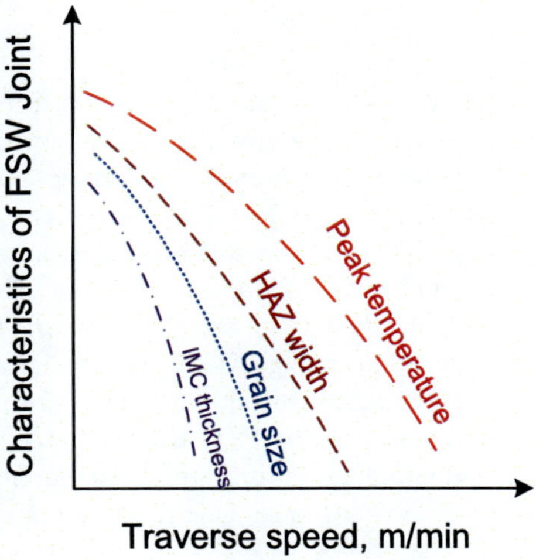

Fig. 4.20 Schematic showing effect of FSW tool rotational speed (please remove tool rotation speed) tool traverse speed on various metallurgical characteristics of a typical dissimilar metal joint developed by FSW

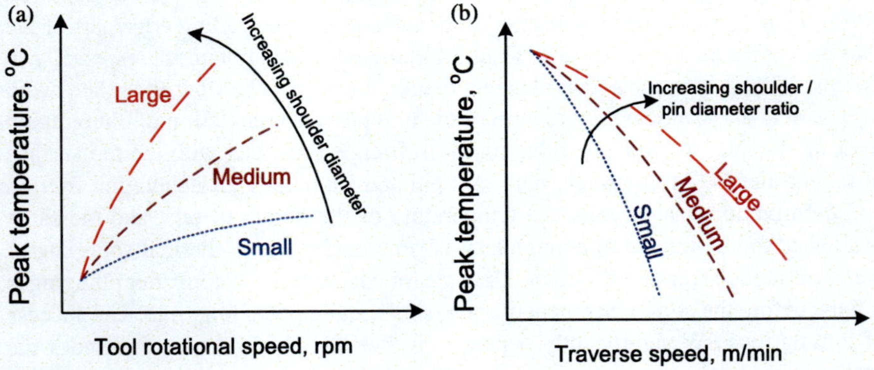

Fig. 4.21 Schematic showing effect of FSW tool **a** rotational speed and **b** traverse speed on peak temperature for varying ratio of shoulder to pin diameter

Tool shoulder rubs the top surface layer of parent metals to generate the frictional heat which in turn causes thermal softening followed by forging action (with the controlled metal flow) under applied normal load to consolidate metal for developing a weld joint. The role of tool shoulder is mostly limited to heat generation and controlling the flow of metal at the top surface during welding. Therefore, the tool shoulder produces wider weld and heat-affected zones at the top than the bottom.

Tool Offset

The tool offset is critical in dissimilar metal joining by FSW. The tool offset means setting the tool pin axis at varying distances away from the weld centreline. Depending upon the difference in yield strength and hardness of the parent metals of dissimilar metal combination, the tool offset can be positive, zero and negative. Increasing difference in hardness/yield strength/thermal softening tendency of parent metals of dissimilar metal combination shifts choice of tool offset from positive to zero to negative. Tool offset affects the tool wear, wear debris formation as inclusions. The debris is formed due to rubbing of tool pin with hard metal, intermixing of metals flowing from the two parent metals (in case of the minor difference in hardness of the two parent metals) during FSW from both the sides which in turn affect the soundness and metallurgical continuity of dissimilar metal joints. The tool offset (0.2–0.5 mm) towards the hard metal side must be optimized for enhanced joint performance.

Tool Plunging Rate and Dwelling Time

The rotating tool in the first stage of FSW is plunged gradually into the parent metals at desired tool offset location leading to penetration of the leading edge/part of the tool pin inside the parent metals. The tool plunging rate is important, especially in the case of high-strength and hard parent metals, and it must be slow enough to avoid overloading and catastrophic fracture of the tool pin. On completion of plunging up to desired depth, the tool shoulder comes in frictional contact with the top surface of parent metals. At this stage, both tool pin and shoulder start generating friction and deformation heat to raise the temperature of the parent metals, and the same is allowed to continue until enough heat is generated to cause thermal softening of metals around the pin and shoulder. The time for which the FSW tool after plunging is rotated before the commencement of traversing is called dwelling time. The success of joining by FSW significantly depends on dwelling time because it decides the extent of thermal softening of metal realized before traversing the tool along the weld line.

Tool Plunge Depth

The tool plunge depth is very important in all three variants of FSW, namely friction stir butt, lap and spot welding because it directly affects the penetration in butt welding and hook length in case of lap and spot welding. The tool plunge depth is about 0.1 to 0.4 mm lesser than the section thickness of plates in friction stir butt welding, while in the case of friction stir spot and lap welding, it must be optimized considering the desired hook length and its orientation and parent metal characteristics (Fig. 4.22). Excessive plunge depth in case of lap/spot welding of hard metals results in weld

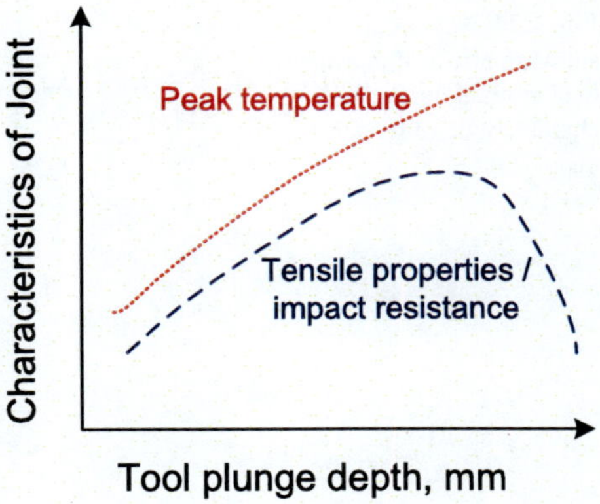

Fig. 4.22 Schematic showing effect of FSW tool plunge depth on characteristics of dissimilar metal joint developed by FSW in lap joint configuration

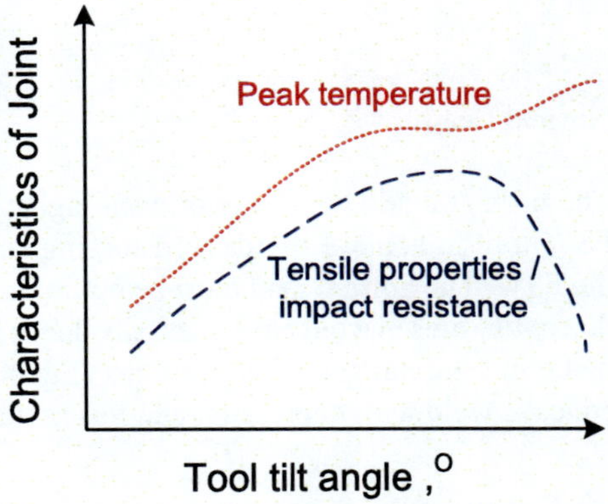

Fig. 4.23 Schematic showing effect of FSW tool tilt angle on characteristic of dissimilar metal joint developed by FSW

defects besides fragmentation of particles leading to inclusions. These inclusions degrade the mechanical performance of the joint.

Tool Tilt Angle

The setup of the FSW tool at a certain angle (1–5°) from the vertical plane along the weld line is called the tool tilt angle. The tilt angle affects the peak temperature and joint strength as shown in Fig. 4.23 suggested that it should be optimized. Tool tilt angle facilitates the flow of metal during FSW from one side to another (mostly front to back) to ensure the metallic continuity. However, the role of tool tilt angle in dissimilar metal welding using zero and negative tool offset is lesser than positive offset due to the limited scope of intermixing of two parent metals.

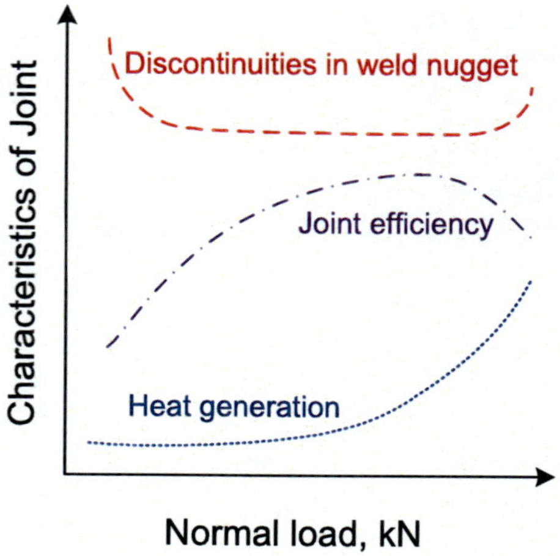

Fig. 4.24 Schematic showing effect of normal load applied during FSW on characteristic of dissimilar metal joint

Normal Load

The normal load (10–50 kN or even higher) is applied through the shoulder of the FSW tool to achieve the desired forging action and consolidation of metal during the FSW. The normal load must be optimum. Less normal load results in many weld discontinuities in weld zone primarily due to limited consolidation/forging of flowing metal in weld nugget during welding (Fig. 4.24). Excessive normal load results in a concave weld with depression and flushing of the weld metal along the weld.

External Heating/Cooling

Dissimilar metal joining using FSW sometimes demands additional heating/cooling to develop a sound weld joint with the desired combination of microstructure and mechanical properties. Localized heating (using a suitable heat source like flame, arc, induction, laser, etc.) of one parent metal of dissimilar metal combination having high yield strength, hardness and thermal softening resistance may be done to achieve desired intermixing and metallic flow to develop the sound weld joint. Application of external heating, however, can cause wider HAZ and coarser grain structure in weld nugget and heat-affected zones. Similarly, external cooling (using forced air, cold water and liquid nitrogen) can also be applied to negate the effect of the weld thermal cycle in HAZ and weld nugget. External cooling reduces the peak temperature and high-temperature retention period and increases the cooling rate which in turn reduces the width of HAZ, grain structure, reversion and grain growth. However, external cooling increases the flow stress and decreases the ductility of parent metals and weld metal during welding. These changes in mechanical properties due to cooling may decrease the flowability of metal and promote the discontinuities in weld nugget.

4.4.3 *FSW Joint Evaluation*

4.4.3.1 Thermo-mechanical Aspect

The weld nugget zone of the friction stir weld joint is subjected to the maximum temperature and strain. As per thermo-physical properties (thermal conductivity and specific heat) of parent metals, both the parent metals experience different weld thermal cycles (Fig. 4.25). For example, during FSW of Al-steel combination, peak temperature and heating rate observed on Al side are around 400 °C and 100 °C/sec, while those for steel are 600 °C and 40 °C/sec, respectively. Deformation coupled with a rise in temperature of metals mostly above recrystallization temperature during FSW produces four zones, namely weld zone (WZ), thermo-mechanically affected zone (TMAW), heat-affected zone (HAZ) and parent metal (PM). However, thermal conductivity and thermal expansion coefficient of the two parent metals of dissimilar combination determine the peak temperature experienced by parent metals and thermal expansion/contraction during welding as unique weld thermal cycles are experienced by two parent metals which in turn accordingly result in different widths of weld nugget, TMAW and heat-affected zones and residual stress conditions. These changes in microstructure and residual stress condition in turn affect mechanical, corrosion and metallurgical characteristics of dissimilar metal FSW joints.

4.4.3.2 Metallurgical Aspect

The FSW butt and lap joining involves severe plastic deformation at a very high strain rate. Most of the time friction and deformation heating increases the temperature of parent metals weld and heat-affected zone above recrystallization temperature. Severe plastic deformation at high strain rate results in not just work hardening but also fracturing and fragmentations of micro-constitutes in weld nugget. Highly strained weld nugget at high temperature during welding experiences dynamic recrystallization leading to very fine and equiaxed grain structure. The typical weld thermal cycle can also cause metallurgical transformation in weld nugget and heat-affected zones (Fig. 4.26). As per the metallurgical compatibility, intermixing of elements from both the parent metals can form a solid solution, phases and intermetallic compound (IMCs).

Metallurgically incompatible parent metals of dissimilar combination during FSW form hard and brittle IMCs. However, the thickness of these IMCs depends on the heat input during welding. Increase of heat input (with the increase of rotational speed/reduction in traverse speed, increase in tool shoulder and pin size) increases the thickness of IMC which in turn deteriorates the mechanical, corrosion performance of FSW joints of dissimilar metals. As per the metal strengthening mechanisms (solid solution strengthening, grain refinement, work hardening, precipitation hardening, dispersion and transformation hardening) of parent metals, both can show completely different responses to mechanical and thermal stress applied during the FSW in form

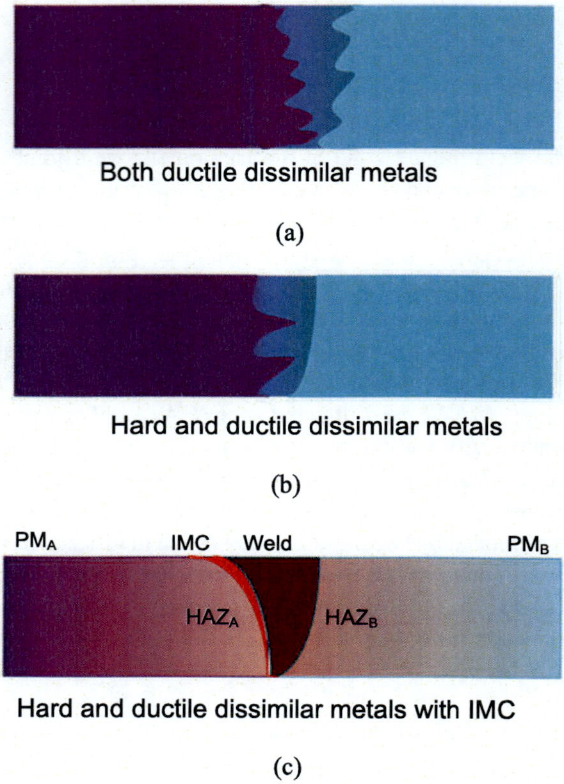

Fig. 4.25 Schematic showing various types of joint morphologies observed in FSW joints of dissimilar metal joint when **a** both metals are soft and compatible; **b** one metal is soft, other is hard still, and both are compatible; and **c** one metal is soft, other is hard, and both are incompatible

of hardening/softening of weld zone, TMAZ and HAZ. The weld nugget of the FSW joint can experience solid solution strengthening, grain refinement, transformation hardening, work hardening and even loss or formation of hardening precipitates (Fig. 4.27). However, the relative impact of these metal strengthening mechanisms on weld nugget properties depends on the parent metal being joined. Similarly, HAZ can be subjected to reversion, recovery and recrystallization, solid solution strengthening, grain coarsening and transformation hardening.

For example, HAZ of FSW joint of steel-aluminium results in hardening on the steel side, while the aluminium side experiences softening. Therefore, metallurgical changes in the weld zone, TMAZ and HAZ in totality determine the joint performance. An increase in normal load, rotational speed, tool shoulder and pin diameter and reduction of transverse speed increases the heat generation, which in turn increases the width of heat-affected zone. Common IMCs reported in friction stir weld joint of various dissimilar metal combinations are Al-steel (FeAl, Fe_3Al, $FeAl_3$, Fe_2Al_5, Fe_4Al_{13}) Al–Cu (CuAl, Cu_3Al_2, Cu_9Al_4, $CuAl_2$, Cu_3Al), Al–Mg ($Al_{13}Mg_2$, $Al_{12}Mg_{17}$) and steel-Mg (No IMC).

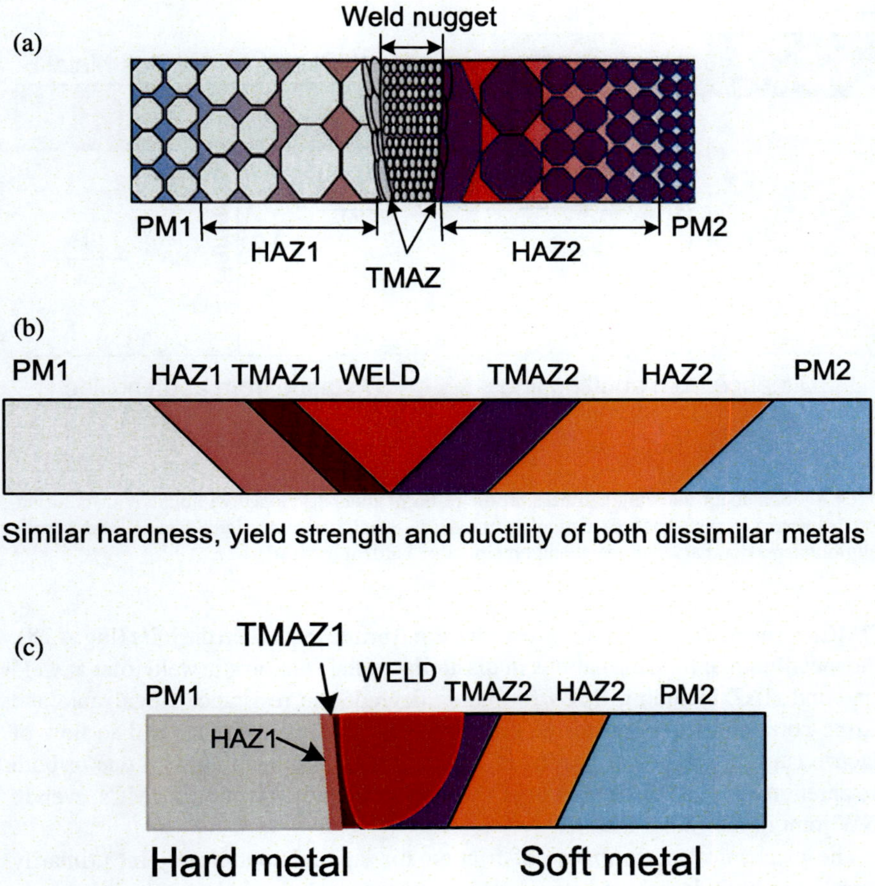

Fig. 4.26 Schematic FSW joints **a** microscopic zone, **b** macroscopic zones formed in a joint made between two dissimilar metals having almost matching mechanical and physical properties and **c** macroscopic zones formed in a joint made between two dissimilar metals having large difference in mechanical and physical properties

4.4.3.3 Mechanical Aspects

The strength of the friction stir butt weld joint depends on the soundness and metallurgical characteristics of weld nugget (WZ), thermo-mechanically affected zone (TMAZ) and heat-affected zone (HAZ) of both side parent metals. The failure under tensile loading can take place from anywhere, i.e. one side parent metal, HAZ1, TMAZ1, WZ, TMAZ2, HAZ2, and other side parent metal as per the response of metal systems to thermo-mechanical stresses imposed during friction stir welding. The formation of thick hard and brittle IMC in the weld zone promotes the weld failure. The hardness profile across the FSW joint of dissimilar metals clearly indicates the minima hardness regions which is expected to fail during the tensile loading

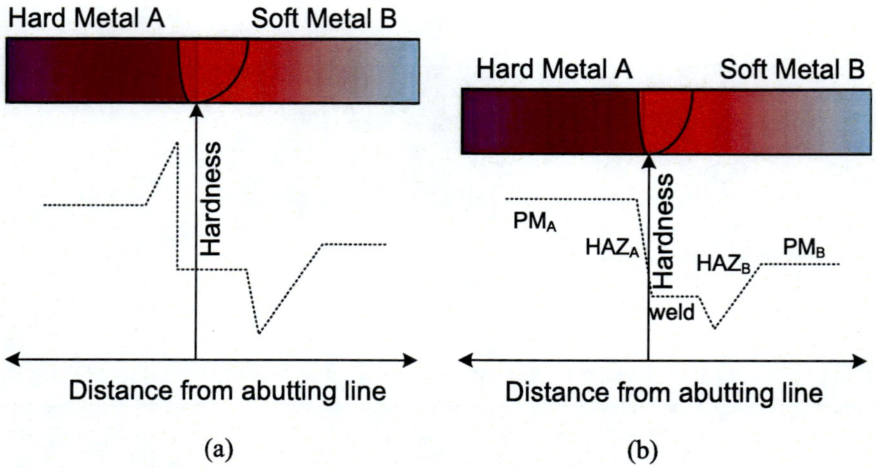

Fig. 4.27 Schematic showing hardness profile of FSW joints developed without any IMC formation between **a** transformation hardening metal and precipitation hardening metal and **b** solid solution/dispersion hardened metal and precipitation hardening metal

if IMCs formed in the weld zone are not hard and brittle to weaken the joint (Fig. 4.28). The metallurgical transformations in dissimilar metal friction stir weld joint at weld zone and HAZ including IMCs formation degrade the resistance to galvanic and pitting corrosion, stress corrosion cracking, electrical conductivity and so flow of electric current. Films and coating of suitable metals compatible with one or both the parent metals can be used as interlayers to develop the favourable IMCs, even in FSW joint of dissimilar metal.

The mechanical performance of friction stir lap and spot weld joint primarily depends on hook length and orientation, stress concertation at/around hook, stress concentration at periphery of joint in spot welding and along the weld in lap weld joint, besides the soundness and metallurgical transformation occurring during FSW on both sides of the parent metals. Effect of FSW process parameters (including rotational speed, plunge depth, normal load, plunge rate, etc. as per the process) on hook geometry and metallurgical transformation should be analysed and obtained for realizing the desired joint performance.

4.4.4 Ultrasonic Welding

Ultrasonic welding (USW) is primarily used for joining of thin-section components in lap joint configuration using in-situ interfacial frictional heating coupled with micro-scale localized plastic deformation achieved through high-frequency (> 20 kHz) ultrasonic vibration applied through a tool called sonotrode. Ultrasonic welding is characterized as a very low heat input solid-state joining process; therefore, it suits

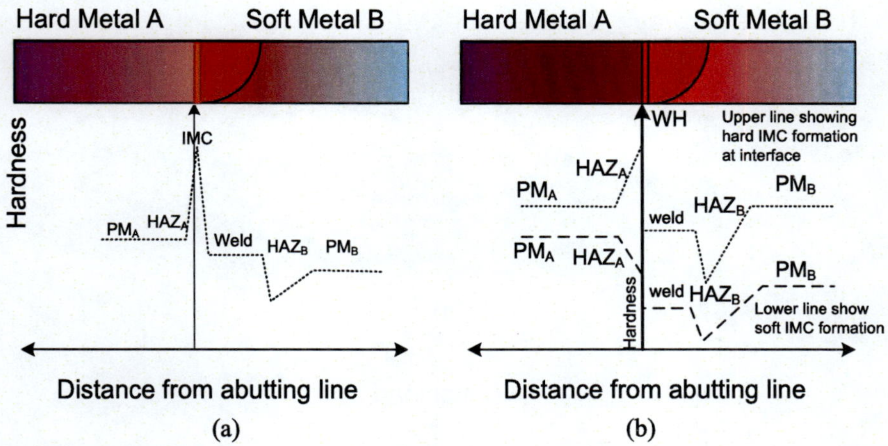

Fig. 4.28 Schematic showing hardness profile of FSW joints developed between **a** solid solution/dispersion hardened metal and precipitation hardening metal leading to hard and brittle IMC formation and **b** metals strengthened by different mechanisms like transformation, precipitation hardening and solid solution strengthened

for joining highly thermal/electrical conducting dissimilar metals in the form of wires and multiple stacked sheets for applications in electrical, electronic and microelectronic industries to join aluminium, copper, magnesium and other electrically conducting metals. Accordingly, two variants of the ultrasonic welding process are uniquely designed for wire bonding and additive manufacturing using a layer-by-layer approach for joining the stacked sheet. The ultrasonic weld joint of dissimilar metals develops a very narrow weld zone, IMC layer (in any) and heat-affected zone primarily due to the low energy input applied for joining.

4.4.4.1 Dissimilar Metal Joining by Ultrasonic Welding

The ultrasonic welding uses a combination of the ultrasonic vibration of frequency greater than 20 kHz and short amplitude (few micrometres to few hundreds of micrometres) coupled with compressive load (few Newton to few hundreds of Newton) for joining of the dissimilar metals. Ultrasonic vibration and compressive force are applied using sonotrode and anvil (Fig. 4.29). Application of the ultrasonic vibrations under compressive load results in interfacial friction conditions between mating surfaces of the components to be joined. The interfacial friction phenomenon is key for ultrasonic welding. The interfacial friction leads to (a) interfacial frictional heating, (b) thermal softening of surface layers of dissimilar metals, (c) mechanical interlocking due to severe interfacial plastic deformation, (d) recrystallization, (e) diffusion across the interface for metallurgical bonding, (f) work hardening and (g) sometimes localized partial melting depending upon the energy applied for ultrasonic welding as per welding parameters, namely compressive load, welding time

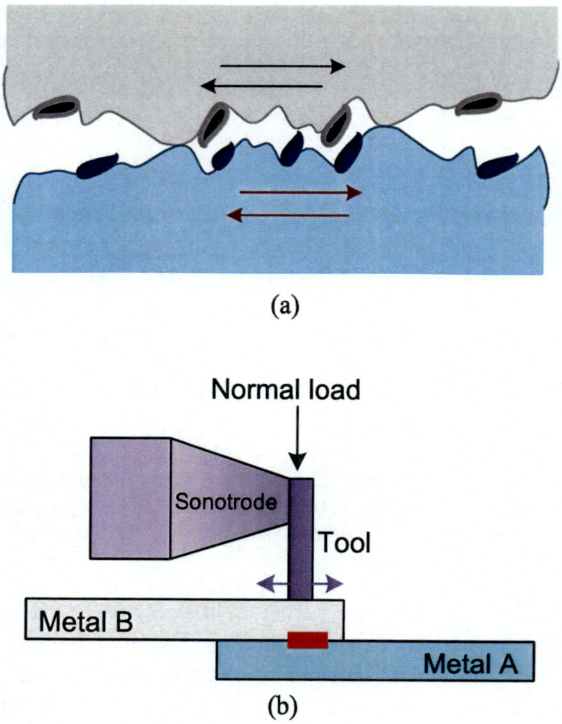

Fig. 4.29 Schematic showing **a** surfaces of dissimilar metals with oxides and impurities and **b** scheme of ultrasonic welding with components and movements

(few micro-second to few hundreds of second), welding power to apply ultrasonic vibrations, frequency and amplitude of vibrations (Fig. 4.30). One of the parent metals of a dissimilar combination should have reasonably good ductility to facilitate the plastic deformation and joining at the mating interface.

Temperature attained at the interface due to interfacial friction and localized micro-scale deformation is a major parameter, which needs to be measured for ensuring the quality of the joint. In general, an increase of energy input (few J to few kJ) increases the interfacial temperature and deformation which in turn increases the metallurgical bonding between dissimilar metals and the width of heat-affected zone. On the other hand, the low energy input can cause interfacial voids and limited metallurgical and mechanical bonding leading to reduced load carrying capacity of the dissimilar metal joint. The energy input, therefore, needs to be optimized considering the yield strength of metals and section thicknesses. Mechanisms in ultrasonic welding of dissimilar metal predominantly responsible for development of the joint include mechanical interlocking, micro-melting, metallic bonding and dynamic recrystallization.

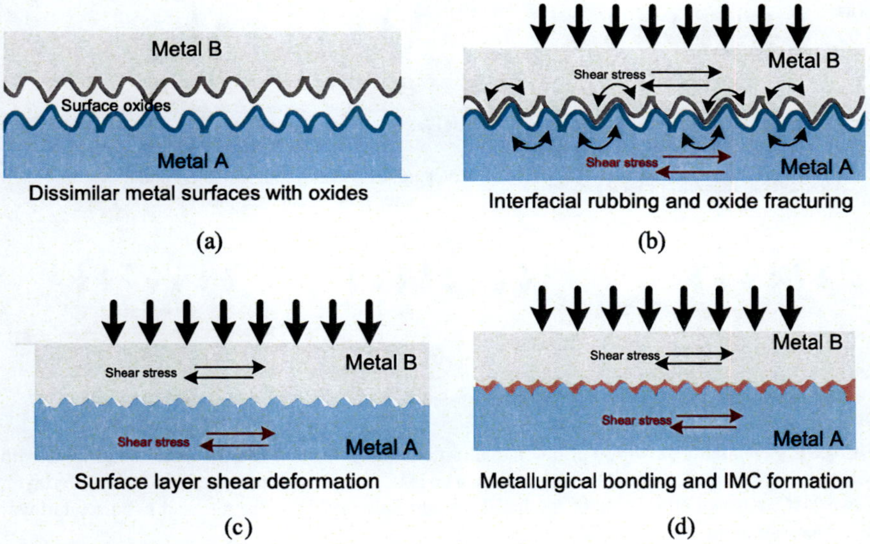

Fig. 4.30 Schematic showing stages of ultrasonic welding of dissimilar metals

4.4.4.2 Characterization of Ultrasonic Dissimilar Metal Weld Joints

The dissimilar metal joint developed using ultrasonic welding exhibits a very narrow weld interface (few micrometres to few tens of micrometres), followed by heat-affected zones on both sides as per the metal system. Energy input, interfacial surface conditions, welding time, normal load and dissimilar metal combination significantly determine the nature, size of weld interface zone and heat-affected zone. The weld interface may exhibit multiple features like a plane, wavy, sinusoidal interface geometry (few micrometres to few hundreds of micrometres), amorphous structure, high-density dislocation and stacking fault zones (few nanometres to few tens of nanometres), localized melting zone, intermetallic compounds and segregation/depletion of alloying elements (Fig. 4.31).

The maximum temperature attained at the interface can be as high solidus temperature of one of the parent metals or eutectic/low melting point constituents formed in presence of interlayer, if any. For example, maximum interface temperature of more than 500 °C and 650 °C has been reported while joining Al-Cu and Cu-steel dissimilar combinations, respectively. An intermetallic compound formed at the interface during dissimilar metal joining is mostly found to be hard and brittle which deteriorates the performance of the joint in terms of load carrying capacity, corrosion resistance and electrical conductivity. The interlayer of suitable metal compatible with one or both parent metals of a dissimilar combination can be used to develop a more favourable IMC and improve the joint performance. For example, interlayers of Zn in Al-steel joining and Al alloy in Al–Cu joining help to form favourable IMCs, which in turn improve the mechanical performance of ultrasonic weld joints. High

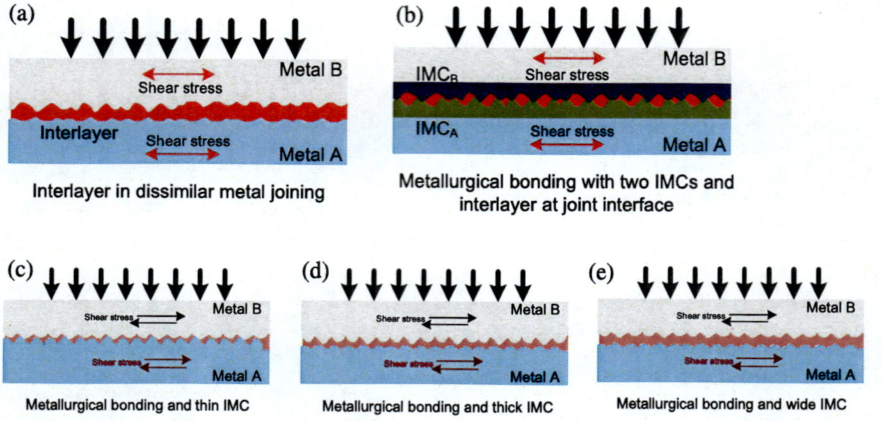

Fig. 4.31 Schematic showing various metallurgical zones formed in dissimilar metal joint with without and with interlayer and IMC formation **a** no IMC formation in presence of suitable interlayer, **b** two IMCs formed either side of the interlayer and interlayer are discrete and **c–e** no interlayer with increasing thickness of IMC

temperature rise at interface results in more metallurgical transformation and heat affected at weld interface and on both the parent metals. Heat-affected zone experiences multiple transformations such as recovery, recrystallization, reversion and grain growth as per the metal system of dissimilar metal combination (Fig. 4.32).

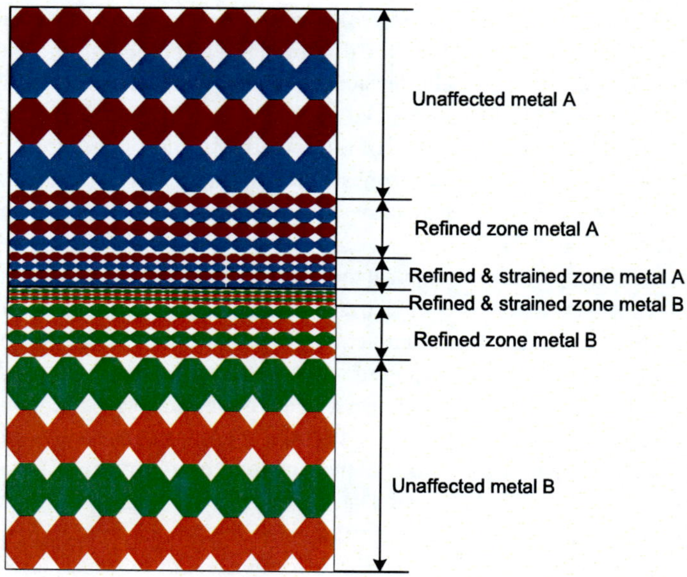

Fig. 4.32 Schematic showing formation of different zones in ultrasonic weld joint of dissimilar metals

4.4.5 Impact Welding

The impact welding is a group of solid-state joining processes like explosive welding and electromagnetic welding mainly used for developing dissimilar metal joints in lap configuration with almost nil or negligible HAZ. In these processes, the kinetic energy of the high-velocity moving (flyer) component is used to develop a metallurgical joint through the impact with another stationary (target) component to be joined (Fig. 4.33). A part of the kinetic energy of the flyer on impact is used to cause interfacial shear deformation at the mating surfaces, and the remaining is converted into heat. Therefore, the velocity of the flyer component becomes a critical parameter in development of the sound metallurgical joint. All the factors related to impact welding processes such as standoff distance, type and amount of explosive, electromagnetic forces, etc. determine the impact velocity and impact angle affecting the localized interfacial deformation which in turn dictate the joint interface morphology, soundness, metallurgical properties (IMC, recrystallization, refinement, AZ, localized melting, etc.) and mechanical performance the joint (Fig. 4.34). Apart from impact (velocity)-related parameters, the desired amount and type of shear deformation (at microscale) influence yield strength and ductility of the joint. One parent metal (usually flyer) of dissimilar metal combination must be comparatively softer and more ductility than other metals. Optimum impact conditions (velocity and angle) of flyer with stationary components result in self (in-situ) surface cleaning (through scouring off surface oxides and other contaminants), nascent surfaces of stationary and flyer component that result in direct metal-to-metal contact, localized shear deformation of mating surfaces and finally metallurgical joint. Inappropriate impact conditions (velocity and angle) cause interfacial inclusion due to improper surface cleaning, voids, poor metallurgical bonding, mostly flat interface due to limited shear deformation, formation of IMC and hard and brittle micro-constituents due to excessive heat generation, localized melting under very severe impact conditions. These undesirable features and discontinuities at the interface degrade the mechanical and corrosion performance of joints.

Explosive welding is a very established and commonly used process for developing dissimilar metal cladding (Al, ASS cladding on mild steel) over large size, thick sections and robust components used in chemical industries for improved corrosion and wear resistance. The joint interface of these claddings with the substrate is mostly stronger than parent metal and cladding due to strain hardening. These claddings of suitable metal on one or both parent metals developed using explosive bonding are therefore used as a buttering layer for joining of dissimilar metals using fusion welding processes also. Electromagnetic welding is a comparatively new solid-state joining process and is primarily used joining of thin-section sheets, tubes and films. Like explosive welding, electromagnetic welding also uses flyer and stationary components to realize the desired impact velocity and angle of impact for developing solid-state joint between dissimilar metals. This is a more sophisticated and easy-to-automate process for the factory environment.

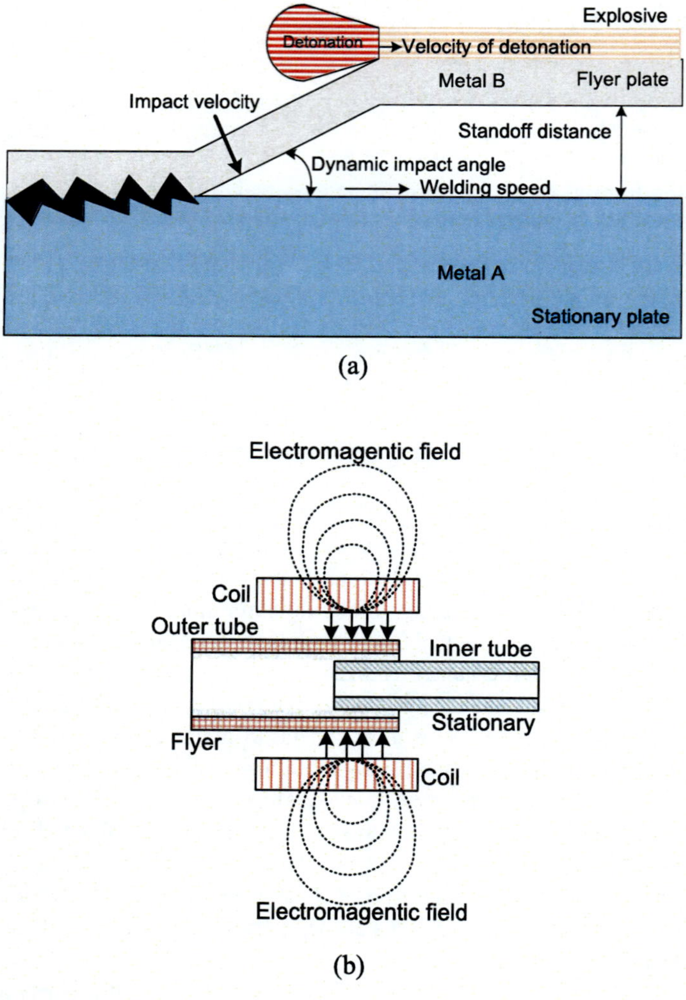

Fig. 4.33 Schematic of the general approach of **a** explosive welding and **b** electromagnetic pulse welding with different component

4.4.5.1 Characterization of Dissimilar Metal Weld Joints by Impact Welding

The mechanical and corrosion performance of dissimilar metal joints developed by the impact welding process depends on interfacial soundness of joint, morphology, metallurgical characteristics and HAZ, if any (Fig. 4.35). The joint interface morphology varies significantly from flat, wavy with jetted metal and wavy/sinusoidal wave depending upon the section thickness, impact angle and velocity of flyer component, ductility and yield strength and difference in hardness of two dissimilar metals being joined (Fig. 4.36).

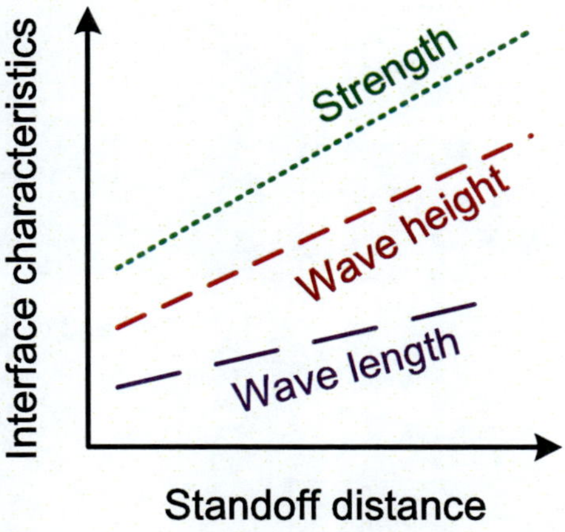

Fig. 4.34 Schematic showing the effect of standoff distance on joint interface characteristics

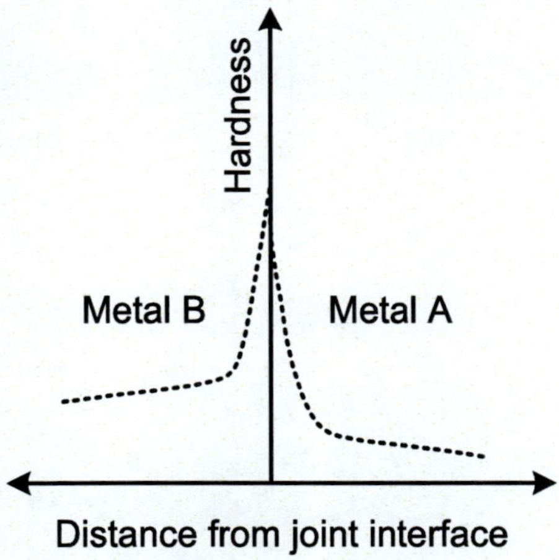

Fig. 4.35 Schematic of hardness profile across the joint interface

Thick section, low velocity and large difference in hardness of parent metals of dissimilar combination promote flat joint interface. An increase in impact velocity in combination with a suitable impact angle results in a transition of joint interface morphology from flat to slightly wavy, wavy and wavy with jetted, melted layer (Fig. 4.37). Joint interface with jetted metal morphology is observed at high impact velocity mainly due to shear instability at the interfacial plastic flow.

A more systematic investigation of the joint interface using a very low magnifying lens (2–10×) to scanning electron microscope (>10 k×) is done to study the interface morphology. Accordingly, the interface can reveal wavy/flat interface, pitch and amplitude of waves, wave with jetted metal, if any, presence of IMC and cracks at

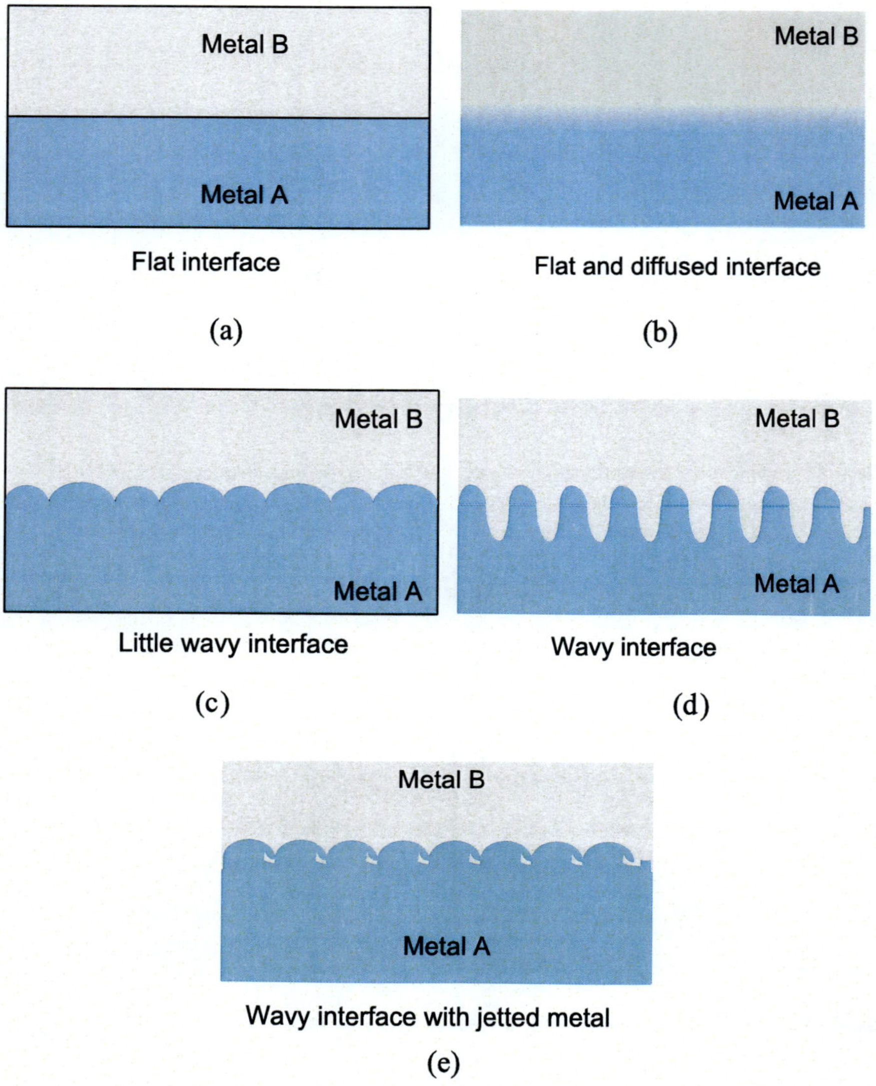

Fig. 4.36 Schematic of different morphologies of the joint interface observed in dissimilar metal joining by impact welding

wavy surfaces and grains/subgrains of metal within waves (Fig. 4.38). These macro-/micro-/nano-scale features can help to explain the behaviour of impact weld joint of dissimilar metals.

However, there is no consensus on the most preferred interface morphology for the maximum mechanical performance of the joint as it has been found sensitive to metals of dissimilar combination. The wavy or sinusoidal interface morphology is

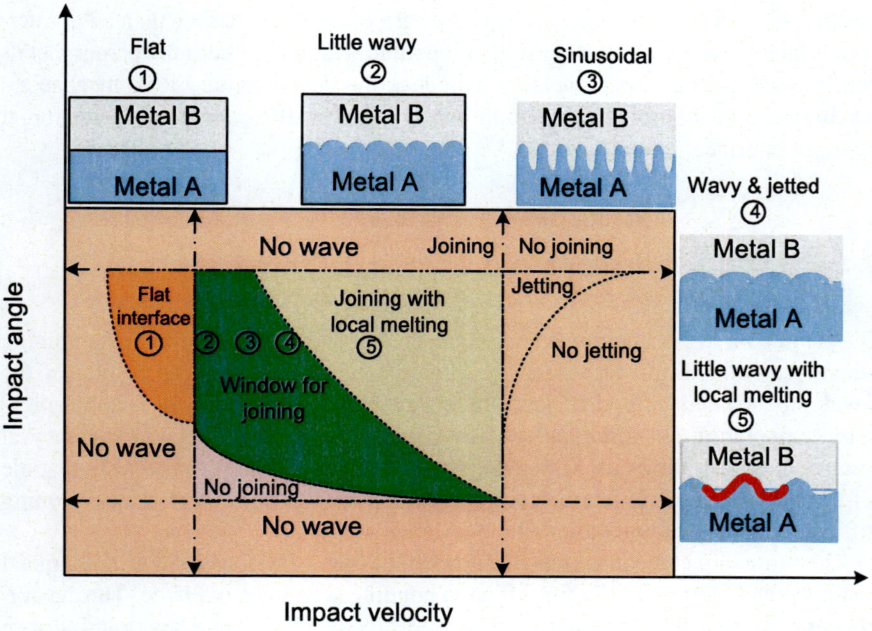

Fig. 4.37 Schematic showing the effect of impact velocity and angle on joint interface characteristics

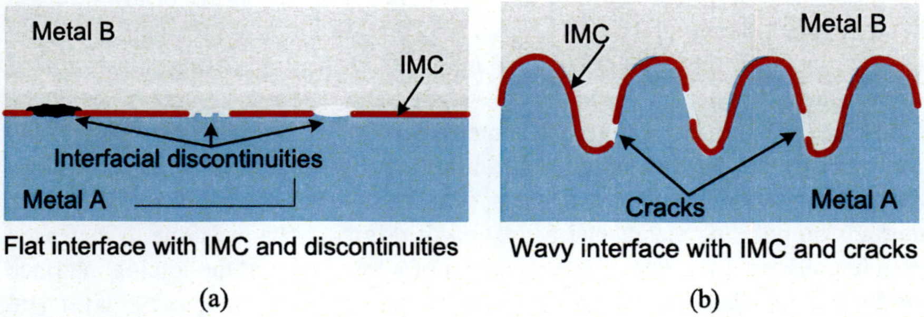

Fig. 4.38 Schematic interface of dissimilar metal joint developed by impact welding with **a** flat and **b** wavy interface with IMC and discontinuities

expected to enhance the mechanical performance in case of the most of the dissimilar metal joints due to two reasons (a) improved mechanical interlocking and (b) tendency of crack growth in weak zones (discontinuities, HAZ and IMC) of joint with wavy interface that is minimized due to zigzag path. The presence of discontinuities voids, oxide inclusion and cracks reduces the load carrying capacity (tensile, shear, fatigue) of joint due to reduction in load-resisting cross-sectional area and stress concentration. Moreover, the extent of degradation of joint performance due to discontinuities is somewhat lower in the case of a wavy interface than a flat interface. The interlayer of suitable metal compatible with one or both the parent metals can be used during impact welding to reduce the deterioration in the mechanical performance of dissimilar metal joint due to intermetallic compound formation at the joint interface.

4.4.6 Diffusion Bonding

Diffusion bonding is a common and established process for dissimilar metal joining wherein atomic diffusion across the interface of mating surfaces results in the development of metallurgical joint under favourable conditions of vacuum, pressure, temperature and time and usually interlayers as well. Metallic intimacy at joining interface, diffusion across the interface and suitable interlayer compatible with one/both the dissimilar metals are three important requirements for developing diffusion bonding of good joint efficiency (Fig. 4.39).

Therefore, direct metal-to-metal contact at the mating surfaces of dissimilar metal components is a prerequisite for diffusion bonding across the interface. The cleanliness of mating surfaces, tendency of parent metals to form passive layers and surface roughness apart from diffusion bonding parameters, namely temperature, pressure and vacuum significantly affect the metallic intimacy and so success of the diffusion bonding process (Fig. 4.40). Thoroughly cleaned and smooth surfaces of dissimilar metals kept under pressure (usually less than the yield stress of low-strength metal of dissimilar combination) at high temperature in vacuum conditions result in good metallic intimacy at the mating interface. Further, increasing surface finish, temperature, pressure and vacuum increases the metal-to-metal contact. An increase in pressure and temperature promotes interfacial metallic intimacy due to creep and localized surficial deformation of peak and valley of a rough surface. This localized plastic deformation, however, facilitates the recrystallization, grain refinement at the interface on both the metals and formation of diffusion channels due to increased grain boundaries. Moreover, the application of a few fluxes at the joining interface during diffusion bonding can also be useful to decompose the passive layer and improve the metallic contact at the interface.

Enough bonding pressure as per soft metal properties (yield strength and ductility) of dissimilar metal combination needed to collapse the surface irregularities and promote the creep at high temperature to enhance the direct metal-to-metal contact. Pressure can be static or pulsating type. The pressure pulsation between the maximum (below the yield strength of soft metal) and minimum levels has been found to be more effective than the static pressure in improving the metallic intimacy. Lack of enough pressure results in limited bonding area, which in turn deteriorates the mechanical performance of diffusion bonds. Increasing pressure generally reduces the interfacial voids and increases the bonding, and so mechanical performance is improved (Fig. 4.41).

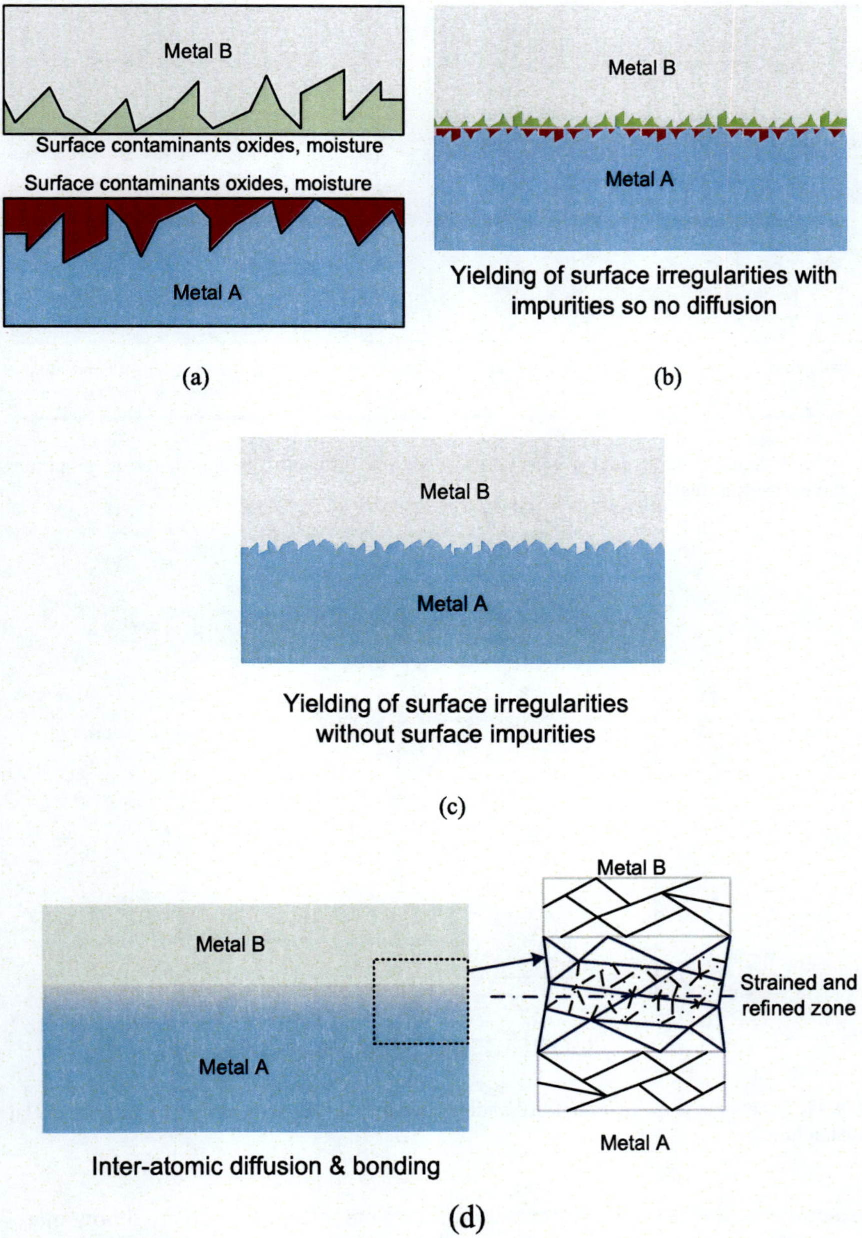

Fig. 4.39 Schematic showing different stages of diffusion bonding **a** dissimilar metal surfaces with surface impurities, **b** plastic deformation of surface irregularities with at the interface leading to no diffusion, **c** plastic deformation of surface irregularities without at the interface leading to direct metal to contact and **d** diffusion across the interface developing metallurgical bond with fine interfacial grain

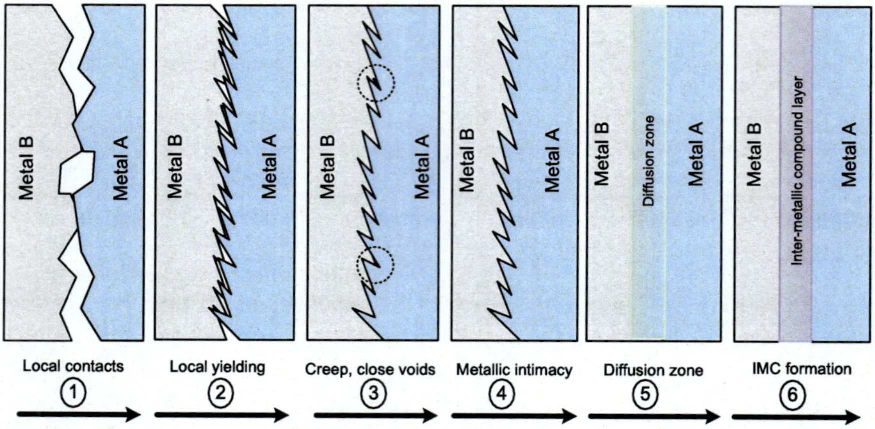

Fig. 4.40 Schematic sequential stages observed during diffusion bonding between properties cleaned surfaces of dissimilar metals

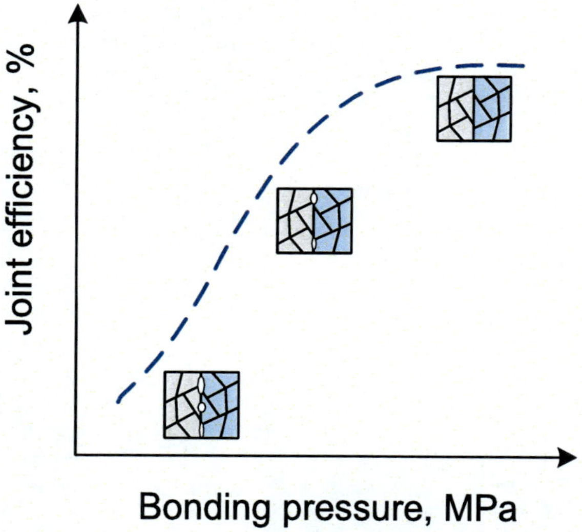

Fig. 4.41 Schematic showing effect of bonding pressure on joint efficiency of dissimilar metal diffusion bonds

Interatomic diffusion of elements across the joint interface (with or without interlayer) needs a high enough temperature (0.5 to 0.9 times of Tm, where Tm is the melting temperature in Kelvin) for a long time to achieve metallurgical continuity through the formation of new grains and phases and even intermetallic compound at the interface (Fig. 4.42). Metallurgical characteristics of joint interface primarily determine the mechanical, electrical and corrosion properties of diffusion bonds of dissimilar metals. The diffusion bonding temperature and pressure for dissimilar

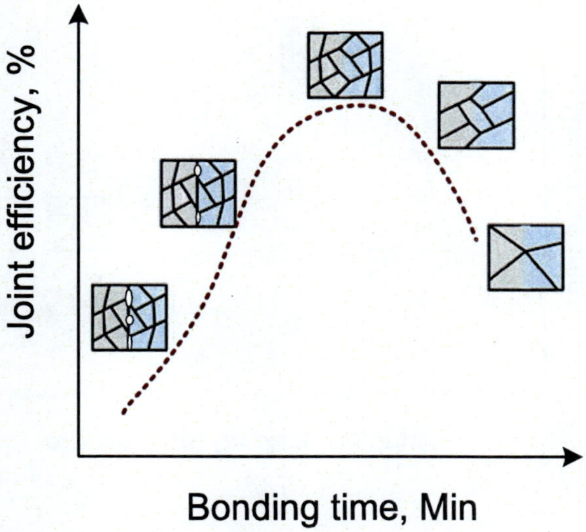

Fig. 4.42 Schematic showing effect of bonding time on joint efficiency of dissimilar metal diffusion bonds

metal joining are selected considering the melting temperature and yield strength of weak metal of a given dissimilar metal combination. Parent metals (like Al, Ti, Cr) that have tendency to form a passive layer (oxides) due to chemical affinity with atmospheric gases need higher vacuum temperature and time for diffusion bonding. Increasing bonding temperature initially increases the joint efficiency due to an increase in metallurgical bonding and reduced voids at the joint interface. After reaching maximum, joint efficiency starts decreasing with an increase in bonding temperature primary due to grain coarsening of parent metals, thickening of IMCs and formation of undesirable metallurgical constituents at the bonding interface (Fig. 4.43).

Interlayer in diffusion bonding of dissimilar metals plays a crucial role in decreasing the issues linked with incompatibilities related to thermos-physical, mechanical, chemical and metallurgical properties. There can be single or multiple interlayers as per the need of controlling IMC or formation of preferred IMC and decreasing residual stress. The interlayer of a suitable metal and thickness for diffusion bonding given a dissimilar metal combination improves metallic intimacy, forms favourable intermetallic compounds, reduces the cracking tendency and residual stress due to mismatch of thermal expansion/contraction behaviour, improves the diffusion across the interface and acts as diffusion accelerator. Thick interlayers (> 100 μm) degrade the mechanical performance of the diffusion bonds because these start taking load due to (its) metallic continuity across the interface, and so it acts as the weak link of the joint. Similarly, very thin interlayers may also not be enough to fill the ups/downs of the rough surface of parent metals thereby lowering metallic intimacy and increasing the interfacial voids. Therefore, the thickness of interlayer

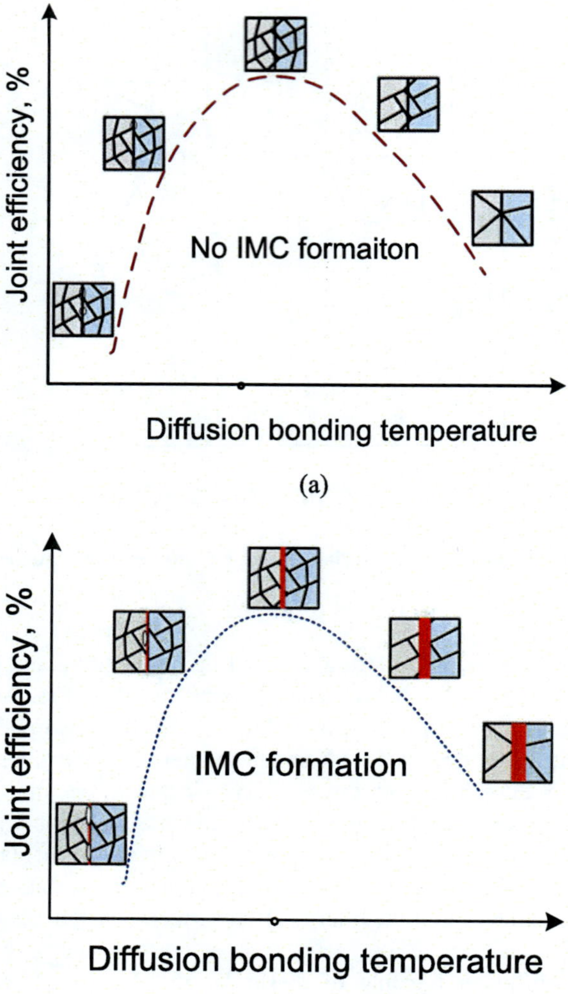

Fig. 4.43 Schematic showing the effect of diffusion bonding temperature on joint efficiency of dissimilar metal diffusion bonds formed **a** without IMC formation and **b** with IMC formation

needs to be optimized for a given set of diffusion bonding parameters of temperature, pressure and time. Increasing diffusion bonding time in general results in thinning of interlayer and widening of diffusion zone (Fig. 4.44). Further, the interaction effect of diffusion bonding parameters, namely temperature, time, pressure and interlayer is very complex and nonlinear depending upon the variant of the diffusion bonding process and dissimilar metal combination in consideration.

There are three common variants of diffusion bonding, namely solid-state diffusion bonding, roll bonding and transient liquid phase bonding. Interfacial melting

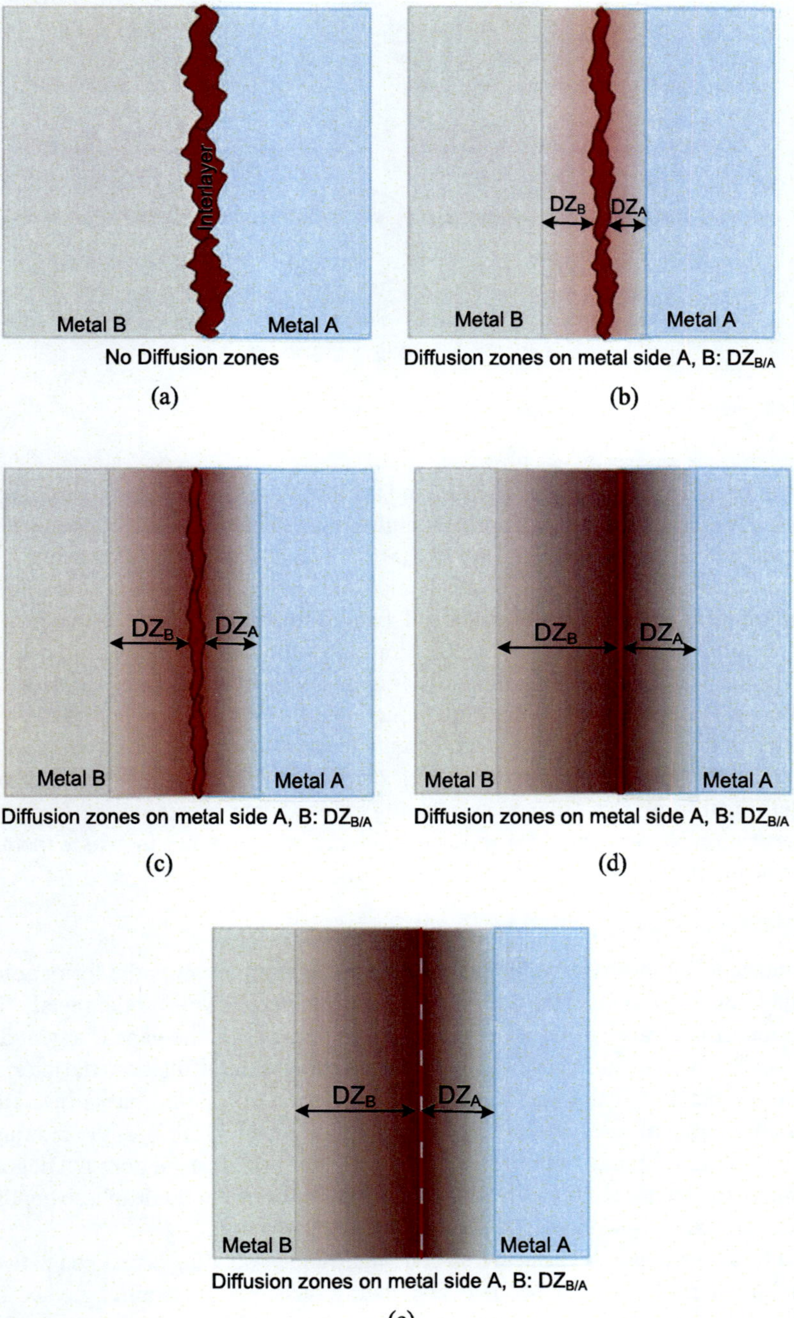

Fig. 4.44 Schematic showing the effect of diffusion bonding time on sequential metallurgical transformation including thinning of interlayer and widening diffusion zone at the dissimilar metal joint interface

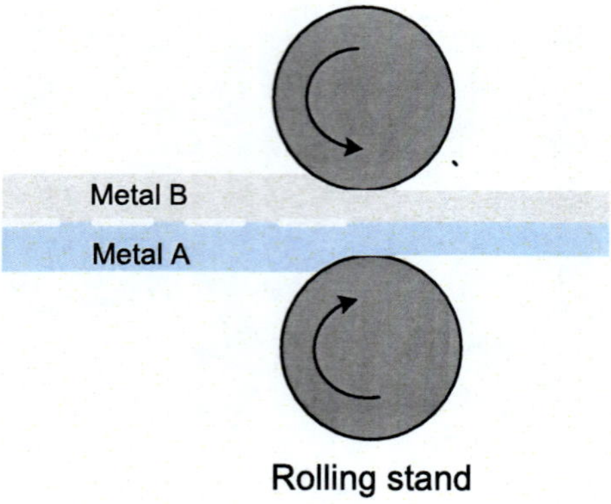

Fig. 4.45 Schematic of roll bonding for dissimilar metal joining

does not occur in the first two variants, but in the third variant, it is there. Solid-state diffusion bonding is usually performed at high temperature > 0.5 Tm in vacuum, while roll bonding can be cold (< 0.5 Tm), hot (> 0.5 Tm) and warm (commencing < 0.5 Tm and completing > 0.5 Tm). The mechanism of dissimilar metal joint development by roll bonding is similar to that of solid-state diffusion bonding, i.e. establishing direct metallic intimacy followed by diffusion across the joint interface. Roll bonding involves the passage of dissimilar metal components/sheet to be joined through a set of rollers with desired contact pressure/reduction ratio (Fig. 4.45). Very cleaned and smooth surface of parent metals is a prerequisite for roll bonding. High temperature accelerates the joining and improves joint characteristics. Localized and differential deformation of dissimilar metals during roll bonding increases the fracturing and separation of surface oxides and contaminants, which in turn increases metallic intimacy and diffusion across the joining interfaces.

Transient liquid-phase bonding

The transient liquid-phase bonding as the name suggests involves the formation of transient liquid to facilitate the diffusion bonding between the dissimilar metals. The formation of the transient liquid at the mating interface fills the gaps if any and so improves the metallic intimacy, which in turn encourages the diffusion at the interface to develop metallurgical bond. However, the presence of a liquid phase invariably forms an intermetallic compound at the joining interface. Therefore, the interlayer for transient liquid-phase bonding is chosen such that IMC formed does not degrade the joint performance appreciably. Different stages of transient liquid-phase bonding are schematically shown in Fig. 4.46.

The localized partial melting in transient liquid-phase bonding is realized with the help of an interlayer sandwiched between mating surfaces of dissimilar metals. The interlayer should contain melting point reducing alloying element. These elements on interaction with mating surfaces form low melting point constituents like eutectic. During the transient liquid-phase bonding, diffusion (out) of such elements from the interlayer/molten phases into mating surfaces of parent metals again raises the

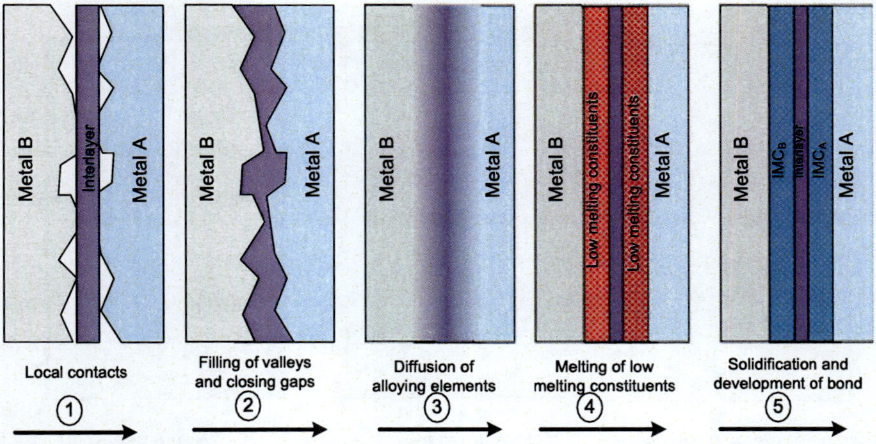

Fig. 4.46 Schematic showing different stages of transient liquid phase bonding of dissimilar metals

melting temperature which in turn causes the solidification of such constituents. Therefore, the bonding temperature for the transient liquid-phase bonding is kept slightly above the expected eutectic temperature during bonding.

Performance and characterization of the diffusion bonds

The interface of the diffusion bonds exhibits a wide range of metallurgical characteristics such as grain and phases structure, grain coarsening/refinement, locally strained and elongated grain structure, segregation/depletion of an alloying element, formation of single or multiple banded zones, diffusion zones, intermetallic compound formation, if any, voids/unbonded zone. However, the presence of these features depends upon the combination of diffusion bonding parameters (surface roughness, temperature, pressure, time and interlayer), variant of diffusion process in consideration and dissimilar metal combination. Solid-state diffusion bonding and roll bonding of dissimilar metals usually result in refined, strained and recrystallized grain structure along with a thin layer of IMCs. The localized surface yielding promotes the grain refinement, recrystallization and diffusion channel through grain boundaries. The formation of IMC (hard and brittle) is the obvious result of interaction between the metallurgically incompatible dissimilar metals through diffusion at elevated temperature. Application of suitable interlayer compatible with one or both the dissimilar metals helps to develop IMCs that are more favourable and improve the bond performance. For example, the use of Ag and Cu as interlayers for diffusion bonding of Al–Mg and Ti–SS dissimilar combination respectively improves the joint performance.

The weak interface of diffusion bonds of dissimilar metals besides IMCs can also be attributed to locked-in strains (residual stress) due to differential thermal expansion and contraction and the presence of discontinuities (voids, cracks) at the bonding interface (Fig. 4.47). Interlayers reduce some of these issues related to diffusion bonding of dissimilar metals.

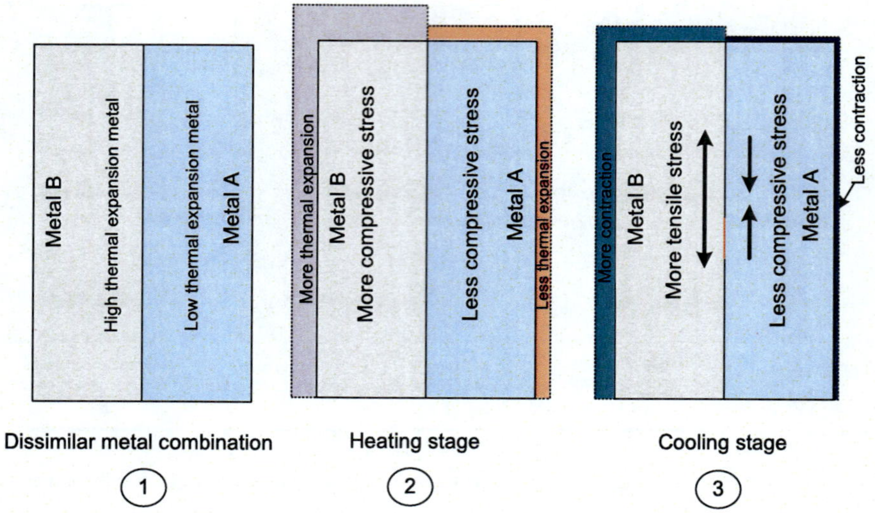

Fig. 4.47 Schematic showing thermal expansion and contraction behaviour of dissimilar metal diffusion bonds leading to the developing of residual stress across the bond interface

Diffusion bonding at a high temperature can degrade the parent metal properties as prolonged high-temperature retention may cause recovery, recrystallization, grain coarsening, tempering, annealing and reversion in one of the parent metals, while other may remain unaffected completely. The degradation in strength of polycrystalline metal occurs more than single-crystal metal with an increase in bonding temperature primarily due to grain coarsening and undesired metallurgical transformation (Fig. 4.48). Therefore, post-diffusion bonding treatment may be useful to restore the parent mental properties of dissimilar metal joints.

The electrical conductivity of diffusion bonds is usually found to be lower than the respective parent metal, and the same is attributed to the presence of interfacial voids reducing the path of flow of current and formation of thick and low conductivity intermetallic compound formation. The suitable interlayer is certainly expected to improve the electrical conductivity of diffusion bonds due to improved metallic intimacy, reduced voids and formation of favourable electrical conducting IMCs (Fig. 4.49). The concept of variation in electrical resistance of the diffusion bonds due to the presence of void has been exploited in the evaluation of fatigue crack growth/stage of crack growth in diffusion bonds. The cumulative fatigue damage has been related to increasing electrical resistance for the flow of current through the metallic bonds/joints.

The corrosion behaviour of diffusion bonds of dissimilar metals is largely influenced by the process parameters (temperature, time, pressure, interlayer) of diffusion bonding affecting the metallurgical characteristics of bond interface such as grain refinement, segregation, formation of the strained zone, banded zone and IMCs. Dissimilar metal joining is particularly sensitive to corrosion due to easy galvanic cell formation caused by different electro-negativity of the different zone of dissimilar

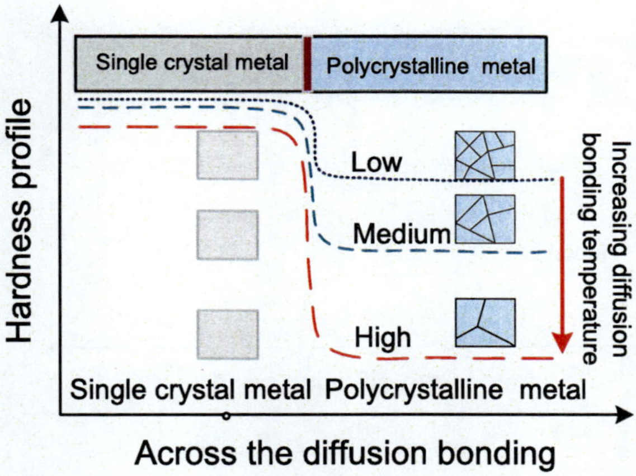

Fig. 4.48 Schematic showing hardness profile of diffusion bonds of dissimilar metal characterized as single and polycrystalline metals

metal bonds/joints. A fine-grain structure and low coincidence of site lattice boundaries at the bond interface (AISI 316 L) result in better intergranular corrosion resistance. An inappropriate combination of diffusion bonding such as bonding temperature, time and interlayer may result in thick and unfavourable IMCs promoting the galvanic corrosion at the interface. For example, diffusion bonding of titanium and Al results in TiAl IMC, which is selectively eaten out in a corrosive environment leading high corrosion rate of the diffusion bond interface (Fig. 4.50).

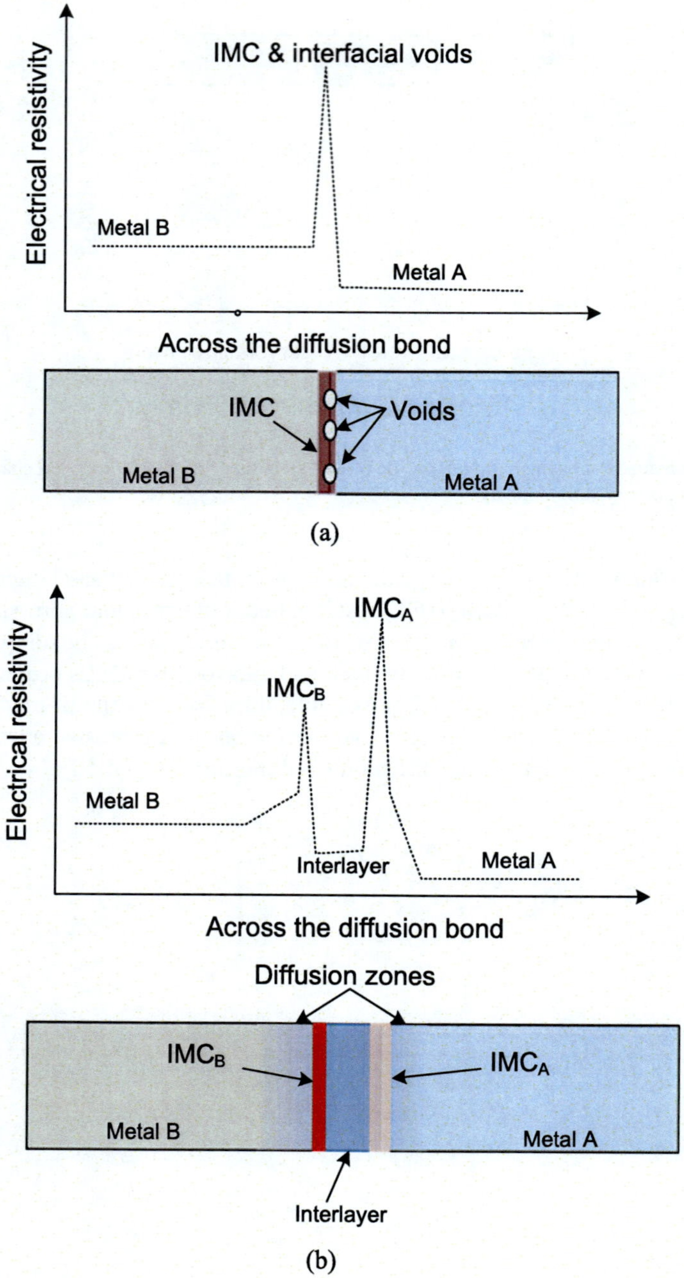

Fig. 4.49 Schematic showing variation in electrical resistivity across the diffusion bonds developed **a** without interlayer and **b** with interlayer between dissimilar metals

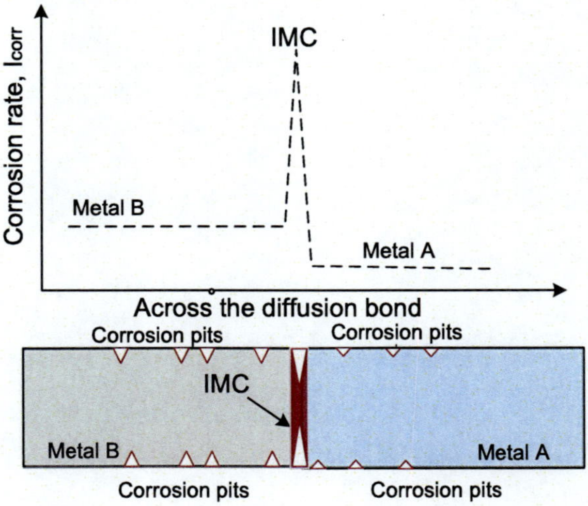

Fig. 4.50 Schematic showing degradation in corrosion resistance due to unfavourable IMC formation in diffusion bond of dissimilar metals leading to deeper pits than parent metals

References

Kaushik P, Dwivedi DK (2020) Selective induction heating in FSW of Al-steel combination. Presented in “28th international conference on processing and fabrication of advanced materials (PFAM 28)” held at VIT University, Chennai during 7th–9th December 2020

Kaushik P, Dwivedi DK (2021a) Effect of tool geometry in dissimilar Al-steel friction stir welding. J Manuf Process 68(Part B):198–208

Kaushik P, Dwivedi DK (2021b) Induction preheating in FSW of Al-Steel combination. Mater Today: Proc 46(1091–1095):2021

Kaushik P, Dwivedi DK (2022), Influence of hook geometry in failure mechanism of Al-Steel dissimilar FSW lap joint. Arch Civ Mech Eng (accepted)

Mohamed HA, Washburn J (1975) Mechanism of solid state pressure welding. Weld J 9:302–209

Sharma G, Dwivedi DK (2017) Microstructure and mechanical properties of dissimilar steel joints developed using friction stir welding. Int J Adv Manuf Technol 88(2017):1299–1307

Chapter 5
Dissimilar Metal Joining Using A-GTAW and HW-GTAW

Check for updates

This chapter presents the fundamental of A-GTAW, the issues and methodologies to deal with challenges of dissimilar metal joining using A-GTAW. Apart from A-GTAW, principles and issues related to dissimilar metal joining and remedies of hot wire GTAW and, further, effect of different factors related to parent metals, fillers, dilution, A-GTAW on properties weld metal, weld geometry and HAZ have been elaborated.

5.1 Introduction

The tungsten inert gas welding (TIGW) process is also known as gas tungsten arc welding (GTAW). This is one of the most preferred fusion welding process to produce high quality weld joint for critical application due to the two most attractive features, i.e. ability to produce (a) clean weld metal with very low oxygen and nitrogen content and (b) that too using very low heat. However, the GTAW suffers from low productivity owing to limited penetration capabilities coupled with requirement of many-many number of passes to complete the weld joint, especially in case of joining of thick sections as it uses non-consumable welding electrode. Therefore, the process is not popular in manufacturing industries where high volume production is needed (Vidyarthi and Dwivedi 2016, 2017a). To overcome above limitations, many advancement and development have taken place, e.g. high-current GTAW, hot wire GTAW and flux-assisted GTAW. The high-current (500–600 A) GTAW helps to achieve high penetration through increased heat input and arc forces leading the molten weld pool flow favourably to produce deeper penetration. Hot wire GTAW helps in welding of thick section in fewer passes by increasing deposition rate without increasing penetration capability while feeding the preheated filler metal in arc zone. Hot wire GTAW can produce deposition rate as high as that of submerged arc welding (without using high heat input of welding arc). Activated GTAW (A-GTAW) is another variant of GTAW wherein application of selected fluxes helps to increase the penetration many folds as compared to conventional GTAW. However, A-GTAW is primarily an

© The Author(s), under exclusive license to Springer Nature Singapore Pte Ltd. 2023
D. K. Dwivedi, *Dissimilar Metal Joining*,
https://doi.org/10.1007/978-981-99-1897-3_5

autogenous welding, which in case of dissimilar metal poses multiple challenges in development of a proper weld joint.

5.2 Fundamentals of Activated Flux-GTAW

5.2.1 Methodology

For A-GTAW, few oxides (TiO_2, SiO_2, Cr_2O_3, ZrO_2) and halide fluxes (as per parent metal) are applied along the weld centreline without any edge preparation. Only square groove geometry is used for A-GTAW. The flux is applied in the form of paste at the top surface of parent metals. After drying the applied flux, GTAW arc is applied along the weld centreline for welding (Fig. 5.1). Application of the flux produces arc constriction and reverse Marangoni convention to cause the deep penetration (Vidyarthi and Dwivedi 2017a, b).

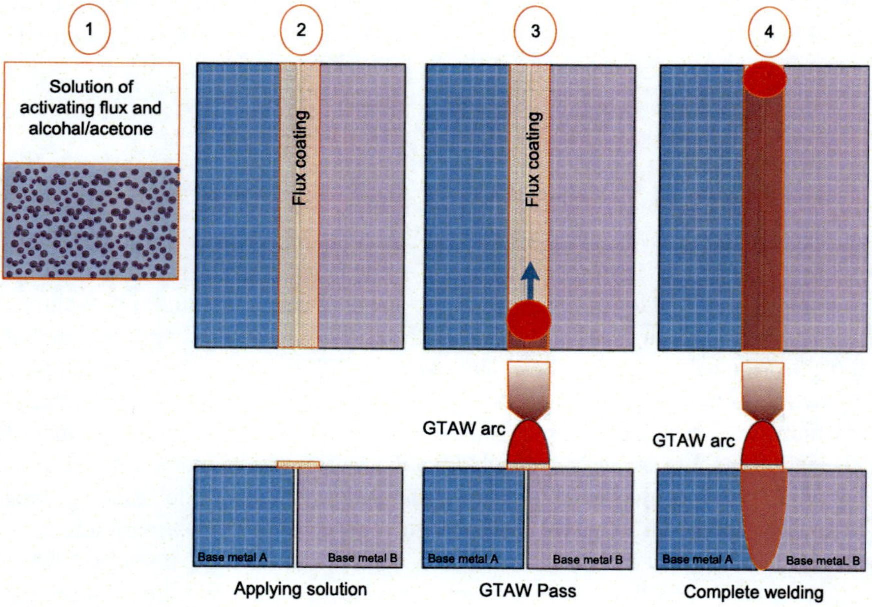

Fig. 5.1 Schematic of steps used A-GTAW process (1) prepare flux paste is acetone/alcohol, (2) applying paste along weld centreline, (3) welding arc passed along the weld centreline and (4) through thickness penetration weld produced

5.2.2 Mechanisms in A-GTAW

Application of activating flux results in many desirable effects during welding by increasing heat generation and heat intensity (localization) both. The increase of heat generation is attributed to higher arc voltage noticed during A-GTAW as compared to conventional GTAW process under identical welding conditions of arc length and welding current (Vidyarthi and Dwivedi 2017c; Vidyarthi et al. 2018). Increase of heat intensity or power density (W/mm^2) is caused by constriction of the welding arc. The constriction of arc reduces the weld width and increases the depth of penetration (Fig. 5.2).

The single-pass through thickness penetration weld joints having high weld depth-to-width ratio (> 0.6) reduces the angular distortion and residual stress. The penetration capability and depth-to-width ratio of the weld in A-GTAW is significantly influenced by welding conditions (welding current, speed and shielding gas governing the heat input) and application of flux and its spread (g/mm^2) (Vidyarthi and Dwivedi 2018a, b). In addition to the arc constriction, the penetration capability and depth-to-width ratio of the weld in A-GTAW strongly depends on the extent up to which heat applied by welding arc at the top surface of molten weld pool is actually used for melting the parent metals at bottom of the pool during the welding due to reversal of the Marangoni convention (Shankar and Dwivedi 2016). According to Marangoni convection, the flow of molten weld metal in normal fusion weld occurs from the centre of weld having highest temperature (so low surface tension) towards fusion

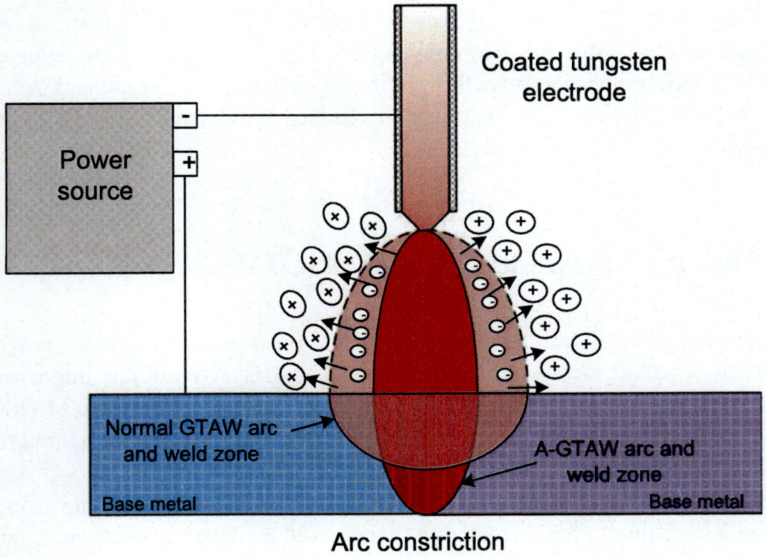

Fig. 5.2 Schematic of A-GTAW showing constriction of normal arc and corresponding change in weld bead geometry in case when two dissimilar metals have identical physical and chemical properties

boundary having minimum molten metal temperature zone (so high surface tension). This flow pattern facilitates the transfer of heat supplied by the arc from the centre to fusion boundary; therefore, in conventional fusion welding a wider weld bead with shallow penetration is realized. Application of activating flux causes reversal of Marangoni convection flow (due to reversal in surface tension versus temperature relationship) in the weld pool so molten metal flows from fusion boundary to weld centre (Fig. 5.3). The reversal in trend of surface tension versus temperature relation has been found to be related with oxygen content in molten metal. Higher the oxygen, greater the reversal in surface tension and temperature relationship. Oxygen concentration in the molten metal introduced during A-GTAW strongly depends on type and amount of oxide fluxes applied (Kulkarni et al. 2018a, 2019a).

How far reversal in Marangoni convention is able to transport the heat from the top of weld pool to bottom determines the depth of penetration which in turn is affected by two factors, namely (a) heat applied (H_{net}) as per welding conditions (welding current, speed, shielding gases, etc.) and (b) solidification time. The time for which the weld pool remains in molten conditions determines penetration due to extent of reversal in Marangoni convention. Longer the time (for solidification), deeper the penetration under identical conditions.

In the light of above, it can be said that A-GTAW (autogenous) welding should result in welding arc constriction and reversal of Marangoni convection both symmetrically even during the dissimilar metal joining to produce a symmetric weld joint. However, the difference in thermo-physical properties (namely thermal conductivity, melting temperature and solidification temperature range), chemical properties (parent metal and its alloying element dilution during the welding affecting pool composition) of the parent metals to be joined results in asymmetric reversal of Marangoni convection and arc constriction. Therefore, single-pass autogenous A-GTAW of dissimilar metals (with significant difference in physical and chemical properties) makes it extremely difficult to produce the desired symmetrical weld (Fig. 5.4).

5.3 Advantages and Limitations of A-GTAW

The development of dissimilar metal joint(s) of thick sections without any edge preparation of parent metals in a single pass using A-GTAW not just improves the economy and productivity but also offers multiple benefits in the form of reduced tendency of angular distortion, and residual stress development, as compared to the conventional GTAW (Fig. 5.5).

The dissimilar metal joining as compared to the same metal joining using A-GTAW is a little more complicated due to increased tendency of asymmetric weld, formation of hard and brittle IMCs, cracking tendency and varying residual stress in vicinity of the weld joint (Fig. 5.6) (Sharma and Dwivedi 2019a, b).

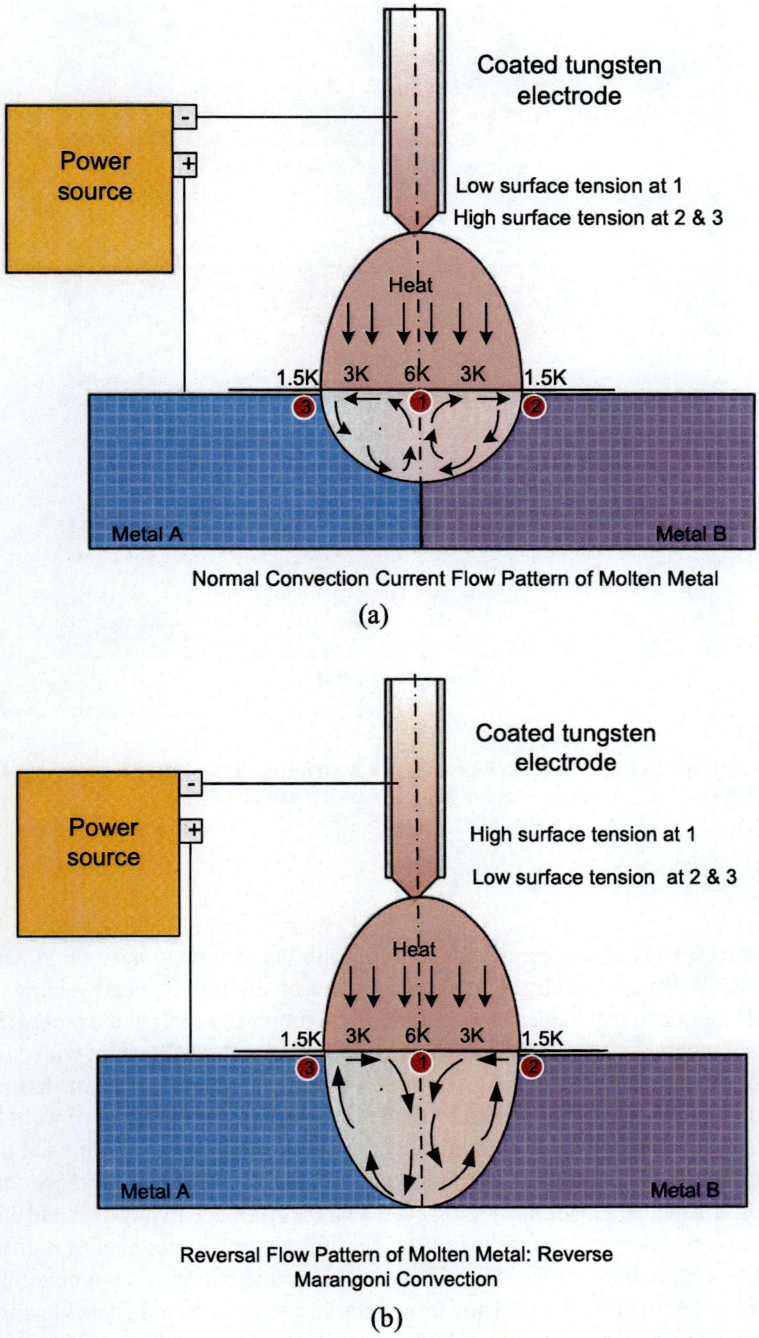

Fig. 5.3 Schematic showing pattern of flow of molten metal **a** centrifugal in conventional GTAW as per Marangoni convection current and **b** centripetal in A-GTAW as per reverse Marangoni convection current

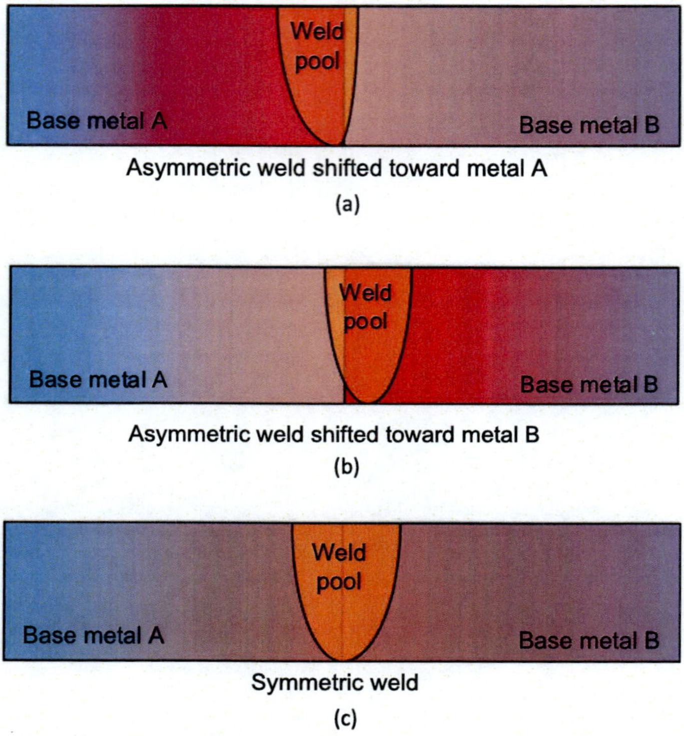

Fig. 5.4 Schematic of A-GTAW bead geometry: **a** asymmetric weld shifted towards metal A, **b** asymmetric weld shifted towards metal B and **c** symmetric weld

5.4 Parent Metals and A-GTAW

The characteristics of the parent metals of dissimilar combination to be joined by using A-GTAW significantly determine the ease of dissimilar metal joining. The physical, chemical, mechanical and metallurgical properties each of the parent metal of dissimilar combination determines how good, sound and reliable joint with desired set of mechanical and corrosion properties it will be developed. Ease of dissimilar metal joining depends on efforts to be made to develop a symmetrical of weld joint with desired combination of mechanical and metallurgical properties of weld metal and HAZs on both parent metals without compromising corrosion resistance for the desired performance of assembly. However, a large difference in physical and chemical characteristics of the parent metals of dissimilar combination makes it difficult to produce a symmetric weld, while the chemical and metallurgical incompatibilities result in formation of hard and brittle intermetallic compounds and phases leading to increased cracking tendency and reduced toughness. A large difference in physical and mechanical properties of dissimilar metal combination leads to asymmetric weld and differential nature and magnitude of residual stress in vicinity of the weld joint.

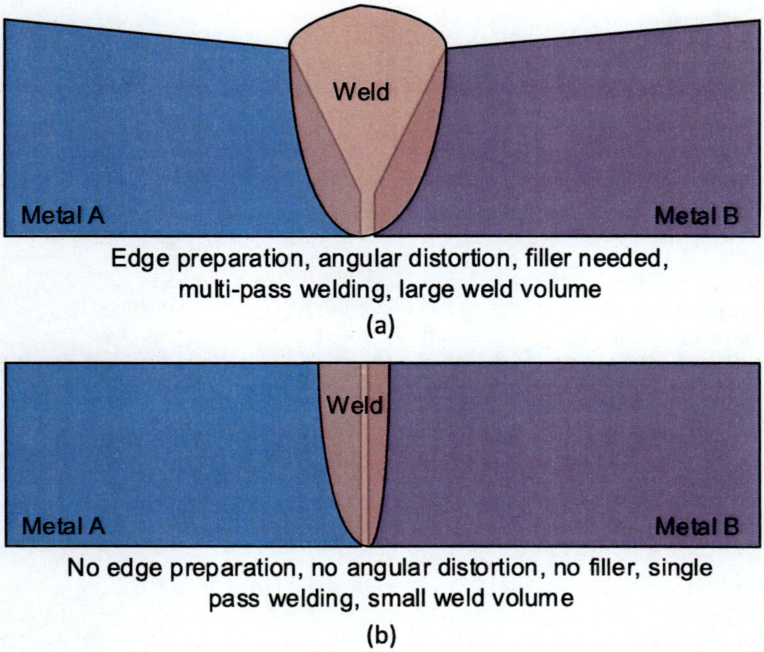

Fig. 5.5 Schematic showing distortion tendency in dissimilar metal weld joint developed using **a** conventional GTAW and **b** A-GTAW

The cracking of dissimilar metal joints in presence of residual tensile stress coupled with hard and brittle phases, and compounds becomes inevitable. Therefore, many precautions and efforts are needed to avoid or minimize such unfavourable conditions. Increased requirement for precautions and efforts to develop a sound dissimilar metal joint using A-GTAW reduces the ease of joining (Kulkarni et al. 2020; Sharma and Dwivedi 2021a).

5.4.1 Physical Properties

The difference of physical properties such as thermal conductivity, thermal expansion coefficient, melting and boiling point, surface tension, solidification temperature range and electromagnetic behaviour of dissimilar metal combination all affect the A-GTAW in different ways in developing sound and symmetric weld joint with desired mechanical and metallurgical characteristics.

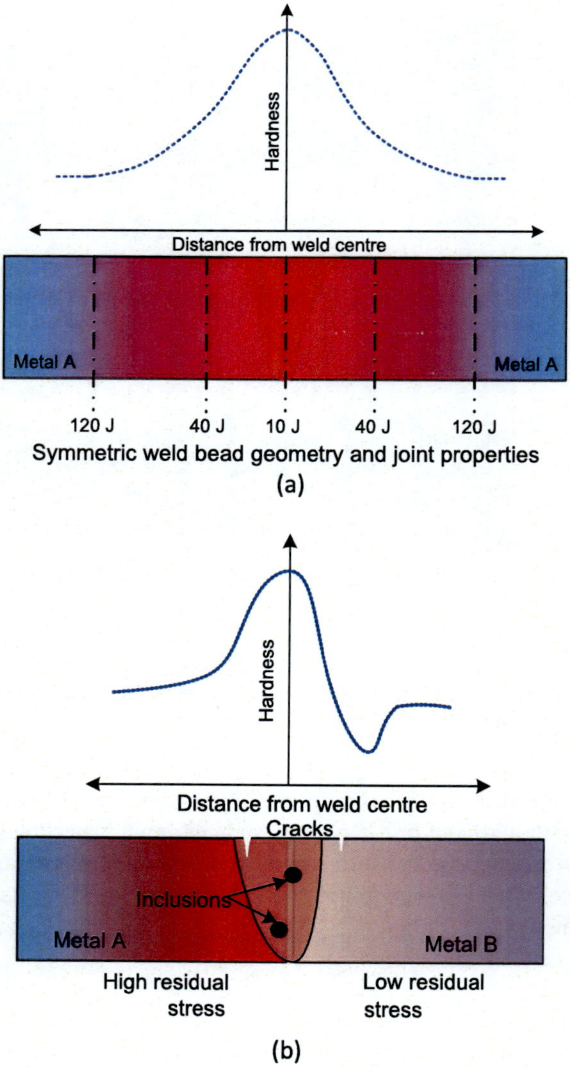

Fig. 5.6 Schematic showing A-GTAW weld joints of **a** similar metals with symmetric and sound weld and **b** dissimilar metal weld joint with asymmetric weld bead geometry and weld discontinuities (Numerical values in figure show impact toughness in J at different locations of weld joint across the weld suggesting embrittlement of weld metal and heat affected zone)

5.4.1.1 Thermal Conductivity

A large difference in thermal conductivity of parent metals primarily affects heat transfer rate from the weld pool on two sides, which in turn affects the weld pool retention time, weld thermal cycle of the weld pool and solidification rate experienced by molten weld pool in vicinity of fusion boundary of two parent metal.

A large difference in molten metal temperature of weld pool near the fusion boundaries on two sides of the parent metals affects the convection current and flow pattern of molten metal (from weld centre towards fusion boundaries or vice versa) as per

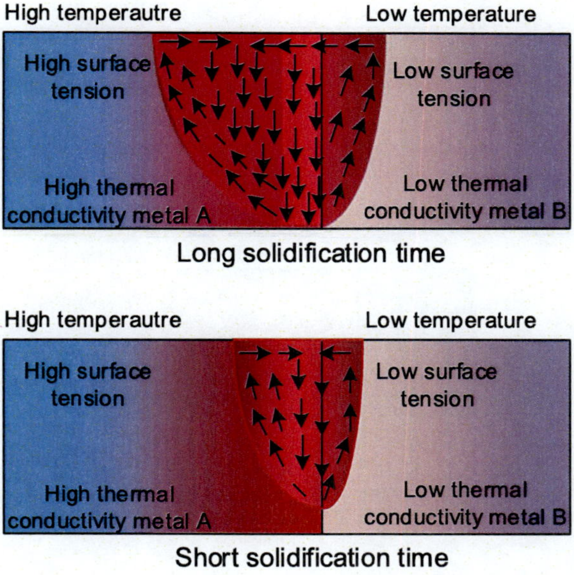

Fig. 5.7 Schematic showing effect of thermal conductivity of parent metals on weld bead geometry A-GTAW weld joints during dissimilar metal joining of **a** low thermal conductivity metals leading long solidification time and **b** high thermal conductivity metals resulting in short solidification time

the surface tension and viscosity of molten metal on two sides. Even in conventional GTAW, the weld pool tends to shift towards the high thermal conductivity metal from the weld centre besides causing wider heat-affected zone than the low thermal conductivity metal of dissimilar combination. Shifting of weld pool towards a particular metal in A-GTAW of dissimilar metal combination is more predominant than conventional GTAW due to the dependence of weld pool formation on reversal of Marangoni convection on dilution and the weld composition (Fig. 5.7).

The weld pool retention time (as per solidification/cooling rate) is very important in A-GTAW in achieving desired depth of penetration by ensuring the transport of the arc heat from the top to bottom of weld pool through centripetal flow of molten metal caused by reversal of Marangoni convection. Therefore, weld pool must have long enough weld pool retention/solidification time else, effectiveness of activated fluxes in A-GTAW to produce deep penetration is reduced. This difference can easily be noticed from the weld pool geometry of GTAW and laser welding produced using identical activated fluxes (Fig. 5.8).

5.4.1.2 Thermal Expansion Coefficient

The development of sound, symmetric and distortion-free weld joint is significantly affected by differential thermal expansion and contraction experienced by two parent metals during A-GTAW (Fig. 5.9). The A-GTAW being a single-pass through thickness penetration fusion welding process with reasonably high heat input invariably produces asymmetric residual stress on the two sides of parent metals of dissimilar combination, which in turn increases distortion and cracking tendency (Fig. 5.10).

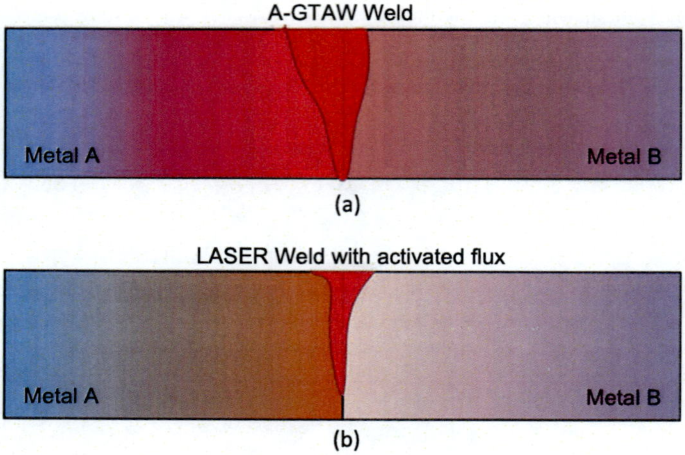

Fig. 5.8 Schematic showing effect of welding process on weld bead geometry of the weld developed with the help of activated fluxes using **a** A-GTAW and **b** laser welding

Increasing difference in thermal expansion behaviour of two parent metals increases above issues in case of A-GTAW without any filler. Moreover, the distortion tendency can somewhat be reduced using A-GTAW with suitable low yield strength and high ductility filler metal addition.

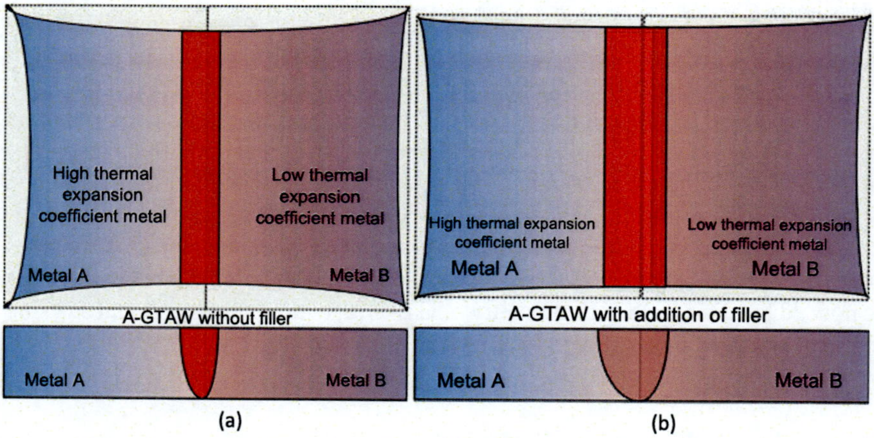

Fig. 5.9 Schematic showing effect of thermal expansion coefficient of metal distortion tendency in dissimilar metal welding using A-GTAW **a** without filler and **b** with filler metal

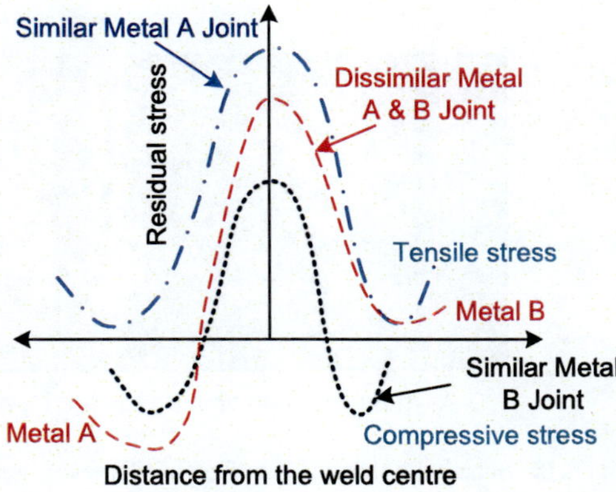

Fig. 5.10 Schematic showing residual stress distribution in transverse direction in similar and dissimilar metal A-GTAW joints

5.4.1.3 Melting and Boiling Temperature

The fusion welding of two parent metals (having a large difference in melting temperature) using A-GTAW becomes difficult primarily due to issues related to poor control over the molten metal and skewed melting of the parent metal. For example, A-GTAW may cause melting of one side parent metal of dissimilar combination while other may still be in solid state and far away from melting temperature (Fig. 5.11). In such cases, the A-GTAW is affected by varying dilution, weld composition and reversal of Marangoni convection. Similarly, a large difference in boiling temperature of parent metals and alloying elements in parent metals of dissimilar metal combination may cause evaporation and loss of alloying elements in one of the parent metals. While the other metal may remain unaffected by any such loss leading to an unexpected change in weld metal composition, penetration, mechanical and metallurgical properties of the weld metal, for example, elements like magnesium and zinc show significant vaporization during the fusion welding.

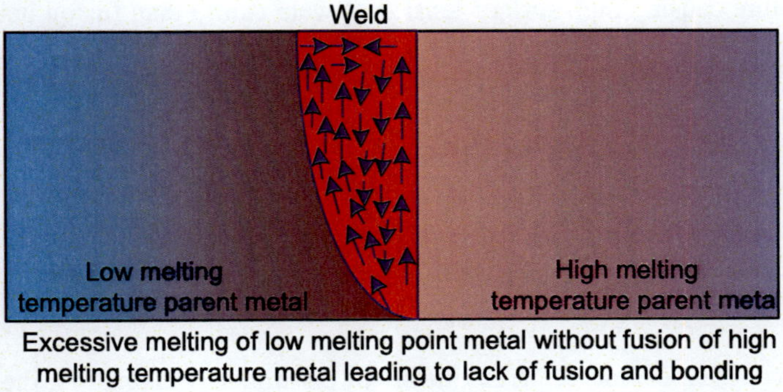

Fig. 5.11 Schematic showing effect of a large difference in melting temperature of two parent metals on weld bead geometry of dissimilar metal A-GTAW joints

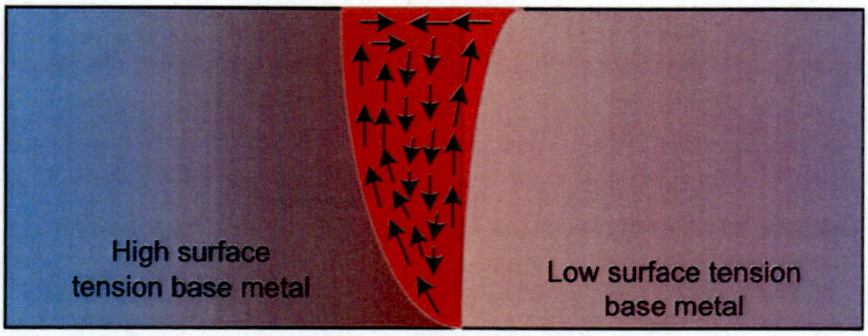

Fig. 5.12 Schematic showing the effect of a large difference in surface tension of two parent metals of dissimilar combination in molten state on weld bead geometry of dissimilar metal A-GTAW joints

5.4.1.4 Surface Tension

The surface tension of the parent metal in molten state at given temperature determines the direction of convection current in the weld pool. The molten metal flows from the low surface towards the high surface tension zone at a given temperature. The surface tension of the metals in general decreases with increase of temperature. Therefore, molten metal in similar metal welding flows from the centre (high-temperature zone) towards the fusion boundary (low-temperature zone) resulting in flow/transfer of arc heat flow from centre to fusion boundary causing wider weld bead and shallow penetration. In dissimilar metal joining, both parent metals may show different surface tensions at a given weld pool temperature; therefore, arc heat with molten metal may flow more from the weld centre towards the fusion boundary of the parent metal having high surface tension (Fig. 5.12). This in turn causes asymmetric weld and shifting of the weld centre away from the weld centreline.

The situation in A-GTAW is just reverse wherein addition of fluxes has opposite effect of temperature on surface tension, i.e. increasing surface tension with increase of temperature causing high surface tension at weld centre than fusion boundaries of both the parent metals. This leads to the flow of the molten metal from the fusion boundary to the weld centre (centripetal flow) which in turn transfers the arc heat from the top to bottom of weld pool (Fig. 5.12). The centripetal flow helps to achieve deeper penetration and narrow weld bead.

5.4.2 *Chemical Properties*

The chemical composition of parent metals of dissimilar combination, the affinity of elements in molten weld pool and those in solid state of parent metals with atmospheric gases and ability of elements to affect the surface tension of the weld at high temperature are few important aspects of chemical properties from A-GTAW point

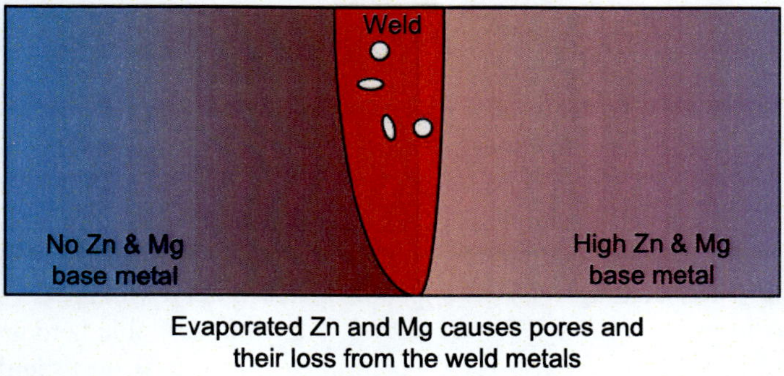

Fig. 5.13 Schematic showing effect of composition and alloying element on soundness and weld bead geometry of dissimilar metal A-GTAW joints

view. Incompatible elements in weld pool coming from two parent metals of dissimilar combination may lead to the formation of hard and brittle micro-constituents, undesirable segregation leading to increased tendency of cracking, embrittlement and corrosion. Elements (Ti, Al, Cr, Mg, Zn, etc.) in weld pool having high affinity with atmospheric gases (oxygen, nitrogen) can form variety of oxides, nitrides, inclusions and other impurities, which in turn deteriorates the mechanical and corrosion performance of the A-GTAW joints (Fig. 5.13). Alloying element and their concentration significantly affect the surface tension of the molten pool. Presence of elements like S and O reduces the surface tension while few others are known to increase the surface tension. The weld pool convention current (due to reversal/normal Marangoni convection) is primarily determined by the surface tension of the molten weld at centre and in vicinity of fusion boundary of the two parent metals of dissimilar combination. Therefore, chemical composition of the parent metals, all the factors affecting the weld pool composition like dilution, welding conditions and filler metal if any affect the A-GTAW penetration, bead geometry, mechanical and metallurgical properties of the weld joint (Sharma and Dwivedi 2021b, c).

Further, precise mechanism contributing deeper penetration by A-GTAW is still debated primarily due to varying effectiveness of fluxes used for A-GTAW for different parent metals. Efforts have been made to identify few fluxes and their suitability for selected engineering metals and understanding the possible mechanisms like arc constriction and reverse Marangoni convection. However, relative contribution of these mechanisms has been found to vary with parent metals. Changing effectiveness of fluxes to produce desired penetration for different parent metals further complicates the A-GTAW to produce a symmetric weld during dissimilar metal joining.

5.4.3 Metallurgical Properties

The metallurgical compatibility of the two parent metals in dissimilar metal joining by A-GTAW without filler is something, which is more important for realizing the sound, crack free, tough weld joint than other arc welding processes like SMAW, GMAW and SAW (Fig. 5.14). The requirement of metallurgical compatibility between two parent metals of dissimilar combination primarily arises due to autogenous nature of A-GTAW. The metallurgically incompatible metals like Fe with Al, Cu and Ti in molten condition during A-GTAW are known to form undesirable hard and brittle intermetallic compounds. Even A-GTAW of ferritic steel and austenitic stainless steel produces hard, brittle and crack-sensitive weld by A-GTAW (Fig. 5.15). High heat input and dilution both increase issues in dissimilar metal joining of metallurgically incompatible metals (Vidyarthy and Dwivedi 2019a).

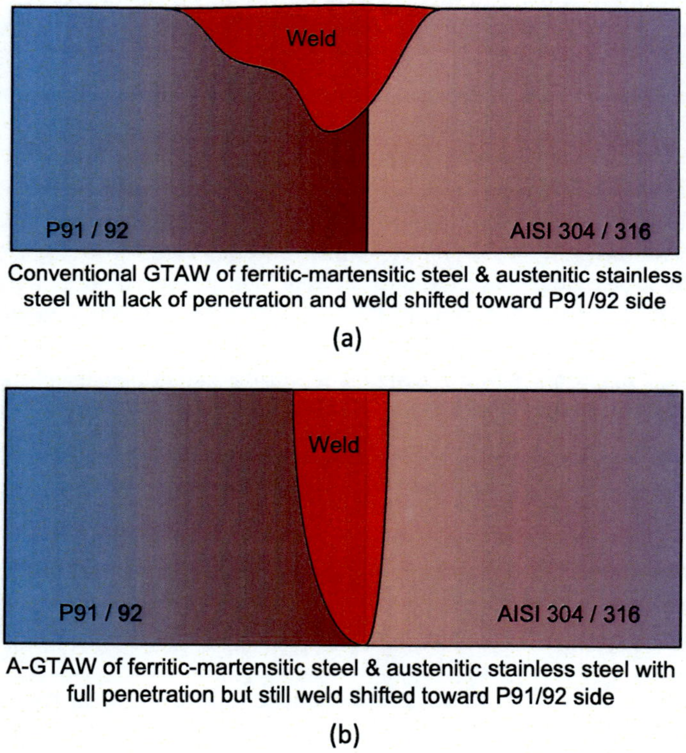

Fig. 5.14 Schematic showing issues related to metallurgical incompatible dissimilar metal welding by **a** conventional GTAW and **b** A-GTAW

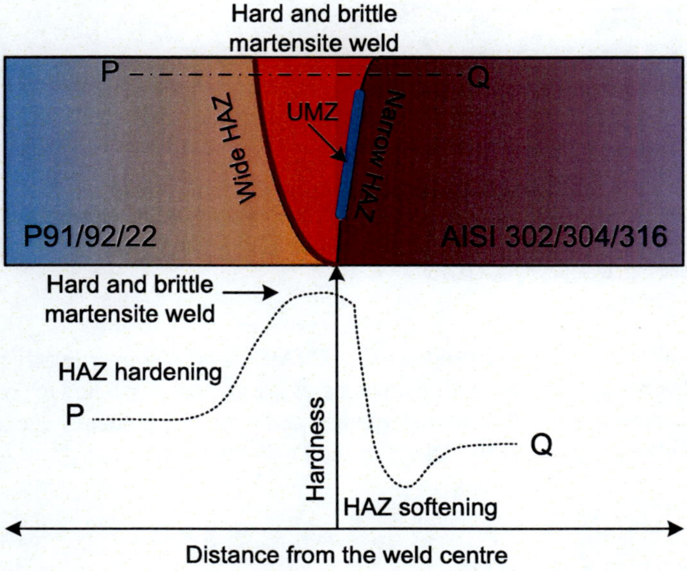

Fig. 5.15 Schematic showing issues related to metallurgical incompatible dissimilar metal welding by A-GTAW

5.4.4 Mechanical Properties

The role of mechanical properties of parent metals in developing sound joint of dissimilar metals using A-GTAW primarily limited to residual stress development and tolerance to the cracking of the HAZs of respective parent metals due to metallurgical transformation (Fig. 5.15). For example, HAZ and weld metal of most of the transformation hardening steels joined using A-GTAW show very low toughness, which is usually not good enough for many engineering applications. Such weld joints need either A-GTAW using filler and or post weld heat treatment to restore the mechanical properties or induce the desired ones in the weld metal. The yield strength of parent metals determines the maximum residual developed in vicinity of fusion boundary of each side and weld metal. The ductility and toughness of parent metals indicate the resistance to cracking in presence of residual tensile stress (Vidyarthy and Dwivedi 2019a, b; Kulkarni et al. 2019b).

5.4.5 Dimensional Properties

The section thickness of parent metals of dissimilar combination is probably one of the significant dimensional properties in A-GTAW as it affects (a) penetration requirement and (b) distortion tendency (Fig. 5.16). Relatively thin Sect. (3–5 mm)

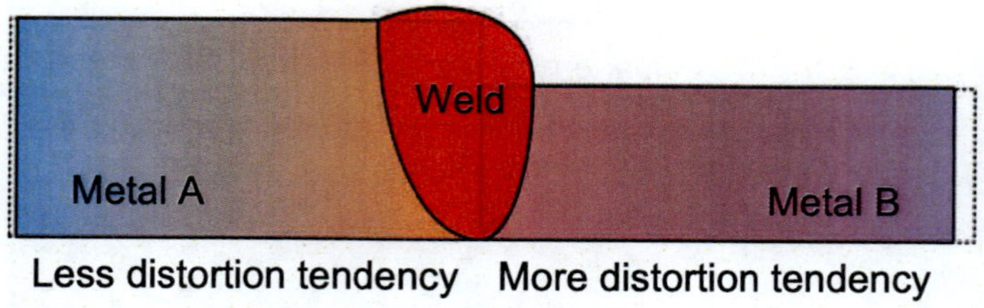

Fig. 5.16 Schematic showing weld bead geometry of different thicknesses of dissimilar metal plated subjected to welding by A-GTAW

parent metals can be directly welded using GTAW. A-GTAW is primarily preferred for deeper penetration when section thicknesses are greater than 6 mm or so. Further, thin section components show more wrapping and distortion tendency due to limited rigidity (Kulkarni et al. 2018b; Anagdha and Dwivedi 2019).

5.5 Flux and A-GTAW

The activated flux used in A-GTAW is one of the primary factors that determine the depth of penetration and weld bead geometry for a given set of the welding conditions. However, many characteristics of activating fluxes namely electrical resistivity, basicity, thermal stability in terms of chemical decomposition, melting and boiling temperature, and electron absorption tendency affect the A-GTAW welding arc constriction and reversal of Marangoni convection which in turn significantly affect weld bead geometry.

Electron absorption tendency of vapours produced by oxide fluxes and electrical resistivity of oxides both affect the welding arc constriction. Oxide fluxes under intense heat of welding arc at high temperature are melted and evaporated. The vapours of oxides fluxes form shroud around the welding arc. Absorption of electrons from the surface of welding arc by vapours shrinks the arc by reducing its cross section. The electron absorption mechanism causing arc constriction is found to be more dominating if the flux has melting point closer or lower than the parent metal. Application of high electrical resistivity oxide coating over the parent metals narrow downs the path of welding electrical current flow which is also expected to cause arc constriction. Since electrical resistivity and electron absorption tendency of different oxides fluxes are different, therefore not all types of fluxes are equally effective in arc constriction and increasing the penetration depth and reducing weld bead width (Sharma and Dwivedi 2019a, b, 2021a, b, c; Vidyarthy and Dwivedi 2019a) (Table 5.1).

Reversal of Marangoni convection depends on oxygen being introduced by fluxes after their decomposition in arc zone. Refractory metallic oxides showing high melting/boiling temperature which do not decompose easily during welding induce very limited oxygen in the weld pool to cause centripetal flow pattern of molten

Table 5.1 Physical properties of common fluxes

Activated flux	Density (g/cm^3)	Melting temperature (K)	Boiling temperature (K)	Electrical resistivity (Ω.m) at 293 K
NiO	6.67	2228	2487	10^{13}
SiO_2	2.19	1986	3223	10^{12}
Al_2O_3	1.93	2345	3340	10^7
CuO	6.3	1599	2273	10
TiO_2	4.23	2116	3245	10^8
Cr_2O_3	5.22	2707	4273	10^4
La_2O_3	3.34	2588	2898	3–8
MoO_3	4.69	1068	1428	10^{11}
SeO_2	3.24	613	805	–

metal as desired for deeper penetration. Therefore, refractory oxides show lower penetration than oxides having low melting temperature. These fundamentals help in explaining the behaviour of many oxides fluxes with many parent metals; however, these do not find undisputed acceptance among the scientific community primarily due to varying effectiveness of fluxes with changes of parent metals.

5.5.1 Factors Affecting the Role of Fluxes

The effectiveness of activating fluxes to produce deep penetration and high depth-to-width ratio weld depends on many multiple factors namely solidification/weld pool retention time, characteristics of the parent metal, dominating mechanism contributing the deeper penetration, section thickness, welding parameters affecting net heat input (welding current, voltage, welding speed, shielding gas, arc length) and interaction between the parent metal and oxygen resulting in oxide/slag formation. To realize the desired effect of activating fluxes through reversal of Marangoni convection, the weld pool retention/solidification time must be long enough. All the factors increasing the cooling rate of the weld pool like low net heat input (as per welding current, voltage, welding speed, shielding gas, arc length) and high section thickness in fact reduce the effectiveness of activating fluxes due to reduced centripetal convection caused by the reversal of Marangoni convection. Decreasing current, voltage, arc length and increasing welding speed lowers the net heat input. Similarly, application of Ar as a shielding gas in place of He, oxygen and hydrogen-modified Ar reduces the net heat input under identical welding conditions.

Parent metals form different types of oxides and slag after interaction with oxygen present in weld pool. These oxides/slag may be thin/thick of low/high viscosity accordingly these interfere/assist in weld pool flow pattern. The flow pattern accordingly affects the weld pool mixing and weld bead geometry.

5.5.2 Selection of Fluxes

The two predominating mechanisms namely arc constriction and centripetal convection current in weld pool contribute to deeper penetration and higher depth-to-width ratio of weld bead. However, effectiveness of activating fluxes to produce above-desired results using A-GTAW is not the same for all engineering metals. For example, few fluxes work effectively with austenitic stainless and other steels but not with non-ferrous metals like aluminium, titanium and copper alloy. Choice of the flux is very metal specific, and therefore, it is required to establish proper welding procedure specification including identification of fluxes, amount of fluxes to be applied (wt./area) and coating patterns in case of dissimilar metal joining besides other welding conditions. Certainly, it is important to consider melting/boiling temperature, electrical resistivity of oxide fluxes and their ability to provide/supply required oxygen to the molten weld pool to achieve desired arc constriction and reversal of Marangoni convection (Fig. 5.17). It therefore might be required to choose different fluxes in varying quantity (wt/area) and different coating patterns on the two parent metals of dissimilar combination to produce a symmetric weld joint. Therefore, different fluxes result in different penetration depths and weld depth-to-width ratio during by A-GTAW. Further, experimental studies have exhibited the deeper penetration and high weld depth-to-width ratio using multi-component fluxes during similar and dissimilar metal joining by A-GTAW (Fig. 5.18) (Shankar and Dwivedi 2016; Kulkarni et al. 2018b, 2019b).

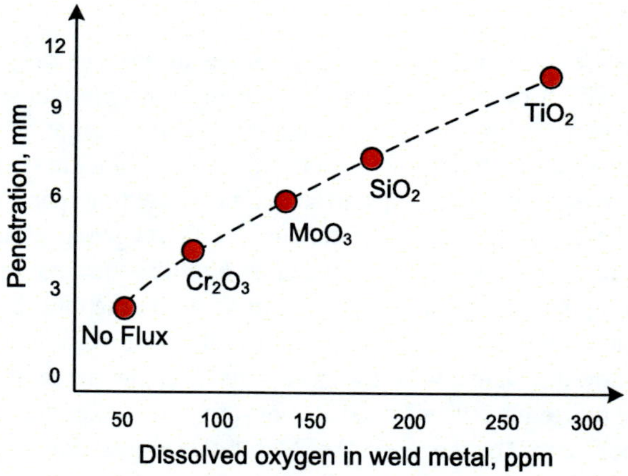

Fig. 5.17 Schematic showing effect of dissolved oxygen in weld metal on penetration achieved through different fluxes during A-GTAW of ferritic and austenitic stainless steel

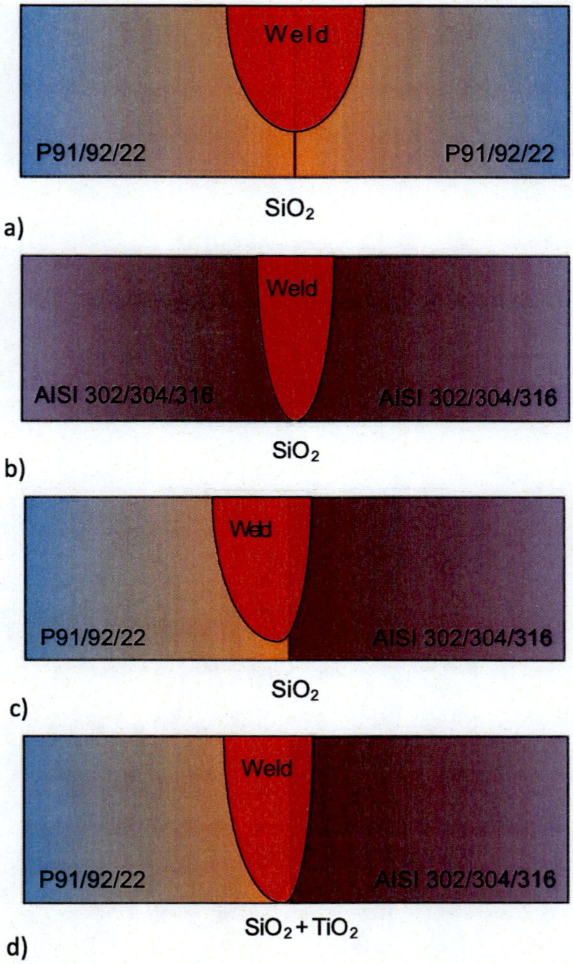

Fig. 5.18 Schematic showing effect of fluxes on weld bead geometry **a** single component flux (SiO_2) on similar ferrite-martensite steel A-GTAW, **b** single component flux (SiO_2) on similar austenitic stainless steel A-GTAW, **c** single component flux (SiO_2) on dissimilar ferrite-martensite steel and austenitic stainless steel A-GTAW and **d** multi-component fluxes ($SiO_2 + TiO_2$) on weld bead geometry during dissimilar metal joining of ferrite-martensite steel and austenitic stainless steel A-GTAW

5.5.3 Welding Parameters

All the factors affecting the net heat input and shielding gas used during the A-GTAW certainly affect the penetration and weld bead geometry. Net heat input during A-GTAW not only affects penetration but weld cross section as well. In general, an increase in heat input even in conventional GTAW simply increases the penetration and weld cross section linearly up to a limit before becoming insignificant. The effect of net heat input in A-GTAW on penetration and weld cross section is more predominant as it directly affects the weld pool solidification/retention time as described earlier (Fig. 5.19). Heat input, however, must be consistent with quality of flux applied (wt/area). Welding current (heat input) and flux quantity (wt/area) must be optimized to maximize penetration. In general, welding current should be

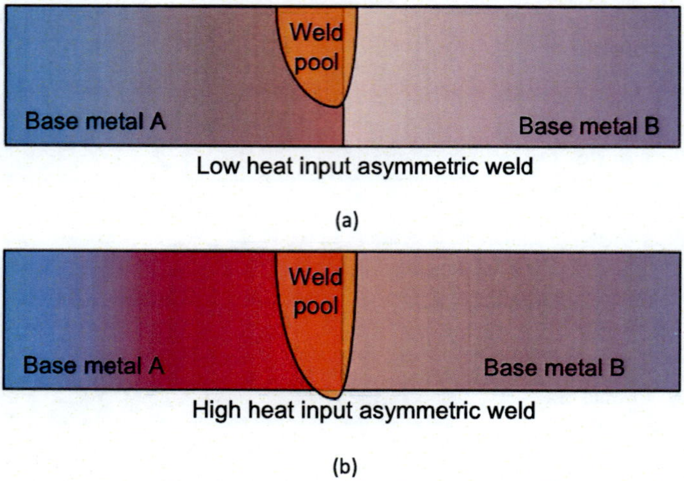

Fig. 5.19 Schematic showing effect of heat input A-GTAW bead geometry **a** low heat input and **b** high heat input

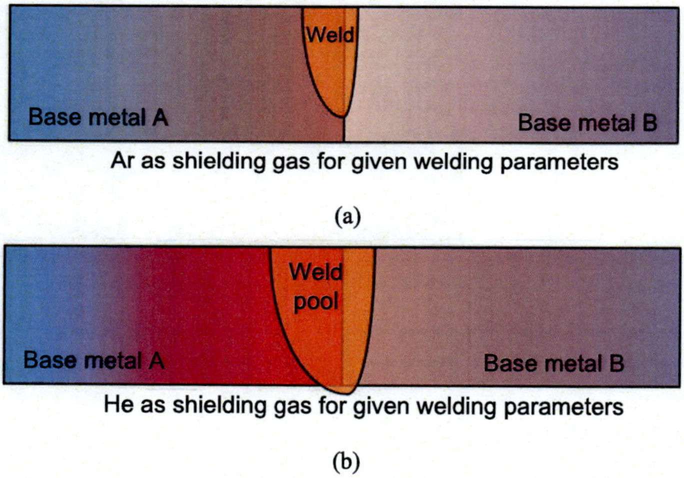

Fig. 5.20 Schematic showing effect of shielding gas on A-GTAW bead geometry **a** Ar and **b** He

increased with quantity of flux to get the desired penetration for joining of thick sections. Shielding gas affects the arc voltage so the arc heat generation, which in turn affects penetration, and weld bead geometry (Fig. 5.20). Further, He and Ar modified with H_2, O_2 and N_2 generate more heat than Ar alone. Argon modified with oxygen further helps to produce deeper penetration due to increased centripetal convection current caused by reverse Marangoni convection.

5.5.4 Flux Coating Patterns

The idea of using different flux coating pattern on two different parent metals to be joined using A-GTAW suits more for dissimilar metal joining. The scope of flux coating pattern may include applying different fluxes in varying quantities at different locations. The flux coating pattern choice needs a consideration of requirement to symmetrical centripetal convection current (through reversal of Marangoni convention) in weld pool even during the dissimilar metal joining (Fig. 5.21). This may require different amount of oxygen in two sides of weld pool (as per the parent metal) to adjust surface tensions of molten metal of respective metals to have symmetric centripetal flow and weld bead geometry. Following are few coating patterns for dissimilar metal joining.

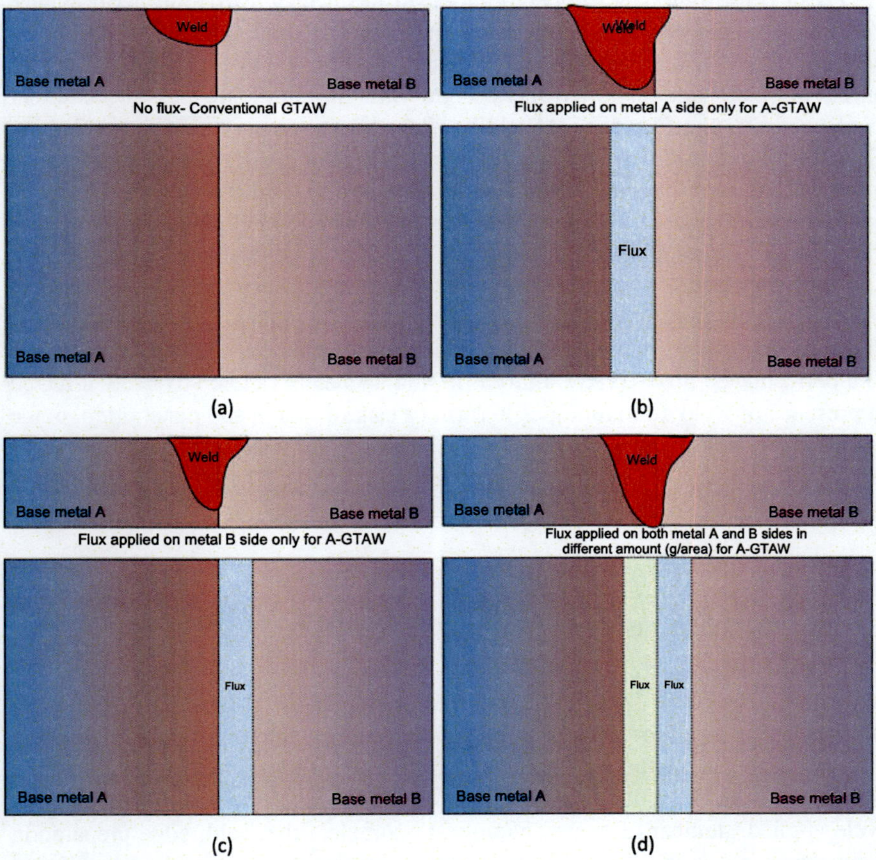

Fig. 5.21 Schematic showing effect of flux coating pattern on A-GTAW bead geometry **a** No flux, **b** coating on metal A, **c** coating on metal B and **d** coating on both metal A and B in varying amounts

5.6 Approaches to Enhance Joint Efficiency

The performance of a sound dissimilar metal joint produced by A-GTAW depends on microstructure, mechanical and corrosion properties of each zone of the joint. There are minimum five zones, i.e. BM1-HAZ1-W-HAZ2-BM2. Additionally, there can be two more zones in form of unmixed zone in weld close to the respective fusion boundaries of two parent metals. The performance of dissimilar metal joints produced by autogenous A-GTAW can be significantly lower than that of filler wire fed A-GTAW primarily due to the scope of the adjusting structure and composition of the weld metal suitably using desired filler metal compatible with both the parent metals. Moreover, the weld thermal cycle (WTC) experienced by HAZ and weld zone close to the fusion boundary on two parent metals can be appreciably different. The WTC, in turn, as per physical metallurgy and metallurgical transformations in weld and HAZ can cause improvement/deterioration in mechanical/corrosion performance. For example, autogenous A-GTAW of ferritic steel (AISI 1020, P22, P91, P92, etc.) with high alloy steel (AISI 302, 304, 316, etc.) produces very hard and brittle HAZ on ferritic steel side and weld metal both while austenitic stainless steels either largely remains unaffected or get slightly softened due to recovery and coarsening. A-GTAW with suitable filler (in form of wire/interlayer) for joining of such combination improves the weld toughness significantly by controlling the weld composition and microstructure (Fig. 5.22). Moreover, effective reversal of Marangoni convection during A-GTAW with/without filler can help to reduce the chemical heterogeneity.

Other approaches to improve the microstructure and mechanical properties of HAZs, and weld zone of A-GTAW dissimilar metal joint include selective preheating, electrode off-setting from the weld centreline, post weld heat treatment (Fig. 5.23). The performance of A-GTAW joint of dissimilar metals can be adversely affected in presence of weld discontinuities in form of lack of fusion and penetration, oxide inclusions in weld zone, solidification crack and liquation crack, under bead cracks and various types of HAZ cracking due to residual stresses. These must be taken care of by establishing suitable welding procedure specification.

5.7 Comparison of A-GTAW and M-GTAW Weld Joints

Activated flux GTAW (A-GTAW) and multi-pass GTAW (M-GTAW) have many similarities in terms of heat source (welding arc established between tungsten electrode and parent metal), short arc length and effective weld pool shielding. However, A-GTAW and M-GTAW differ in many ways including weld thermal cycle applied, number of passes required to complete the weld, edge preparation requirement, productivity, filler metal, dilution, weld metal composition, etc. Therefore, the microstructure mechanical properties of the weld and heat-affected zones of both similar and dissimilar metal joint produced by A-GTAW and M-GTAW differ significantly. However, a lot will depend on physical metallurgy and strengthening

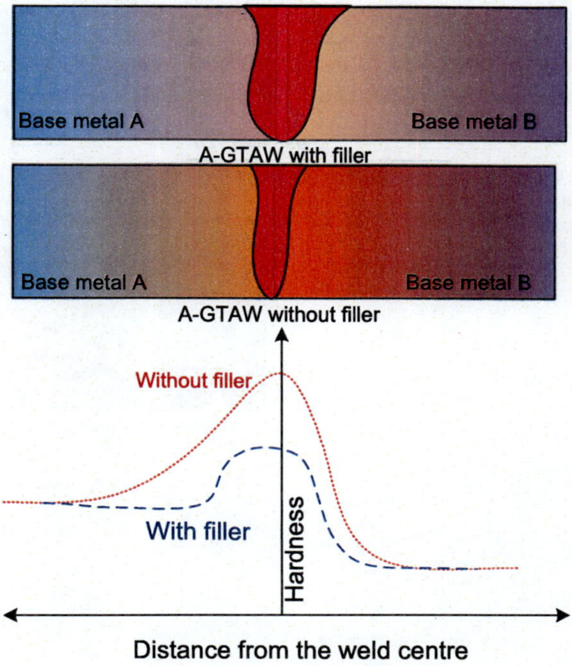

Fig. 5.22 Schematic showing effect of suitable filler addition on A-GTAW bead geometry and hardness distribution

mechanisms of two parent metals being joined (Shankar and Dwivedi 2017a; Sharma and Dwivedi 2019a).

5.7.1 *Preheat and Post-weld Heating (Bead Tempering)*

The A-GTAW is a single-pass welding and so it imposes just one weld thermal cycle to parent metals while multi-pass GTAW imposes many weld thermal cycles. The metallurgical transformation in weld and heat-affected zone realized after one weld thermal cycle of A-GTAW may lead to significant hardening or softening and residual stress development. On the other hand, the M-GTAW imposes one thermal cycle on each pass of welding which in turn results in many beneficial effects like preheating, tempering of already deposited bead and post weld heat treatment HAZs (Fig. 5.24).

5.7.2 *Weld Metal Composition*

The weld metal in case of autogenous A-GTAW is determined by dilution from two parent metals. A-GTAW with use of filler/interlayer allows the adjustment of weld metal composition even in single weld pass (thermal cycle) while M-GTAW

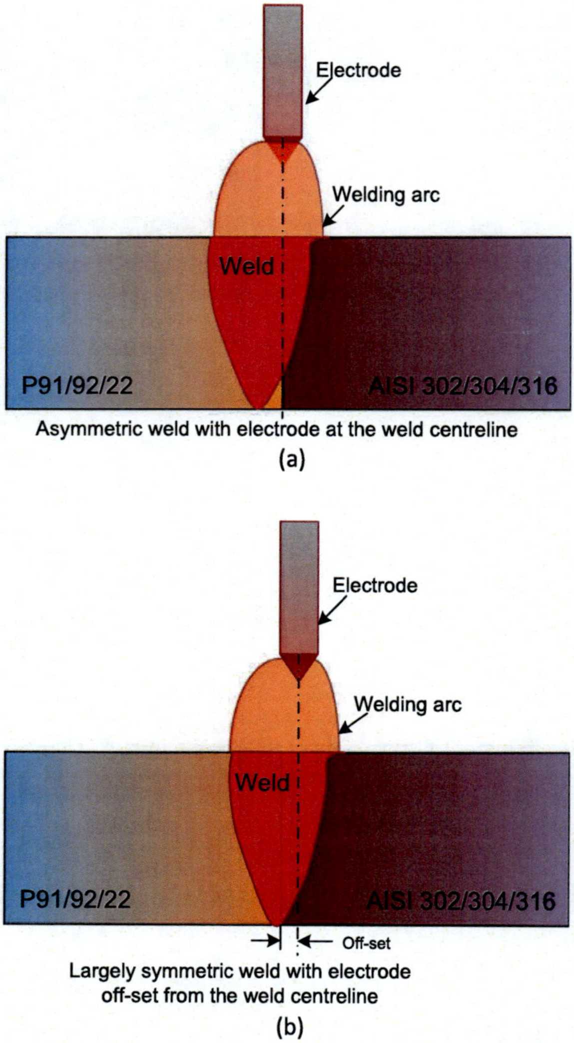

Fig. 5.23 Schematic showing effect of electrode off-setting on weld bead geometry during A-GTAW of dissimilar metal joints **a** weld electrode and arc at the weld centreline and **b** weld electrode and arc off-set from the weld centreline

develops the weld joint using suitable filler in multiple passes (thermal cycles). M-GTAW, therefore, can allow developing weld metal of desired composition and microstructure while autogenous A-GTAW suffers with this limitation specially in dissimilar metal joining. Therefore, A-GTAW with filler variants has been developed to realize not just deeper penetration, high depth-to-width ratio weld but also weld metal of the desired composition and properties.

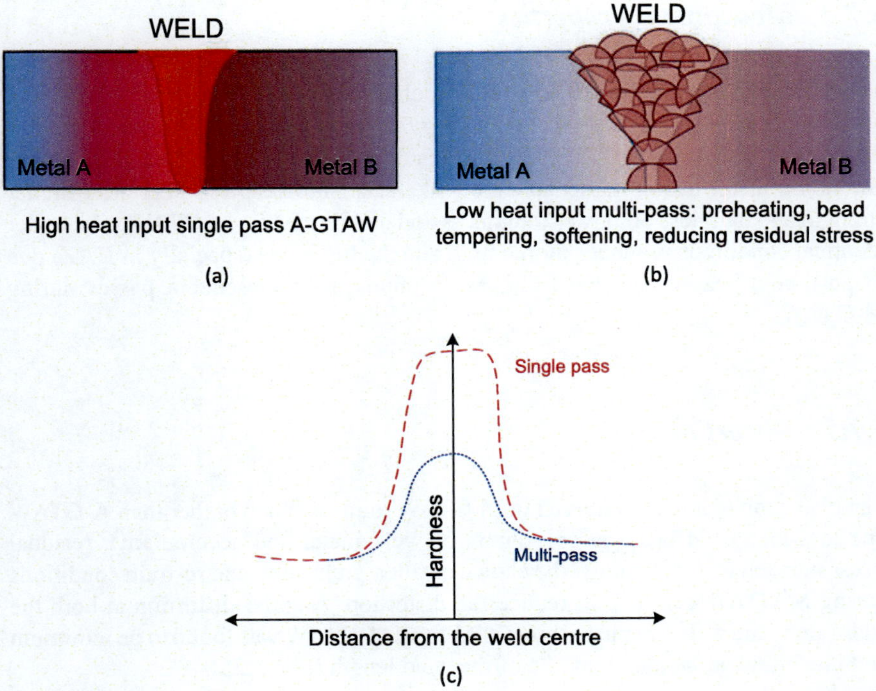

Fig. 5.24 Schematic showing features of weld joint developed using **a** single-pass A-GTAW, **b** multi-pass GTAW and **c** typical hardness profile of hardenable steel-weld joint developed using single-pass A-GTAW and multi-pass GTAW

5.7.3 Microstructure

The microstructure of the HAZ and weld metal developed using A-GTAW and M-GTAW joints differ significantly. A-GTAW although develops weld joint in a single pass and the heat input (kJ/mm) applied is usually much higher (2–3 kJ/mm) than M-GTAW (< 1 kJ/mm). High heat input causes wider heat-affected and coarser grain structure and increased tendency of untempered martensite formation in HAZ of hardenable steel while strain and precipitation hardenable metals show HAZ softening due to recrystallization, recovery, reversion and grain growth. These changes in M-GTAW may be very limited due to narrow width of HAZ. The microstructure of the weld metal on the other hand is primarily determined by composition of the two parent metals, dilutions and weld thermal cycle occurring in A-GTAW. In case of M-GTAW, composition of the filler apart for the two parent metals and welding parameters, weld thermal cycle, fluxes, etc., also play an important role in determining the microstructure of the weld metal.

5.7.4 Mechanical Properties

The mechanical properties of dissimilar metal joints developed by A-GTAW and M-GTAW depend on the physical metallurgy of metal systems of dissimilar combination. Autogenous A-GTAW joints of transformation hardenable metals like steels and non-transformation hardenable steel like austenitic stainless steel offer higher strength and hardness but lower toughness and ductility while M-GTAW joint under identical conditions produces more tough and ductile joints (specially in HAZ) due to post-weld heating and bead tempering imposed by subsequent passes during M-GTAW.

5.7.5 Distortion

The distortion tendency observed in M-GTAW joints is much higher than A-GTAW due to wider weld bead, large volume of weld metal and accumulative residual stress in reach pass. Certainly, the choice of filler, preheating and restrain conditions during M-GTAW can help to reduce the distortion. Angular distortion at both the ends (along the weld line) in both A-GTAW and M-GTAW was found to be minimum and maximum at the about middle of the weld length (Fig. 5.25).

5.7.6 Economics

An economic analysis of A-GTAW and M-GTAW considering labour, filler, power and shielding gas indicates that A-GTAW is more cost effective than M-GTAW. In case of M-GTAW, filler metal and labour costs are major components while in case of A-GTAW cost the flux is a major factor.

5.8 Characteristics of a Typical Ferrite-Martensite and Austenitic Steel Weld of A-GTAW

The dissimilar metal joining of ferrite-martensite steel (P91/92) with austenitic stainless steel (AISI 304/316) using A-GTAW with and without filler metal addition has been studied. The direct A-GTAW of ferrite-martensite steel (P91/92) with austenitic stainless steel (AISI 304/316) without any filler addition results in very hard and brittle weld and HAZ on ferrite-martensite steel side and the same is attributed on martensitic transformation in HAZ and weld metal both. The heat-affected zone of high carbon equivalent (CE) ferrite-martensite steel (P91/92) invariably causes embrittlement due to the typical weld thermal cycle imposed during fusion welding

Fig. 5.25 Schematic showing distortion tendency in dissimilar metal weld joint developed using **a** multi-pass GTAW and **b** single-pass A-GTAW

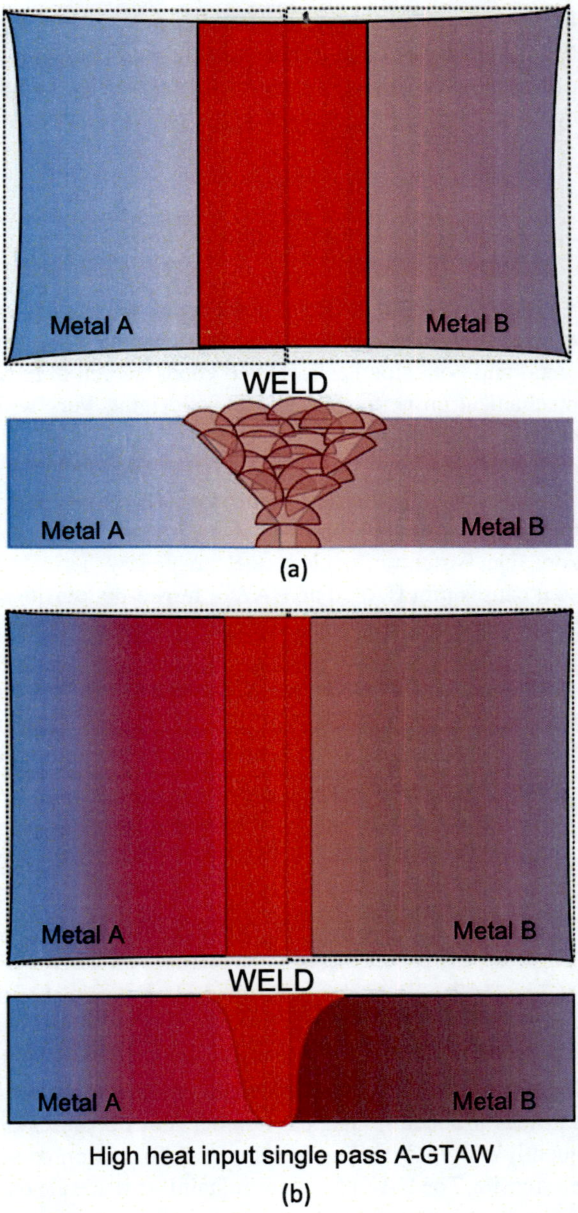

while the hard and brittle weld metal of such a dissimilar metal combination is primarily due to dilution from both the parent metals leading to reasonably high CE of the weld metal, which in turn produces hard and brittle weld metal with extremely low toughness (5–10 J). For the development of A-GTAW joint with reasonably acceptable toughness (> 40 J), multiple approaches have been suggested/explored like post weld heat treatment (tempering), A-GTAW with suitable filler metal, A-GTAW coupled with induction heating, etc. These approaches work on the principle of realizing the desired microstructure using either application of controlled thermal cycle or changing the chemistry of weld metal using suitable filler.

5.8.1 Metallurgical Characteristics

The A-GTAW joint of ferrite-martensite steel and austenitic stainless steel dissimilar combination develops significantly different metallurgical characteristics in weld metal and both side heat-affected zones, which in turn leads to a large variation in mechanical properties across the weld joint. Very wide HAZ of more than 3000–4000 μm is formed on ferrite-martensite steel while HAZ on austenitic stainless is comparatively very thin 100 to 200 μm depending upon the heat input. High thermal conductivity of ferrite-martensite steel (~29 $Wm^{-1} K^{-1}$) rapidly dissipates heat from the weld metal to the parent metal to a greater distance resulting in wider heat-affected zone than austenitic stainless steel having low thermal conductivity (~17 $Wm^{-1} K^{-1}$) so it causes thin HAZ. The HAZ of ferrite-martensite steel shows both coarse-grain heat-affected zone (in vicinity of fusion boundary) and fine-grain heat-affected zone while HAZ of austenitic stainless steel reveals coarse-grain heat-affected zone only.

Chemical analysis of the weld metal can be used to estimate the possible microstructural constituents of the weld metal such as martensite, martensite ferrite, ferrite, ferrite–austenite and austenite with help of Schaeffler diagram using Cr and Ni equivalent.

$$C_{req} = Cr + 2Si + 1.5Mo + 5V + 1.75Nb + 0.75W$$

$$Ni_{eq} = Ni + 0.5Mn + 30C + 25N + 0.3Cu$$

The HAZ of ferrite-martensite steel primarily shows the untempered martensitic structure due to typical weld thermal cycle (high temperature and high cooling rate) experienced by coarse-grain heat-affected zone in vicinity of the fusion boundary. High-temperature exposure results in complete dissolution of phases, compounds and carbides producing high CE homogeneous coarse grain of austenite which subsequently on rapid cooling leads to transformation of the austenite into untempered martensite. The HAZ of austenitic stainless steel side on contrary shows coarse grains coupled with recovery and recrystallization, which in turn cause the softening of the HAZ.

5.8.2 Mechanical Properties

The mechanical properties of ferrite-martensite steel and austenitic stainless steel dissimilar metal A-GTAW joint show huge variation across the weld joint in terms of tensile strength, ductility, toughness and hardness. All such variations in mechanical properties of the dissimilar metal weld joint can easily be captured and understood using micro-hardness distribution of transverse section of the weld joint, which can indirectly indicate the tensile strength, ductility and toughness of a particular location. In general, increase of hardness is coupled with increase of tensile strength and decrease of ductility and toughness.

On approaching from ferrite-martensite side parent metal to weld metal to austenitic stainless steel side, the hardness profile of the weld joint typically suggests (a) hardening of HAZ of ferrite-martensite steel, (b) very hard weld metal formation and (c) softening of HAZ of austenite steel (Fig. 5.26). During the tensile test, such dissimilar metal weld joint invariably fails from either softened HAZ or parent metal of austenite stainless steel side as suggested by minimum hardness region. Toughness of weld metal and HAZ of the ferrite-martensite steel is very badly compromised (5–25 J) as compared to that of austenite stainless steel (100–150 J). Hardening of HAZ of ferrite-martensite steel and that of weld metal is attributed to untempered martensitic transformation while softening of HAZ on austenite is due to coarsening, recovery, recrystallization and grain growth.

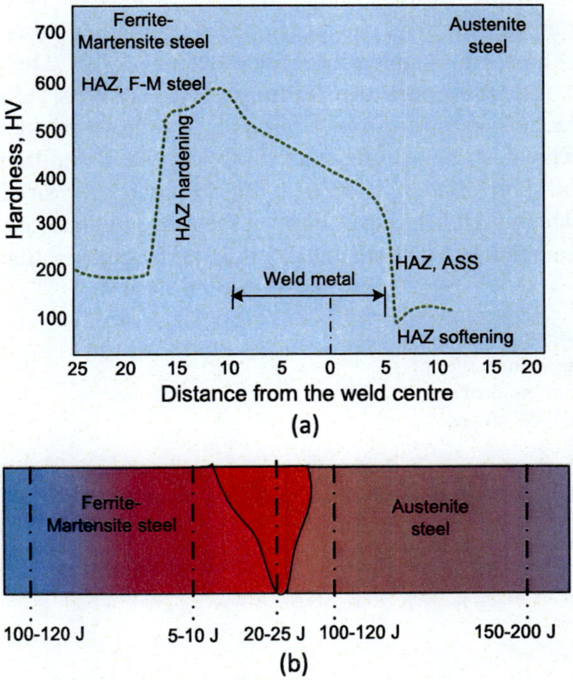

Fig. 5.26 Mechanical properties of dissimilar ferrite-martensite and austenitic stainless steel joint developed by A-GTAW **a** hardness profile cross the weld joint and **b** toughness of different locations across the weld joint

5.9 Hot Wire Gas Tungsten Arc Welding

The hot wire gas tungsten arc welding (HW-GTAW) process is based on the principle of using preheated filler wire during GTAW. This process is primarily designed to reduce heat input to the parent metals while realizing higher deposition rate for welding of thick sections (Fig. 5.27). Preheating of the filler wire reduces the heat used from arc for its melting as some of the sensible heat is provided to the filler during preheating. Therefore, preheating of filler increases melting rate and so welding speed even when using low net heat input. Increased welding speed due to high deposition rate even at low heat input results in high productivity. Preheating of the filler can be done using a suitable external source of heat. AC is commonly used to preheat the filler wire by electrical resistance (R_w) heating principle (Joule heating) (Fig. 5.28). Application of DC for preheating of wire can interfere with welding arc due to interaction of electromagnetic fields around welding arc and filler wire to cause arc blow. This process can be effectively used for welding of ferrous metals and Ni alloys. Welding of aluminium and copper by this process is somewhat limited mainly due to difficulties associated with preheating of Al and Cu fillers by electrical resistance heating as such a high (thermal and electrical) conductivity metals need heavy current (I_w) for electrical resistive heating of filler wire owing to their low electrical resistivity.

The heat input calculations for HW-GTAW need consideration of heat generated by weld arc as per welding current, arc voltage and welding speed (VI/S) and that of electric resistive heating ($I_w{}^2R_w/V_w$) due to the flow of current through filler wire for preheating. R_w can be calculated using filler cross-sectional area (e.g. resistivity of austenitic stainless steel filler ρ is 6.47×10^{-7} Ωm^9).

The most attractive feature of HW-GTAW is the ability to join the very thick sections in fewer passes while using low heat input almost like conventional GTAW, which can be extremely useful to join the dissimilar metal without causing issues like high dilution, wider heat-affected zone, excessive hardening/softening of HAZs due to high heat input. Further, the scope of choosing suitable filler wire (which is compatible with both metals of dissimilar combination) to develop the weld joint in

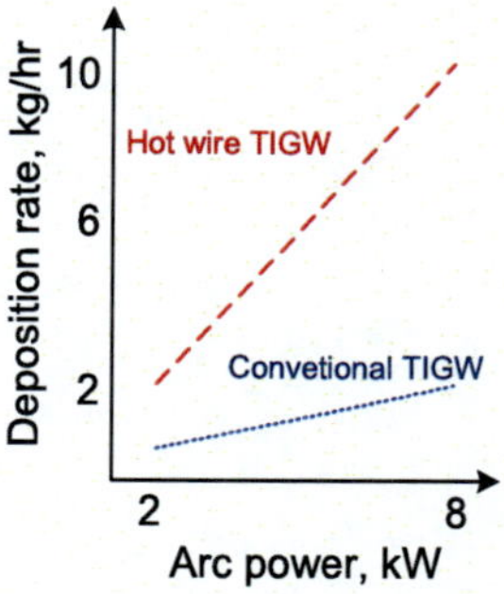

Fig. 5.27 Comparative deposition rates of conventional and hot wire GYAW process

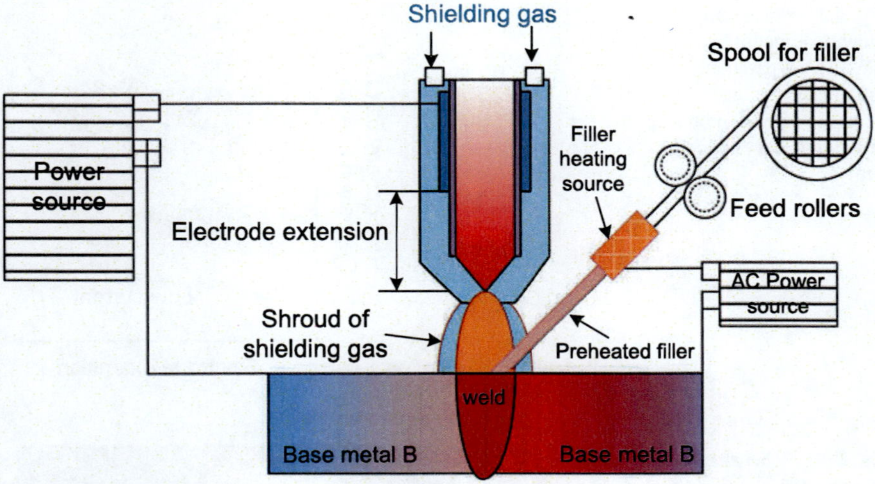

Fig. 5.28 Schematic showing the principle of hot wire GTAW process

few passes not just improves the productivity but also mechanical performance and other joint characteristics.

5.9.1 HW-GTAW Parameters and Weld Joints

Hot wire GTAW parameters (apart from those of conventional GTAW like welding current, arc voltage and welding speed) are current for wire preheating, extension of filler wire, wire feed rate, filler wire temperature, diameter of filler wire and its electrical resistivity. These parameters affect the electrical resistive heat of filler wire and so total heat input to the weld which in turn governs the weld bead geometry, deposition rate, microstructure and mechanical properties of the dissimilar metal joint (Fig. 5.29). The welding parameters namely high wire heating current, high electrical resistivity, long extension of filler wire and small filler wire diameter increase the heat generation (so the filler wire temperature) due to electrical resistive heating. High heat generation in turn increases the weld deposition rate, weld bead width, coarse-grain structure, while reducing the bead angle, bead reinforcement and impact resistance of the weld metal (Fig. 5.30).

In view of above, it is safe to assume that HW-GTAW can be used effectively to develop dissimilar metal joints just like GTAW while realizing high productivity and deposition rate. The scope for choice of filler metal (for HW-GTAW) compatible to both parent metals can certainly be an advantage over A-GTAW to development of the dissimilar metal joint.

Further, HW-GTAW in dissimilar metal joining of metallurgical incompatible metals needs buttering of suitable metal before applying a closing weld. The buttering

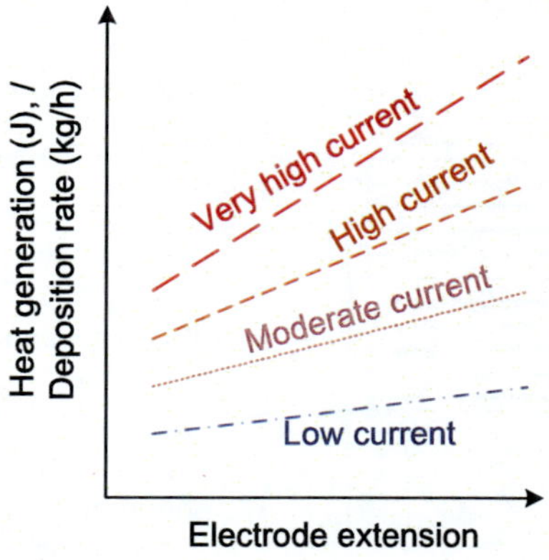

Fig. 5.29 Schematic showing effect of HW-GTAW parameters namely current, and electrode extension on heat generation and deposition rate

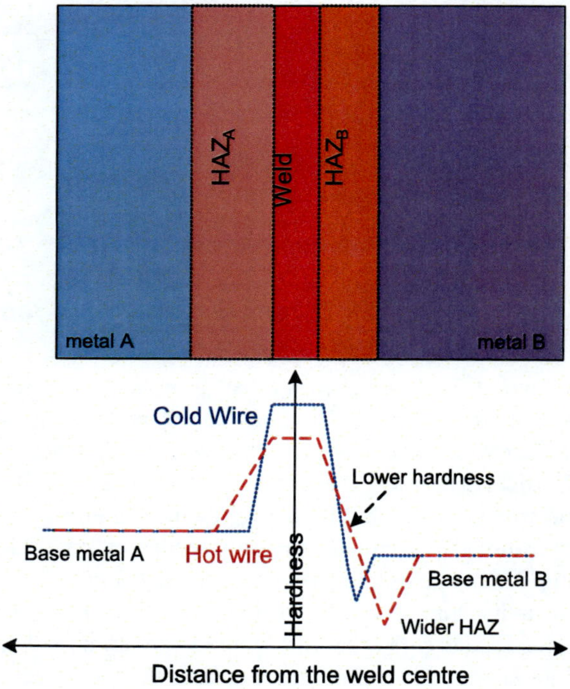

Fig. 5.30 Schematic showing comparative effect of cold and hot wire GTAW on HAZ and hardness profile across the dissimilar metal weld joint

helps to isolate the parent metal from the weld (and other parent metal) which can be problematic for weld properties and long-term performance of the weld during service. For example, HW-GTAW of alloy steel (F22/8630 forging steels) with high-strength steel (X65/F65) using a buttering layer of alloy 625 and low alloy steel result in lot of chemical and metallurgical heterogeneity in weld zone primarily due to large difference in chemistry of the metals coming from filler metal, buttering layer and the parent metal (Fig. 5.31). Additionally, convection current in weld pool and

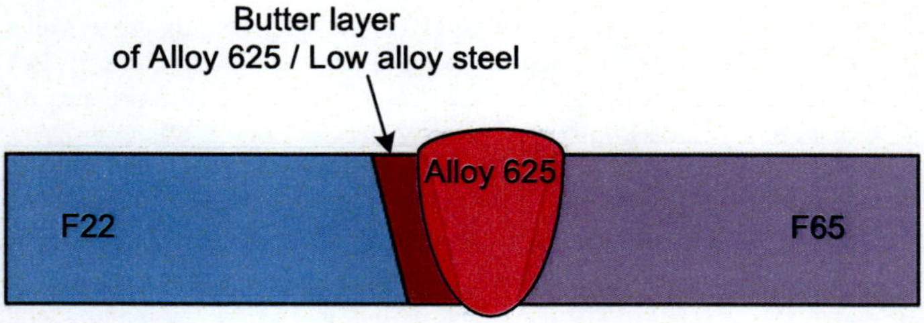

Fig. 5.31 Typical approach used for joining of dissimilar metals namely F22 and F65 using buttering layer appropriate filler metal using HW-GTAW

solidification behaviour of weld metals also affect the compositional homogeneity of the weld metal, which in turn promotes the cracking of weld, embrittlement and tendency of galvanic corrosion.

References

Anagdha S, Dwivedi DK (2019) Effect of developed activated flux on properties of gas tungsten arc welded thick gauge section of steels. In: International conference on nano, advanced materials, Tokyo, Japan, March 7–8, 2019

Kulkarni A, Dwivedi DK, Vasudevan M (2018a) Study of mechanism, microstructure and mechanical properties of activated flux TIG welded P91 Steel-P22 steel dissimilar metal joint. Mater Sci Eng A, 731:309–323

Kulkarni A, Dwivedi DK, Vasudevan M (2018b) Effect of oxide fluxes on activated TIG welding of P91 steel. Presented in international conference on advanced materials, Saint Petersburg, July 9–10, 2018b

Kulkarni A, Dwivedi DK, Vasudevan M (2019a) Dissimilar metal welding of P91 steel-AISI 316L SS with Incoloy 800 and Inconel 600 interlayers by using activated TIG welding process and its effect on the microstructure. J Mater Process Technol 274:116280

Kulkarni A, Dwivedi DK, Vasudevan M (2019b) Effect of oxide fluxes on activated TIG welding of AISI 316L austenitic stainless steel. Mater Today: Proc 18:4695–4702

Kulkarni A, Dwivedi DK, Vasudevan M (2020) Microstructure and mechanical properties of A-TIG welded AISI 316L SS-Alloy 800 dissimilar metal joint. Mater Sci Eng: A 790(2020):139685

Sharma P, Dwivedi DK (2019a) Comparative study of activated flux-GTAW and multipass-GTAW dissimilar P92 steel-304H ASS joints. Mater Manuf Processes 34:1195–1204

Sharma P, Dwivedi DK (2019b) A-TIG welding of dissimilar P92 steel and 304H austenitic stainless steel: mechanisms, microstructure and mechanical properties. J Manuf Process 44:166–178

Sharma P, Dwivedi DK (2021a) Wire-feed assisted A-TIG welding of dissimilar steels. Arch Civ Mech Eng 21(2021a):1–20

Sharma P, Dwivedi DK (2021b) Improving the strength-ductility synergy and impact toughness of dissimilar martensitic-austenitic steel joints by A-TIG welding with wire feed. Mater Lett 285:129063

Sharma P, Dwivedi DK (2021c) Study on Flux assisted tungsten inert gas welding of bimetallic P92 martensitic steel-304H austenitic stainless steel using SiO_2–TiO_2 binary flux: Welding arc/pool behaviour, microstructure and mechanical properties. Int J Pr Vessel Piping 192:104423

Vidyarthi RS, Dwivedi DK (2016) Activating flux tungsten inert gas welding for enhanced weld penetration. J Manuf Process 22(2016):211–228

Vidyarthi RS, Dwivedi DK (2017a) Influence of M-TIG and A-TIG welding process on microstructure and mechanical behaviour of 409 ferritic stainless steel. Mater Eng Perform 26:1391–1403
Vidyarthi RS, Dwivedi DK (2017b) Study of microstructure and mechanical property relationships of A-TIG welded P91-316L dissimilar steel joint. Mater Sci Engi A, 695:249–257
Vidyarthi R, Dwivedi DK (2017c) Analysis of the corrosion behaviour of an ATIG welded SS 409 weld fusion zone. J Mater Eng Perform 26:5375–5384
Vidyarthi R, Dwivedi DK (2018a) Microstructural and mechanical properties assessment of the P91 A-TIG weld joints. J Manuf Process 31(2018a):523–535
Vidyarthi R, Dwivedi DK (2018b) Microstructure evolution and Charpy toughness relationship of A-TIG weld fusion zone for varying tempering time. Trans Ind Inst Metals 71:1287–1300
Vidyarthi RS, Dwivedi DK, Vasudevan M (2018) Optimization of A-TIG process parameters using response surface methodology. Mater Manuf Process 33:709–717
Vidyarthy RS, Dwivedi DK (2019a) Effect of shielding gas composition and activating flux on the weld bead morphology of the P91 ferritic/martensitic steel. Mater Res Exp 6(8):0865f7
Vidyarthy RS, Dwivedi DK (2019b) Weldability evaluation of 409 FSS with A-TIG welding process. Mater Today: Proc 18:3052–3060

Chapter 6
Dissimilar Metal Joining Using Filler Wire Fed A-GTAW

This chapter presents the need, approach and underlying principle of using filler during dissimilar metal joining by A-GTAW. The factors to be considered for selection of suitable filler considering composition and physical metallurgy of parent metals, dilution and welding conditions have been described using Graville and Schaeffler diagrams. Case studies on dissimilar steel welding using A-GTAW with filler wire have also been presented.

6.1 Introduction

The activated flux GTAW is primarily an autogenous welding process involving melting of edges of components of dissimilar metals to be joined followed by solidification of the weld metal leading to metallurgical continuity to a develop weld joint. Dilution and intermixing of the parent metals from chemically and metallurgically incompatible metals during the welding frequently form the hard and brittle micro-constituents in the weld metal. Therefore, A-GTAW may produce a weld joint with very low strength, ductility and toughness making it unsuitable for engineering applications. Issues that are related to chemical and metallurgical incompatibilities to some extent can be minimized using butter layer to isolate one or both the parent metals from each other by applying suitable metal as buttering layer/cladding followed by A-GTAW. However, application of isolating buttering/cladding layer makes the process time consuming and uneconomical due to requirement of edge preparation, buttering the entire faying surface of one or both the parent metals as per need. This problem related to A-GTAW needs to be addressed (Fig. 6.1) (Shankar and Dwivedi 2017; Kulkarni et al. 2018, 2019, 2020; Sharma and Dwivedi 2019a, b, 2021a, b, c; Vidyarthy and Dwivedi 2019).

© The Author(s), under exclusive license to Springer Nature Singapore Pte Ltd. 2023

D. K. Dwivedi, *Dissimilar Metal Joining*,
https://doi.org/10.1007/978-981-99-1897-3_6

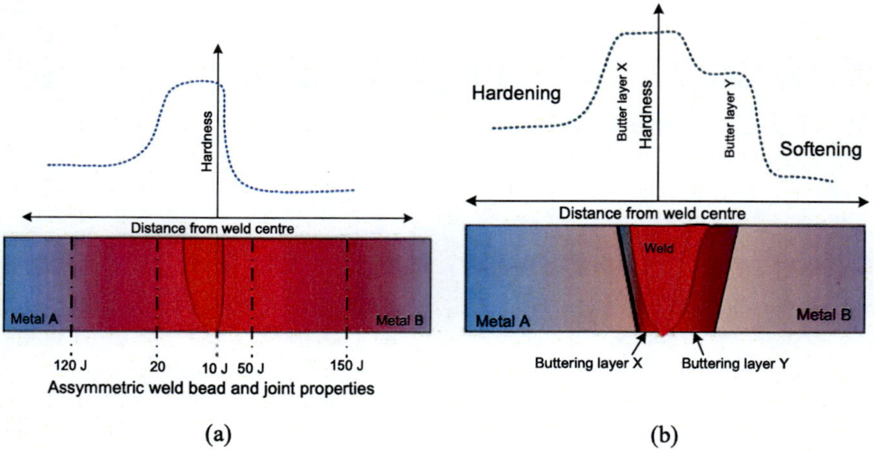

Fig. 6.1 Schematic of showing typical hardness distribution and weld cross section of dissimilar metal joint developed using A-GTAW **a** without buttering layer on to the parent metals and **b** with buttering layers on to the parent metals (Numerical values in figure show the different impact toughness values in different zones of two dissimilar metals showing different hardening behaviours due to weld thermal cycle)

6.2 Filler Wire Fed A-GTAW and Other Variants

Filler wire fed A-GTAW is comparatively newer approach for developing a reasonably good weld joint of metallurgically incompatible metals, wherein A-GTAW is combined with feeding of the suitable compatible filler wire into the arc zone to regulate/adjust the composition, microstructure and mechanical properties of weld joint. A well-calculated and controlled dilution from parent metals and filler helps in achieving the desired mechanical properties of the weld joint. However, the chemical homogeneity of the weld metal is crucial in realizing the desired composition and properties. Therefore, parameters related to wire feeding and A-GTAW must be optimized to achieve chemical and metallurgical homogeneity in dissimilar metal weld (Sharma and Dwivedi 2021a; b).

The approach of filler wire feeding in A-GTAW is similar to that hot wire GTAW or conventional GTAW with filler wire feeding in the arc zone. However, objectives achieved from these variants of GTAW differ significantly. Filler wire feeding in A-GTAW helps to realize development of weld joint of thick sections of dissimilar combination in a single pass with high depth-to-width ratio. On the other hand, the hot wire GTAW offers advantage of high deposition rate but with requirement of multiple passes and low weld depth-to-width ratio, while conventional GTAW offers very low deposition rate, low penetration, a large number of passes and low depth-to-width ratio of the weld (Fig. 6.2).

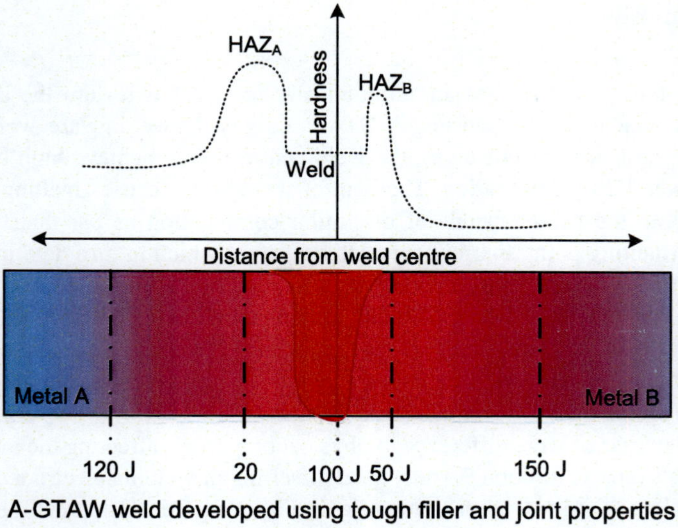

(a)

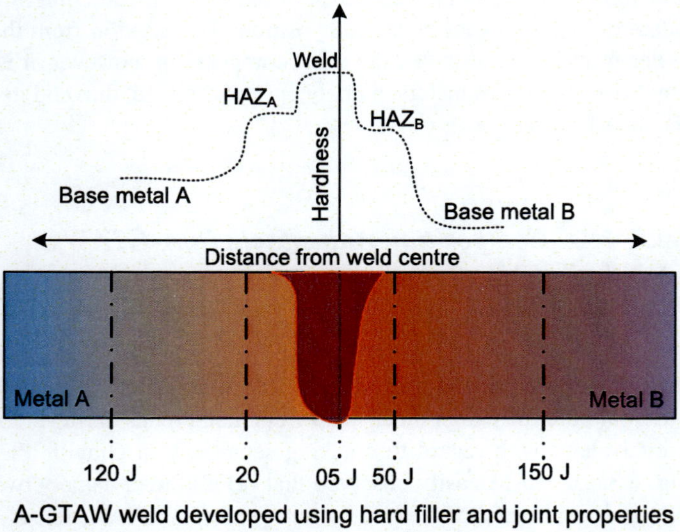

(b)

Fig. 6.2 Schematic showing the effect of filler on hardness distribution and weld cross section of dissimilar metal joint developed using A-GTAW **a** soft and low-strength filler metal and **b** hard and high-strength filler metal (Numerical values in figure **a** shows the different impact toughness values in different zones of two dissimilar metals showing different hardening behaviours due to weld thermal cycle)

6.3 Principle

Filler wire fed A-GTAW uses an automatically fed filler wire into the arc zone, wherein the weld is developed using A-GTAW, i.e. passing welding arc over the flux applied on the parent metals along the weld centre line to achieve high depth-to-width ratio and deep penetration. The heat of welding arc causes melting of filler wire and both the parent metals of dissimilar combination as per their physical properties (melting point, thermal conductivity, etc.), which in turn determines the % contributions of each of the parent metal and filler wire in development of the weld metal. To allow the desired compositional modification using filler in A-GTAW, the root gap between the parent metals is kept little wider (1–3 mm) with suitable backing plate so as to adjust the weld metal composition, microstructure and mechanical properties (Sharma and Dwivedi 2021a).

Further, in case of large difference in physical properties, affecting the melting of parent metals (and so dilution %) may need offsetting of welding electrode from the weld centreline to redirect arc heat more/less to/from a particular parent metal so to increase/decrease the dilution from the specific parent metal as per requirement and achieve symmetric weld.

The following section elaborates influence of base/filler metal composition and dilution in determining the weld metal composition. The dilution from the parent metal and filler primarily determines the weld composition, however, it is further influenced by the welding parameters, namely current, voltage, flux and its amount per unit area and shielding gas.

6.3.1 *Weld Metal Composition Adjustment in A-GTAW of Dissimilar Metals*

A case study on autogenous A-GTAW joint developed without filler assuming 30% contribution from parent metal A and balance 70% from parent metal B is given below. Location of both the parent metals and weld metal (considering carbon wt% and carbon equivalent) with regard to cracking sensitivity in Graville diagram is shown in Fig. 6.3a. It can be easily observed that (a) all three, namely two parent metals (A, B) and weld metal developed by autogenous A-GTAW, fall in zone II, and (b) weld metal composition is lying somewhere in between two parent metals as per dilution. The weld metal is still in crack-sensitive zone wherein cracking can be controlled using suitable welding procedure including filler, preheat, etc. The composition of weld metal can be estimated as given in following table.

Autogenous A-GTAW without filler (30% dilution for A and 70% dilution for B)							
wt%	C	Mn	Si	Cr	Ni	Mo	CE
Parent metal A	0.14	0.6	0.4	0.2	0.5	0.1	0.33

(continued)

(continued)

Autogenous A-GTAW without filler (30% dilution for A and 70% dilution for B)							
wt%	C	Mn	Si	Cr	Ni	Mo	CE
Parent metal B	0.19	1.1	0.8	0.1	0.2	0.05	0.42
Weld composition	0.175	0.95	0.68	0.13	0.29	0.07	0.39

Sample calculation for carbon (C) in weld metal: $0.3 * 0.14 + 0.7 * 0.19 = 0.175$

Case study of the same dissimilar metal combination joined by A-GTAW using appropriately chosen filler to modify the weld metal composition in order to minimize the cracking tendency of weld metal is given below. Assuming the contribution of the parent metal A and parent metal B and filler metal C in form of dilution (%) is 20, 30 and 50, respectively. Location of both the parent metals, filler metal and weld metal (considering carbon wt% and carbon equivalent) with regard to cracking sensitivity in Graville diagram is shown in Fig. 6.3b. The figure shows that (a) both parent metals are in zone II, while filler falls in crack safe zone I, and (b) weld metal (developed by A-GTAW) is lying in crack safe zone I. This suggests that the crack safe A-GTAW joint of two crack-sensitive parent metals can be developed by adjusting the weld composition using suitable filler. The composition of the weld metal can be calculated as given below.

Correct filler for A-GTAW							
wt%	C	Mn	Si	Cr	Ni	Mo	CE
Parent metal A	0.14	0.6	0.4	0.2	0.5	0.1	0.33
Parent metal B	0.19	1.1	0.8	0.1	0.2	0.05	0.42
Filler metal X	0.01	0.2	0.2	0.3	0.4	0.4	0.21
Weld composition	0.09	0.55	0.42	0.22	0.36	0.24	0.3

Sample calculation for carbon (C) in weld metal: $0.14 * 2 + 0.19 * 0.3 + 0.01 * 0.5 = 0.09$

Case study of same dissimilar metal combination welded using different filler metal and welding condition leading to completely different dilution (%) from the two parent metals and filler. Assuming the contribution of the parent metal A and parent metal B and filler metal C in form of dilution (%) is 50, 10 and 40, respectively. Location of both the parent metals, filler metal and weld metal (considering carbon wt% and carbon equivalent) with regard to cracking sensitivity in Graville diagram is shown in Fig. 6.3c. The figure shows that (a) both parent metals are in zone II, while filler falls in crack safe zone I, and (b) but the weld metal (developed by A-GTAW) falls in highly crack-sensitive zone III. This suggests that the inappropriate selection of filler metal and welding parameters causing unfavourable dilutions from two parent metals can result in a highly crack-sensitive weld metal. The composition of the weld metal can be calculated as given below:

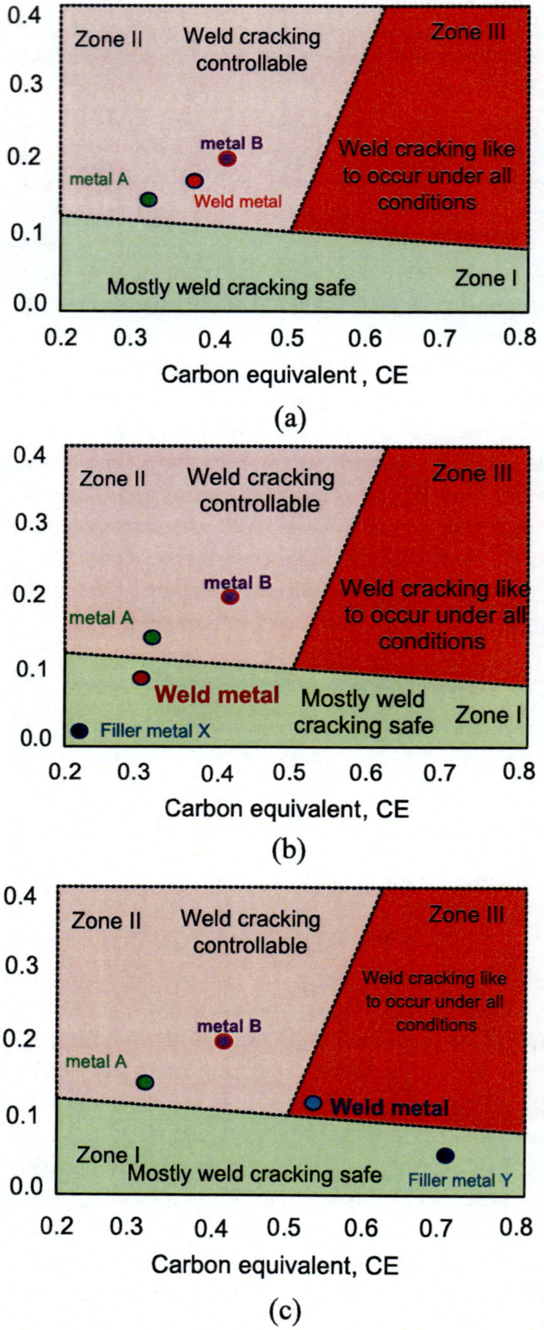

Fig. 6.3 Graville diagram showing cracking sensitivity of different parent metals (varying C wt.% of steel as shown in ordinate and carbon equivalent in abscissa), fillers and weld metal developed using A-GTAW **a** autogenous weld without any filler, **b** with correctly chosen filler X and **c** with incorrectly chosen filler Y

A-GTAW with filler Y							
wt%	C	Mn	Si	Cr	Ni	Mo	CE
Parent metal A	0.14	0.6	0.4	0.2	0.5	0.1	0.33
Parent metal B	0.19	1.1	0.8	0.1	0.2	0.05	0.42
Filler metal Y	0.05	0.5	0.2	2.5	0.6	0.3	0.73
Weld composition	0.109	0.61	0.36	1.11	0.51	0.18	0.5

Sample calculation for carbon (C) in weld metal: 0.14 * 0.5 + 0.19 * 0.1 + 0.05 * 0.4 = 0.109

Case study of autogenous A-GTAW joint of two high alloy steel (A and B) developed without filler assuming 30% contribution from parent metal A and balance 70% from parent metal B. Location of both the parent metals and weld metal (considering Ni and Cr equivalent) in Schaeffler diagram is shown in Fig. 6.4a. It can be easily observed that (a) all three, namely both parent metals and weld metal developed by autogenous A-GTAW, fall in the hard and brittle martensite zone, and (b) weld metal composition is lying somewhere in between two parent metals as per dilution from the two parent metals. The weld metal is in brittle and crack-sensitive zone. The composition of weld metal can be estimated as given in following table.

Autogenous A-GTAW dissimilar metal welding of high alloy steels								
wt%	C	Mn	Si	Cr	Ni	Mo	Cr equivalent	Ni equivalent
Parent metal A	0.14	0.6	0.4	2.25	0.5	0.1	2.95	5
Parent metal B	0.19	1.1	0.8	9.2	0.2	0.05	10.45	6.45
Weld composition	0.175	0.95	0.68	7.115	0.29	0.065	8.2	6.015

Sample calculation for chromium (Cr) in weld metal = 2.95*0.3 + 10.45*0.7 = 8.2

Case study of A-GTAW joint of two high alloy steel (A and B) developed suitable filler P to avoid cracking. Assuming the contribution of the parent metal A and parent metal B and filler metal P in form of dilution (%) is 20, 30 and 50, respectively. The location of both the parent metals (A and B), filler metal (P) and weld metal (considering Ni and Cr equivalent) in Schaeffler diagram is shown in Fig. 6.4b. It can be noticed that (a) both the parent metals correspond to the hard and brittle martensite zone, and filler metal "P" falls in austenite-ferrite zone, and (b) weld metal developed by A-GTAW using filler P falls in the austenite-5%ferrite zone as per dilution from the two parent metals and filler. The austenite-5%ferrite weld metal is known to resist solidification cracking and embrittlement during welding of stainless steel. The composition of weld metal can be estimated as given in following table.

Fig. 6.4 Schaeffler diagram showing phases of different parent metals, fillers and weld metal developed using A-GTAW **a** autogenous weld without any filler causing weld embrittlement, **b** with correctly chosen filler P avoiding solidification cracking and **c** with incorrectly chosen filler Q leading to solidification cracking

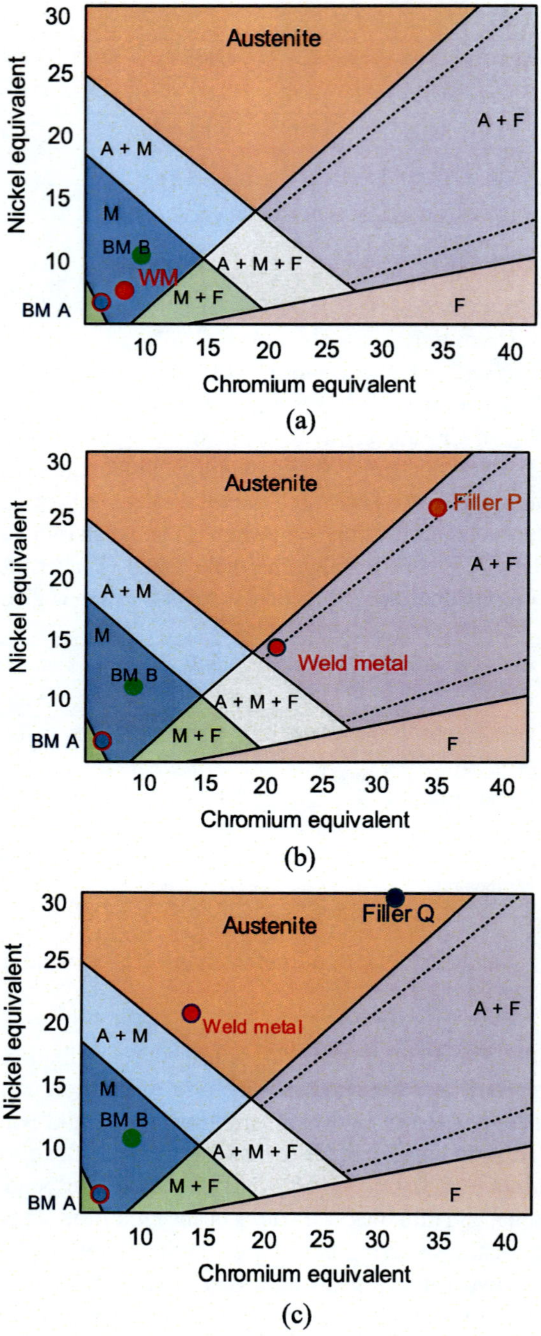

A-GTAW dissimilar metal welding of high alloy steel with filler P

wt%	C	Mn	Si	Cr	Ni	Mo	Cr equivalent	Ni equivalent
Parent metal A	0.14	0.6	0.4	2.25	0.5	0.1	2.95	5
Parent metal B	0.19	1.1	0.8	9.2	0.2	0.05	10.45	6.45
Filler metal P	0.01	0.2	0.2	36	24	0.5	36.8	24.4
Weld composition	0.09	0.55	0.42	21.21	12.2	0.285	22.125	15.135

Sample calculation for nickel (Ni) in weld metal: 0.5 * 0.2 + 0.05 * 0.3 + 24 * 0.5 = 12.2

Case study of A-GTAW joint of two high alloy steel (A and B) developed using incorrectly chosen filler "Q". Assuming the contribution of the parent metal A and parent metal B and filler metal P in form of dilution (%) is 50, 10 and 40, respectively. The location of both the parent metals (A and B), filler metal (Q) and weld metal (considering Ni and Cr equivalent) in Schaeffler diagram is shown in Fig. 6.4 (c). It can be noticed that both the parent metals correspond to the hard and brittle martensite zone, and filler metal "Q" falls in completely austenite zone (beyond the range of diagram with regard to high Ni equivalent). Weld metal developed by A-GTAW using incorrectly chosen filler Q falls in fully austenite zone as per dilution from the two parent metals and filler. The weld metal therefore becomes very crack sensitive. The fully austenite weld shows solidification cracking tendency during the welding of stainless steel. The composition of weld metal can be estimated as given in following table.

A-GTAW dissimilar metal welding of high alloy steel with filler Q

wt%	C	Mn	Si	Cr	Ni	Mo	Cr equivalent	Ni equivalent
Parent metal A	0.3	0.6	0.4	1.2	2.5	2	3.8	11.8
Parent metal B	0.4	1.6	0.8	0	0	0	1.2	12.8
Filler metal Q	0.01	0.2	0.2	32	40	0.5	32.8	40.4
Weld composition	0.194	0.54	0.36	13.4	17.3	1.2	15.14	23.34

Sample calculation for molybdenum (Mo) in weld metal = 2 * 0.5 + 0 * 0.1 + 0.5 * 0.4 = 1.2

6.4 Choice of External Filler Wire

The choice of filler metal for A-GTAW for dissimilar metal joining needs consideration of chemical and metallurgical compatibility of both the parent metals with filler metal, dilution expected from each of the parent metals and filler metal, welding parameters and extent of the change in weld metal composition needed to achieve the desired properties of the weld metal. For example, cracking tendency (due to embrittlement) of weld metal developed by autogenous A-GTAW to join dissimilar steels is significantly affected by choice of filler metal. Proper filler metal can result in crack safe weld metal, while poor selection can produce a very crack-sensitive weld (Fig. 6.5a).

Similarly, to minimize the hardening, embrittlement and related cold cracking in weld metal of dissimilar hardenable steels, the selection of filler metal plays an important role. Autogenous A-GTAW produces very crack-sensitive martensitic weld. Application of a suitable filler "P" helps to develop a solidification/cold crack safe weld metal, while a wrong choice of filler Q results in solidification crack-sensitive weld metal (Fig. 6.5b).

In the same way, solidification cracking in weld metal of dissimilar combination of AISI 316 and AISI 304 parent metal, the filler is selected in such way that weld metal composition corresponds to the formation of weld metal with 3–5% delta ferrite. Similarly, to avoid the development of hard and brittle weld metal during welding of P22 and AISI 304, the filler metal is selected such that it results in either

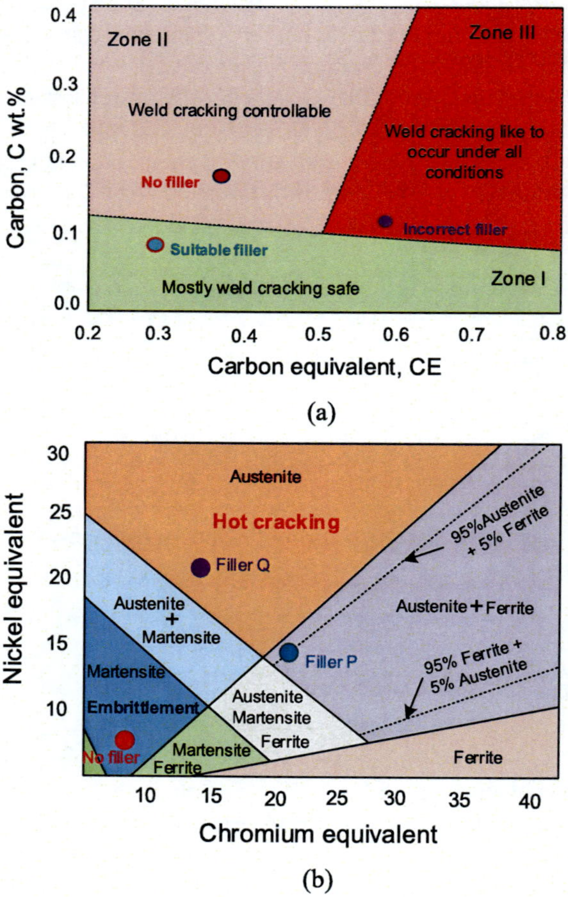

Fig. 6.5 Effect of filler addition/composition on **a** cracking sensitivity of the weld metal as per Graville diagram and **b** formation of phases and cracking tendency of weld metal developed using A-GTAW as per Schaeffler diagram

microstructure comprising ferrite and austenite or austenite only so the martensitic transformation of the weld metal is minimized or avoided through the adjustment of Cr and Ni equivalents. For given set of welding conditions and weld joint design, dilution from each of the parent metals and filler metal needs to be estimated so that using the above sample calculation, suitable filler metal compositions can be identified.

6.5 Weld Composition Homogeneity

Homogeneity of the composition in wire fed A-GTAW is critical for the soundness, microstructure and mechanical properties of the weld joint. It is influenced by not just welding conditions (current, voltage and speed), filler metal, physical and chemical properties of the elements present in the weld pool but also position of filler wire tip in arc zone and its feed rate. A large difference in physical properties specifically density, melting temperature, surface tension and viscosity of molten metal produced by filler and both the parent metals increases the chemical heterogeneity of the weld metal due to limited/poor mixing. Limited convection current in weld pool particularly near the fusion boundary of the parent metals promotes the unmixed zone formation leading to significant chemical heterogeneity. High heat input realized through high welding current facilitates the increased convection current in the weld pool and long enough weld solidification time which in turn helps to achieve chemical homogeneity in weld metal. Additionally, a continuous supply of molten metal from the optimally positioned filler wire results in good chemical homogeneity in the weld metal to offer the desired microstructure and mechanical properties (Fig. 6.6) (Sharma and Dwivedi 2021a).

6.6 Welding Parameters of Filler Wire Fed A-GTAW

The A-GTAW welding parameters including type of flux, flux mixture, amount of flux applied (g/area), welding current, voltage, shielding gas and its flow rate, welding speed, filler wire metal, diameter of filler, feed rate of filler, positon of filler wire tip in arc zone all affect the chemical composition, microstructure and mechanical properties in one or other way. These parameters can categorized under flux, filler and welding conditions.

Flux-related parameters affect the arc voltage (so heat generation), arc constriction (energy intensity), reversal of Marangoni convection and so bead geometry which in turn affect the microstructure and mechanical properties. The strong convection current in the weld pool results in high depth-to-width ratio weld with greater chemical homogeneity.

Filler-related factors include the type of filler metal, its compatibility with both the parent metals and homogeneity of the weld metal composition, kind of change in

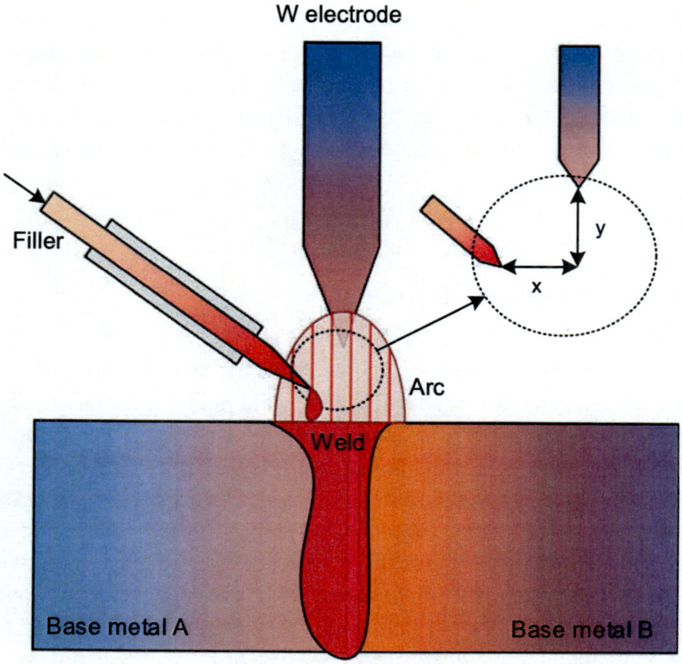

Fig. 6.6 Schematic showing the scheme of A-GTAW with filler with regard to location of tip of filler and electrode and weld pool

chemistry of weld metal brought in by filler metal, which in turn affects the metallurgical transformation in the weld so the microstructure and mechanical properties of the weld metal are influenced accordingly. Filler metal wire position in arc zone affects the homogeneity of the weld chemistry significantly. An optimum filler wire position in arc zone resulting in a continuous feeding of the molten metal from the filler wire to weld pool produces more chemical homogeneity than other positions leading to interrupted supply of the molten metal from the filler into weld pool.

The welding conditions like current, voltage, shielding gas, current pulsation if any, welding speed affecting the heat input and cooling rate and solidification time in turn affect (a) depth-to-width ratio of weld pool, (b) the time available for bring in chemical homogeneity through convection current specially in case of filler wire fed A-GTAW (Fig. 6.7). These factors in turn affect the compositional uniformity, microstructure and mechanical properties of the weld joint.

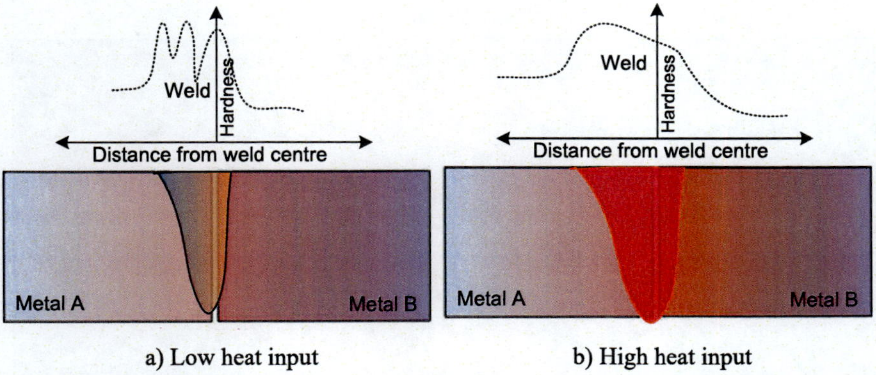

Fig. 6.7 Schematic showing typical hardness distribution and compositional homogeneity in weld cross section of dissimilar metal joint developed using A-GTAW **a** with low heat input and **b** high heat input

6.7 Flux Coating Patterns for Symmetric Weld Bead Geometry

The flux coating pattern in dissimilar metal joining by autogenous A-GTAW is more important than that of A-GTAW with filler wire variant to develop the symmetric weld because addition of filler due to compositional modification of the weld pool influences the strength of reversal in Marangoni convection. Still, a large difference in physical and chemical properties of the two metals of dissimilar combination may lead to skewed weld geometry or even lack of penetration. Therefore, even A-GTAW with filler wire feeding variant may also need little optimization of flux coating patterns to achieve largely symmetric weld through identification and optimization of (a) suitable type of flux for each parent metal, (b) amount of flux (g/area) to be applied on each side and (c) the pattern of flux coating on each side (separately). Thereafter, filler wire fed A-GTAW is carried out. The optimization should be done considering the filler wire composition, dilution from the parent metals and filler for developing symmetric weld in a single pass (Shankar and Dwivedi 2017; Vidyarthy and Dwivedi 2019).

6.8 Filler Wire Fed A-GTAW and Mechanical Properties

Certainly, the filler wire addition in dissimilar metal weld joint developed by A-GTAW brings in many desired changes in the microstructure and mechanical properties despite the possibility of chemical heterogeneity. However, the nature of changes in the weld zone is precisely influenced by the chemical composition of each of the parent metals and filler metal applied, dilution, solidification time and cooling rates

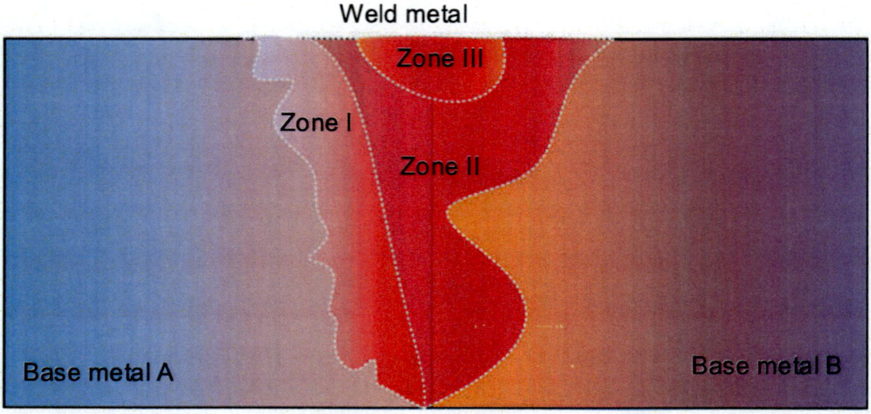

Fig. 6.8 Schematic showing formation of different zones in weld cross section of dissimilar metal joint developed using A-GTAW

in weld as per A-GTAW conditions including current, voltage, speed, shielding gas and fluxes. The chemical heterogeneity in weld metal composition especially near the fusion boundaries due to addition of filler metal is expected to adversely affect the response to heat treatment, resistance to corrosion and even mechanical properties like toughness (Fig. 6.8).

Further, an appropriately selected filler metal certainly improves the microstructure and mechanical properties of the weld metal. For example, direct A-GTAW of transformation hardenable steel like P91 with non-transformation hardenable steel like AISI 304 without any filler results in hard and brittle martensitic structure in weld metal. This leads to embrittlement of the weld joint in form of very low toughness (10–30 J), while A-GTAW of same dissimilar metal combination with filler metal like high Ni–Cr filler produces high toughness weld joint (100 J) with 100% joint efficiency (Sharma and Dwivedi 2021a; Sharma 2021).

Considering the (a) Cr and Ni equivalents of a given stainless steel and alloy steel and (b) expected/estimated dilution from each of the parent metals for a given set of the welding conditions, it is possible to predict the microstructure of the weld zone with the help of Schaeffer diagram. Now, filler metal composition is selected in such a way that composition point of the weld metal (considering Cr and Ni equivalent) is shifted to the desired zone/location in Schaeffer diagram to develop the weld metal of desired microstructure and mechanical properties.

6.9 A-TIG Welding with Wire Feed of Dissimilar P92 Steel-316L ASS

In this section, activated flux-tungsten inert gas (A-TIG) welding with wire feed is proposed as an approach to achieve improved mechanical properties (ductility and impact toughness) of weld joints via modification of the chemical composition and microstructure of weld zone. In this method, the ErNiCrMo-3 filler wire is directly fed into the weld pool during welding (Fig. 6.9a). Welding was carried out using 220 A welding current, 3 mm arc gap and Ar as shielding gas with 10 lpm flow rate. Wire feeding was carried out at 100 mm/min feed rate, 30° feeding angle and horizontal and vertical distance between electrode tip and filler wire as 5.5–6.5 mm and 2–2.5 mm, respectively.

The weld appearance and macro-graphs of the weld bead cross section are given in Fig. 6.9a, b. The top side of welded plate shows the continuous weld seam without spatters. The rear side of welded plate shows the through thickness penetration achieved in single pass. The chemical composition of the weld zone developed using A-TIG welding with wire feed is presented in Table 6.1. For comparison, the chemical composition of weld zone developed without wire feeding is also given. Wire feeding during A-TIG welding showed a significant increase in the content of Cr and Ni. The content of Mn and Mo also increased slightly in the weld zone with wire feed. To understand the effect of chemical composition modification on the phase structure of the weld zone, Ni equivalent (Ni_{eq}) and chromium equivalent (Cr_{eq}) were calculated for weld zone developed with wire feed and without wire feed using Eqs. (6.1) and (6.2), respectively.

$$Cr_{eq} = Cr + Mo + 1.5Si + 0.5Nb \tag{6.1}$$

$$Ni_{eq} = Ni + 30C + 0.5Mn \tag{6.2}$$

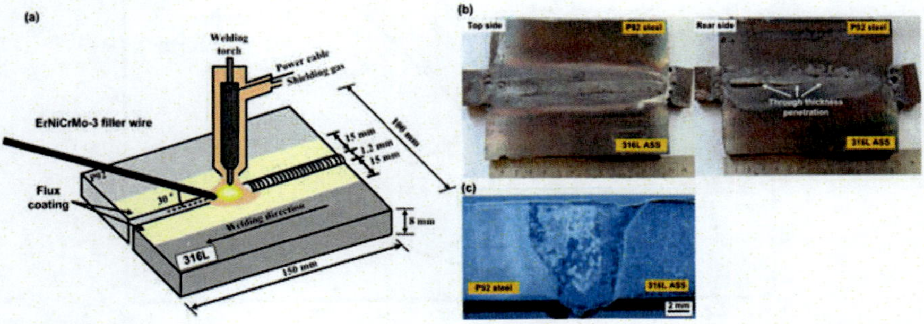

Fig. 6.9 Schematic illustration of A-TIG welding with process, **a** photographs of welded plates (top side and rear side) of obtained using A-TIG welding with wire feed and **b** macro-graph showing the weld bead cross section of developed weld joint (Sharma and Dwivedi 2021a)

Table 6.1 Chemical compositions of weld zones developed using A-TIG welding with and without external wire feeding

	C	Nb	Mo	Mn	Cr	Ni	Si	Fe	Cr_{eq}	Ni_{eq}
A-TIG welding without wire feeding	0.05	0.31	0.40	0.41	9.33	4.89	0.41	Bal	10.5	6.5
A-TIG welding with external wire feed	0.03	1.22	1.10	1.8	18.62	26.44	0.55	Bal	21.15	28.24

where symbol of elements represents their weight percentage. Based on the calculation, the points corresponding Cr_{eq} and Ni_{eq} were located on Schaeffler diagram, as shown in Fig. 6.10. It was observed that the addition of ErNiCrMo-3 wire using optimized wire feeding parameters shifted Cr_{eq} and Ni_{eq} of the weld zone from martensite to austenite region. This is attributed to the enrichment of Ni and Cr with the addition of wire feeding.

To study the effect of wire feeding on microstructure of weld zone, the optical micro-graphs of weld zones are given in Fig. 6.11a, b. The micro-graphs indicated a completely austenitic structure in the weld zone developed with A-TIG welding with wire feed. Further, analysis of the weld zone microstructure reveals columnar/cellular dendrites near the fusion boundaries and equiaxed dendrites at the centre of the weld zone. Whereas, the weld zone of joint developed using A-TIG welding without wire feed showed completely martensitic structure (Fig. 6.11b).

To evaluate the effectiveness of wire feeding during the A-TIG welding process, the mechanical properties of the weld joint were assessed in terms of hardness, tensile properties and impact toughness. The micro-hardness distribution curve across the various zones of weld joints developed using A-TIG welding with and without wire

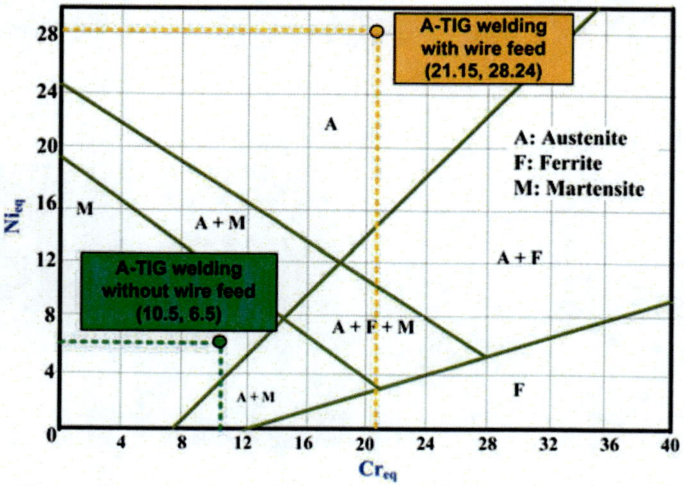

Fig. 6.10 Schaeffler diagram depiction of weld zones developed using A-TIG welding with and without wire feed

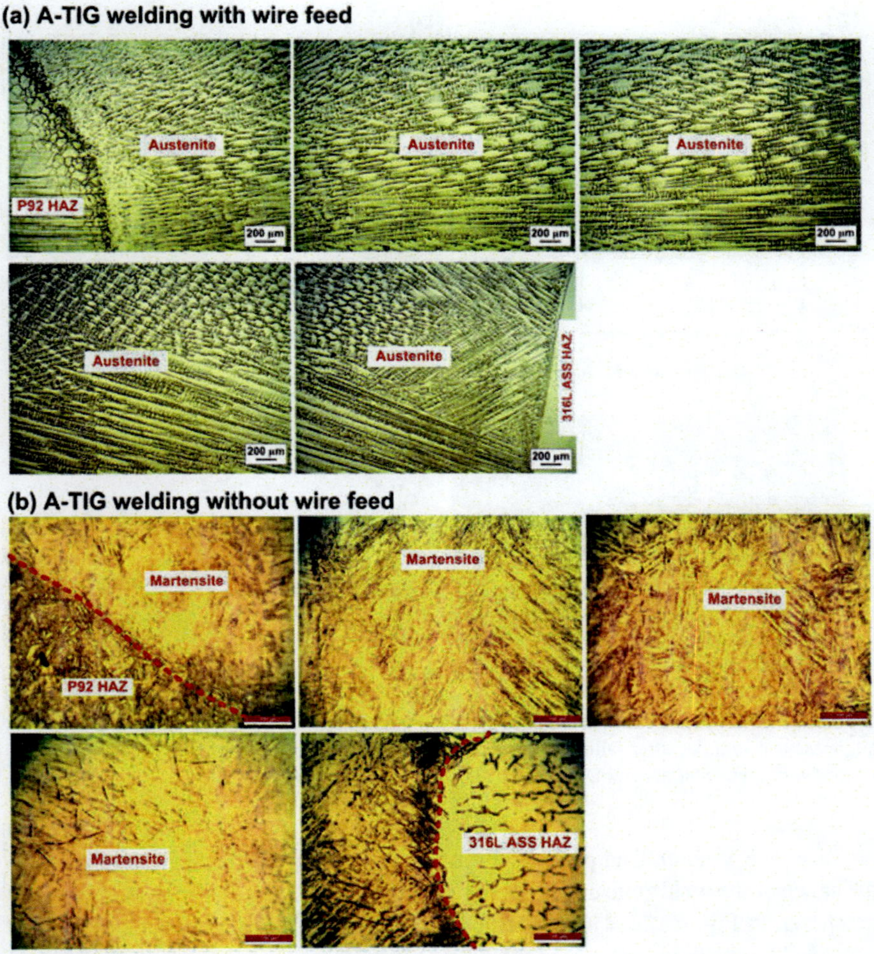

Fig. 6.11 Optical micro-graph showing the weld zone structure of weld joint developed using: **a** A-TIG welding with wire feed and **b** A-TIG welding without wire feed

feed is given in Fig. 6.12a. The hardness of weld zone of A-TIG weld joint without wire feed was 390 ± 10 HV; whereas the A-TIG weld joint with wire feed exhibited the hardness of the weld zone 183 ± 13 HV. This substantial reduction in the hardness of the weld zone is attributed to Fe–Ni–Cr-rich austenitic microstructure. The tensile testing of the transverse section sample of A-TIG weld joint with and without wire feed was performed. The resultant tensile stress-tensile strain curve for both the weld joints is given (Fig. 6.12b). The A-TIG welding with wire feed results in comparatively higher ultimate tensile strength and ductility as compared to A-TIG welding without wire feed. Fracture of the dissimilar metal joint from weak parent metal side indicates that weld joint is stronger. Therefore, feeding the suitable filler wire during A-TIG welding can help to achieve a good combination of strength and

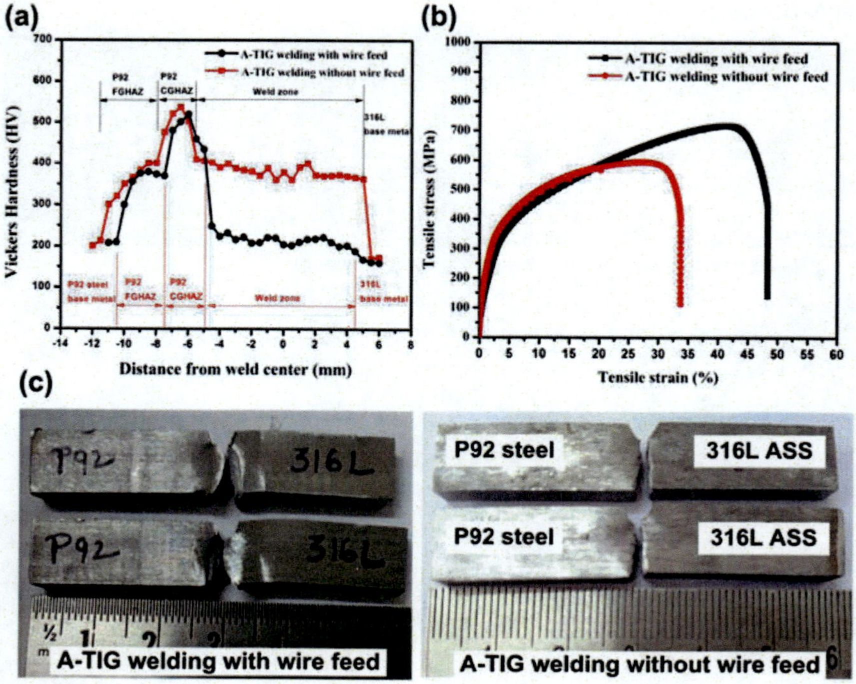

Fig. 6.12 **a** Micro-hardness distribution curve, **b** tensile stress-tensile train curve and **c** impact toughness test samples after failure for the weld joints developed using A-TIG welding with and without wire feed (Sharma and Dwivedi 2021a, b)

ductility. The V-notch Charpy impact toughness test was performed to examine the resistance of the weld zone under impact loading. The impact toughness test samples are shown in Fig. 6.12c. The mean impact toughness of the weld zone developed using A-TIG welding with wire feed 92 ± 2 J; whereas the mean impact toughness of weld zone of joint developed using A-TIG welding without wire feed was 10 ± 2 J. The superior impact energy of A-TIG weld joint with wire feeding is due to the formation of fully austenitic structure in the weld zone (Sharma and Dwivedi 2021a, b).

References

Kulkarni A, Dwivedi DK, Vasudevan M (2018) Study of mechanism, microstructure and mechanical properties of activated flux TIG welded P91 Steel-P22 steel dissimilar metal joint. Mater Sci Eng A, 731:309–323

Kulkarni A, Dwivedi DK, Vasudevan M (2019) Dissimilar metal welding of P91 steel-AISI 316L SS with Incoloy 800 and Inconel 600 interlayers by using activated TIG welding process and its effect on the microstructure. J Mater Process Technol 274:116280

Kulkarni A, Dwivedi DK, Vasudevan M (2020) Microstructure and mechanical properties of A-TIG welded AISI 316L SS-Alloy 800 dissimilar metal joint. Mater Sci Eng: A, 790:139685

Shankar R, Dwivedi DK (2017) Study of microstructure and mechanical property relationships of A-TIG welded P91–316L dissimilar steel joint. Mater Sci Eng, A 695(2017):249–257

Sharma P, Dwivedi DK (2019a) Comparative study of activated flux-GTAW and multipass-GTAW dissimilar P92 steel-304H ASS joints. Mater Manuf Processes 34(2019):1195–1204

Sharma P, Dwivedi DK (2019b) A-TIG welding of dissimilar P92 steel and 304H austenitic stainless steel: mechanisms, microstructure and mechanical properties. J Manuf Process 44(2019):166–178

Sharma P, Dwivedi DK (2021a) Wire-feed assisted A-TIG welding of dissimilar steels. Arch Civ Mech Eng 21(2021):1–20

Sharma P, Dwivedi DK (2021b) Improving the strength-ductility synergy and impact toughness of dissimilar martensitic-austenitic steel joints by A-TIG welding with wire feed. Mater Lett 285:129063

Sharma P, Dwivedi DK (2021c) Study on flux assisted tungsten inert gas welding of bimetallic P92 martensitic steel-304H austenitic stainless steel using SiO_2–TiO_2 binary flux: Welding arc/pool behaviour, microstructure and mechanical properties. Int J Pr Vessel Piping 192:104423

Vidyarthy RS, Dwivedi DK (2019) Effect of shielding gas composition and activating flux on the weld bead morphology of the P91 ferritic/martensitic steel. Mater Res Exp 6(8):0865f7

Chapter 7
Dissimilar Metal Joining by A-TIG Welding Using Interlayers

This chapter presents the approach to develop dissimilar metal A-TIG weld joints using interlayers. The need and selection of interlayer materials in A-TIG welding of dissimilar steels are discussed first, followed by the discussion on the effect of interlayer size. Further, the effect of interlayers on metallurgical and mechanical properties is discussed. Finally, the development of functionally graded joint using the interlayers in A-TIG welding is discussed.

7.1 Background

A-TIG welding process improves the productivity of the conventional TIG welding process by improving the depth of penetration and eliminating the need for groove preparation. Additionally, A-TIG weld joints show a good combination of hardness, tensile strength and metallurgical stability. Moreover, due to the completion of welding in a single pass, residual stress and angular distortion are reduced in A-TIG weld joints. However, the autogenous nature of the process does not facilitate the microstructural modifications in the fusion zone. In dissimilar metal welding, due to dilutions from only parent metals (as no filler metal is used), the fusion zone of the A-TIG weld joint exhibits chemical composition, which is a mixture of the parent metals. As microstructure ultimately controls the mechanical performance of any component, undesirable microstructure might result in the poor mechanical properties of the weld zone. One such example is the fusion welding of dissimilar combination of ferritic steel and austenitic stainless steel (SS). A-TIG welding of dissimilar ferritic steel-austenitic stainless steel couple results in a predominant martensitic structure in the fusion zone (Vidyarthy et al. 2017). Such A-TIG weldments show good tensile strength due to the dominant martensitic microstructure of hard and brittle nature. However, impact toughness and ductility in such joints are found to be very low. Additionally, the weld zone might become susceptible to hydrogen embrittlement-related failures (Kurt et al. 2009). Therefore, in order to

© The Author(s), under exclusive license to Springer Nature Singapore Pte Ltd. 2023
D. K. Dwivedi, *Dissimilar Metal Joining*,
https://doi.org/10.1007/978-981-99-1897-3_7

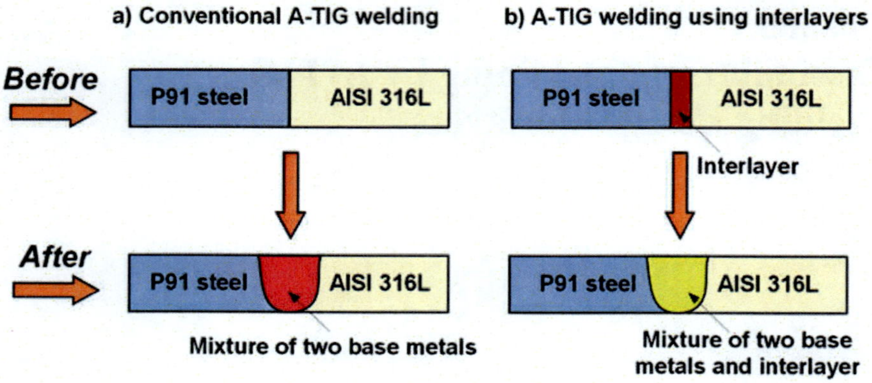

Fig. 7.1 Schematic of **a** conventional A-TIG welding and **b** A-TIG welding using interlayers

improve the usefulness of A-TIG welding for developing dissimilar steel transition joints, the approach to modify the microstructure and mechanical properties of the weld zone is necessary.

The fraction of the martensite phase can be lowered by increasing the stability of austenite, which can be achieved by increasing the content of austenitic stabilizers such as Ni, Mn, C and N in the fusion zone. Approaches of increasing austenitic stability include the use of filler metals, the use of interlayers and use of nitrogen shielding gas. Additionally, preheating and post-heating can also alleviate the detrimental effects of martensitic transformation.

In this chapter, the use of interlayers during A-TIG welding has been presented. The approach can be schematically represented in Fig. 7.1a, b. The selection of suitable interlayer material and its size is discussed initially. Furthermore, the beneficial effect of interlayers on metallurgical and mechanical characteristics is presented.

With an increase in the interlayer width, the volume fraction of austenite increased at the expense of martensite. The presence of austenite improves ductility and impact toughness. Additionally, higher austenite content is desired due to its lower carbon diffusivity coefficient than martensite, which would reduce the tendency of carbon migration.

7.2 Selection of Interlayer Material

The selection of interlayer material is critical for achieving the desired chemical composition and microstructure of the fusion zone. Some of the primary aspects to look into for choosing the interlayer material are discussed here. The desirable properties of the potential interlayer material are summarized in Fig. 7.2. The interlayer material should have the following characteristics (a) metallurgical compatibility with the both parent metals, (b) melting point close to that of the parent metals, (c)

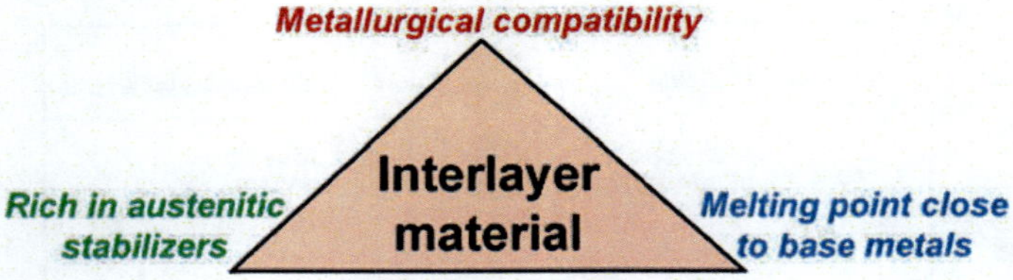

Fig. 7.2 Schematic showing the key requirements of the interlayer material

high content of austenitic stabilizing elements and low fraction of elements that can form low melting point compounds (and subsequently cause solidification cracking during welding), etc. Additionally, the interlayer material should be available in the form of strips or plates to be stacked between parent metals prior to A-TIG welding.

Considering the requirements of the interlayer properties, nickel-based alloys and austenitic stainless steels are found to be suitable for this purpose. The welding between steels and nickel-based alloys is commonly carried out. The possible interlayer materials could be nickel-based alloys such as Inconel 600, Incoloy 800, Inconel 625, Inconel 617, Incoloy 825 and Inconel 718. Moreover, Incoloy 800 (a ferrous-based alloy enriched in Ni content), which is commonly used as a transition piece in the trimetallic configuration, can also be used as an interlayer material (Bhaduri et al. 1994). The chemical compositions of parent metals and candidate interlayer materials are provided in Table 7.1. The effect of interlayers on the chemical composition of the fusion zone can be schematically represented in the Schaeffler diagram (Fig. 7.3). The family of ferritic steels and austenitic stainless steels are also represented.

The following section presents an attempt to alter the chemical composition and microstructure of the A-TIG welded DMW joint between ferritic steel (P91 steel) and austenitic stainless steel (AISI 316L) by introducing the ferrous-based (Incoloy/Alloy 800) and nickel-based (Inconel 600) interlayers. The interlayer materials, i.e. Incoloy 800 and Inconel 600, are enriched in nickel content. Additionally, these alloys are metallurgically compatible with the parent metals, i.e. P91 steel and AISI 316L.

Table 7.1 Typical chemical compositions (wt%) of parent metals, interlayers and filler metal

Element (wt%)	C	N	Cr	Mo	Si	Nb	Ti	V	Ni	Mn	Fe
P91 steel	0.10	–	8.33	0.94	0.20	0.08	–	0.26	0.33	0.37	Bal
P22 steel	0.11	–	2.11	0.91	0.20	–	–	0.09	0.03	0.51	Bal
AISI 316L SS	0.03	0.01	16.33	2.17	0.21	0.25	0.02	0.08	10.0	1.32	Bal
AISI 304L SS	0.03	0.01	18.50	–	0.50	–	–	–	8.50	1.00	Bal
Incoloy 800	0.09	0.12	22.68	0.27	0.25	0.39	0.48	0.03	30.80	1.06	Bal
Inconel 600	0.12	–	16.35	0.21	0.27	0.02	–	–	71.80	1.02	Bal
Inconel 718	0.08	–	19.00	3.05	0.35	–	0.90	–	52.50	0.35	Bal
Incoloy 825	0.05	–	21.00	3.00	0.50	–	0.90	–	42.00	1.00	Bal

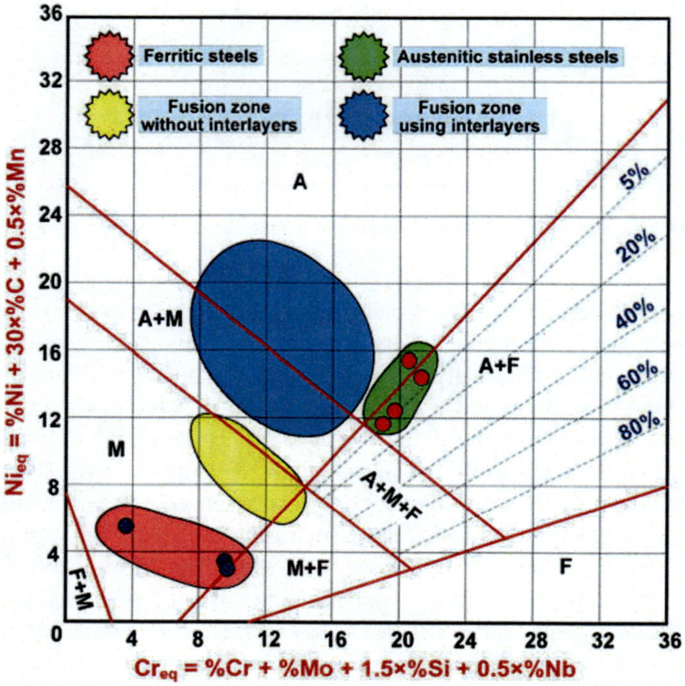

Fig. 7.3 Schaeffler diagram representation of parent metals, interlayers and weld zone

7.3 Weld Joint Development

For the development of A-TIG weld joints between ferritic steel and austenitic stainless steel, the substrates are kept in a closed square butt configuration. Due to differences in surface-active elements and thermo-physical properties on both sides, the weld bead shifting may occur towards one of the parent metal side, having a higher surface tension value (Mills et al. 1998). Typically, the weld bead shifting occurs towards the ferritic steel side parent metal in between ferritic steel and austenitic stainless steel combination. In order to mitigate the weld bead shifting and asymmetric weld bead profile in A-TIG welding, either torch offsetting towards austenitic stainless steel side or flux coating patterns or both these techniques are employed.

For developing A-TIG weld joints using interlayers, the interlayers are sandwiched between the parent metals. It is made sure that the faying surfaces are clean and contaminant free. Tack welds are developed to hold the parent metals and interlayer in their places temporarily. Afterwards, the flux is applied using a paintbrush. The fluxes are typically multi-component fluxes comprising multiple oxides, e.g. 30–40% TiO_2 + 25–35% SiO_2 + 20–30% MoO_3 + 5–15% CuO. The flux coating patterns can be developed by removing some part of the flux from the parent metal side with low surface tension to balance out surface tension forces. The torch offset is done towards the parent metal side with lower surface tension, which generally is an austenitic stainless steel side. The torch offset is measured as the perpendicular distance between the weld centreline and the vertical axis of the tungsten electrode for A-TIG welding. This is followed by autogenous A-TIG welding. The welding current and welding speed need to be optimized for a metal combination and section thickness. The weld joints are further sectioned and characterized.

7.4 Role of Interlayer Width on Weld Joint Characteristics

The primary role of interlayers is to modify the chemistry of the fusion zone and yield desirable mechanical properties of the weld zone. During A-TIG welding using interlayers, complete melting of the interlayer is mandatory. Different sized interlayers can be stacked between the parent metals, and A-TIG welding can be performed (Fig. 7.4a–e). An increase in width of interlayer also increases the dilution from interlayer. However, the probability of the formation of an unmelted region also increases due to increased possibility of partial melting of a wide interlayer.

7.4.1 *Finalization of Interlayer Size*

The interlayer width needs to be optimized by considering the complete mixing of the interlayer and parent metals, desired microstructure and mechanical properties. With an increase in interlayer width, the complete mixing becomes difficult to achieve in a single-pass A-TIG welding, as some region of the interlayer at the root side of the weld joint remains unmelted. For these reasons, the width of the interlayers is limited up to 1.5 mm.

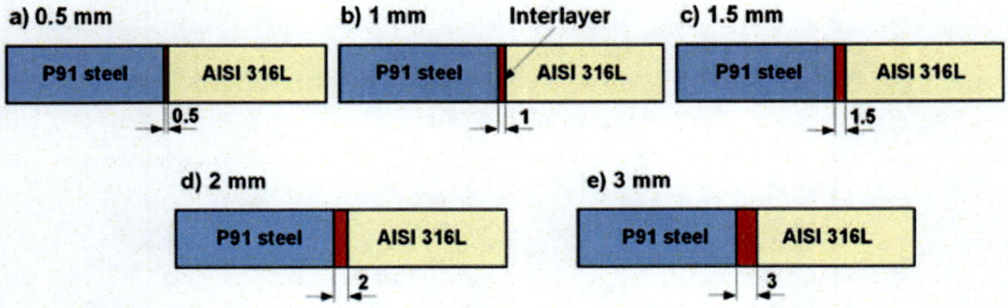

Fig. 7.4 Schematic representation of A-TIG welding using different interlayer sizes **a** 0.5 mm, **b** 1 mm, **c** 1.5 mm, **d** 2 mm and **e** 3 mm

7.5 Calculation of Dilution Levels

In the weld joint developed without interlayers, the dilution from each parent metal in the weld zone can be estimated using the lever rule. On the Schaeffler diagram, the corresponding points for the parent metals and weld zone can be fixed using their chemical compositions (Fig. 7.5). The lever rule can be applied for the weldments without interlayers as given by Eqs. (7.1) and (7.2).

$$\text{Dilution from parent Metal A} = \frac{\text{Length of segment BW}}{\text{Length of line AB}} \tag{7.1}$$

$$\text{Dilution from parent Metal B} = \frac{\text{Length of segment AW}}{\text{Length of line AB}} \tag{7.2}$$

The dilution from the interlayers can be predicted using the cross-sectional area of the weld zone and the interlayer. The macroscopic images showing weld bead geometries can be captured using stereoscopy. The cross-sectional area of the weld bead can be calculated using image analysis techniques. The interlayer cross-sectional area is the product of its width and thickness. The area fraction of the interlayer with respect to the total weld bead area can be assumed as the dilution from the interlayer. The fundamental assumption here is that the parent metals and the interlayer are completely mixed in the fusion zone.

For example, if an interlayer of 100(L) × 1(W) × 10(T) mm^3 was utilized during A-TIG welding of 10 mm thick plates, the cross-sectional area of the interlayer would be 1 × 10 = 10 mm^2. If the measured cross-sectional area of the weld zone is 80 mm^2, the area fraction of the interlayer would be 10/80 = 12.5%, which is the dilution from the interlayer. On the other hand, the area-wise dilution from the parent

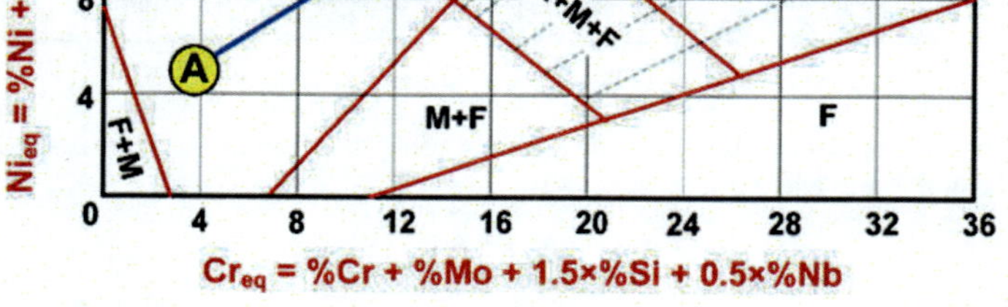

Fig. 7.5 Dilution calculation through the graphical method

metals will be 80–10 = 70 mm^2, which is 87.5%. The same approach can be used for calculating dilutions of the interlayers of any dimensions.

Using a similar approach based on the cross-sectional areas of weld zone and interlayers, the dilution levels were calculated for interlayers of different sizes. The dilution levels from the interlayer increase with an increase in the width of the interlayer.

7.6 Metallography

The microstructure of the as-received parent metals depends on their primary processing and heat treatment conditions. The ferritic-martensitic steels in the as-received form typically exhibit tempered martensitic structure (P91/P92 steel) or ferritic-bainitic structure (P22 steel). The ferritic steels also show precipitates such as carbides and nitrides (Sudha et al. 2002; Kulkarni et al. 2018). The austenitic stainless steels (AISI 316/304) show an austenitic structure of equiaxed morphology. Some delta ferrite stringers are also observed (Mortezaie and Shamanian (2014).

The fusion zone of the autogenous A-TIG weldment developed without the use of interlayer showed an untempered martensitic structure. The structure is characterized by prior austenitic grain boundaries (PAGBs), martensitic laths and packets of laths. The secondary precipitates are not observed in the as-welded condition (Vidyarthy et al. 2017).

With the insertion of nickel-rich interlayers, the retained austenitic content increases at the expense of martensite. A-TIG welding of P91 steel and AISI 316L developed using Incoloy 800 interlayer revealed an austenitic-martensitic structure. With the use of Inconel 600 interlayer, a fully austenitic structure was realized (Kulkarni et al. 2019).

The rationale behind the retention of austenite in the martensitic matrix is explained further. The alloying elements such as nickel, manganese, carbon and nitrogen act as austenite stabilizers. With the use of ferrous-based interlayers enriched in austenitic stabilizers, the higher content of these elements stabilizes austenite during the weld cooling cycle. The stabilized austenite gets mingled with martensite at room temperature. As a result, a two-phase mixture of austenite–martensite is observed. With further increase in nickel content in the weld zone, such as in the case of nickel-based alloy interlayers, austenitic stability is very high, which results in a fully austenitic structure of dendritic morphology. The retention of austenite can be related using martensitic start temperature (M_s). The M_s temperature depends upon the content of alloying elements, which can be estimated by using the empirical Eq. (7.3) (DuPont 2012).

$$M_s(°C) = 540 - 497\%\,C - 46.6\%\,Mo - 36.3\%\,Ni - 10.8\%\,Cr - 6.3\%\,Mn \quad (7.3)$$

It can be seen that a higher amount of alloying elements reduce the M_s temperature. The martensitic finish temperature (M_f) is typically 100–150 °C below the M_s temperature. If M_s temperature is very high (above 150–200 °C), it can be assumed that the weld zone would exhibit a dominant martensitic structure. For the lower M_s temperature values, some amount of austenite is retained, as the M_f temperature is expected to be below the ambient temperature. If the M_s temperature is lower than the room temperature, the driving force (Gibbs free energy) will not be available for nucleation of martensitic grains. Therefore, a full austenite structure can be observed.

An empirical relationship to predict the volume fraction of retained austenite (γ) based on M_s temperature, as given by Eq. (7.4) (Wu et al. 2000). The theoretical value of the volume fraction of retained austenite can be calculated using this equation.

$$\gamma(\%) = 36.412\text{–}0.126 \times \text{Ms}(^\circ\text{C}) \tag{7.4}$$

The retained austenite can be experimentally estimated using image analysis techniques such as image thresholding of the micrograph. The error in the estimation will be minimum if the two-phase boundaries are clearly contrasted in the micrograph.

The quantification of the retained austenite in the weld zone can also be done using X-ray diffraction (XRD) analysis by using the ASTM E975-13 standard (ASTM E975-13 2013). As per the standard, the retained austenite volume fraction (V_γ) in ferrite or martensite matrix (V_θ) is given by Eq. (7.5).

$$V_\gamma = \frac{\frac{1}{q}\sum_{j=1}^{q}\frac{I_\gamma}{R_\gamma}}{\frac{1}{p}\sum_{j=1}^{p}\frac{I_\theta}{R_\theta} + \frac{1}{q}\sum_{j=1}^{q}\frac{I_\gamma}{R_\gamma}} \tag{7.5}$$

In this equation, the integrated intensities of the peaks of austenite and martensite are represented by I_γ and $I_{\theta,}$ respectively. The parameters (R_θ and R_γ) are derived from the Bragg angle (θ), interplanar spacing, Miller indices (hkl), crystal structure and composition of the phase, which can be expressed using Eq. (7.6).

$$R = \frac{1}{V^2}\left\{F^2 P\left(\frac{1+\cos^2 2\theta}{\sin^2\theta * \cos\theta}\right)\right\}e^{-2M} \tag{7.6}$$

Here, V represents the volume of the unit cell. The structure factor is expressed as F^2 ($4f^2$ for BCC and $16f^2$ for FCC structure for probable reflections). The other terms such as P, LP and e^{-2M} represent the multiplicity factor, Lorentz Polarization factor $\left(\frac{1+\cos^2 2\theta}{\sin^2\theta*\cos\theta}\right)$ and the temperature factor, respectively. For better estimation of retained austenite using XRD analysis, slow scanning speeds are recommended.

7.7 Mechanical Properties

The mechanical properties are evaluated through micro-hardness testing, tensile testing and impact toughness testing. The weld zone without interlayer exhibited a very high hardness in the range of 400–500 HV. There is an increase in the micro-hardness of heat-affected zone (HAZ) on the ferritic steel side. The addition of interlayers of nickel-based or ferrous-based superalloys reduces the micro-hardness of the weld zone (Fig. 7.6). The micro-hardness of the HAZs remains unaffected after the insertion of interlayers.

The hard and brittle untempered martensitic structure results in high hardness of weld zone developed without interlayer. The reduction in the micro-hardness after the addition of interlayers is attributed to the formation of retained austenite at the expense of martensite. Austenite being a softer phase than martensite lowers the overall hardness of the weld zone. The increase in micro-hardness of HAZ on the ferritic steel side is due to the transformation hardening that takes place during the weld thermal cycle.

The tensile testing results are typically in line with the micro-hardness studies. The tensile test sample failure takes place from the zone exhibiting the lowest micro-hardness. The weld joint developed without interlayer fails from the parent metal side (depending upon the weakest parent metal region), showing an overmatching strength and poor ductility. The interfacial failures from austenitic stainless steel-weld zone interface may also occur, giving rise to quasi-cleavage type fracture. On the other hand, the weld joints developed with interlayers show improved ductility without much compromise of tensile strength.

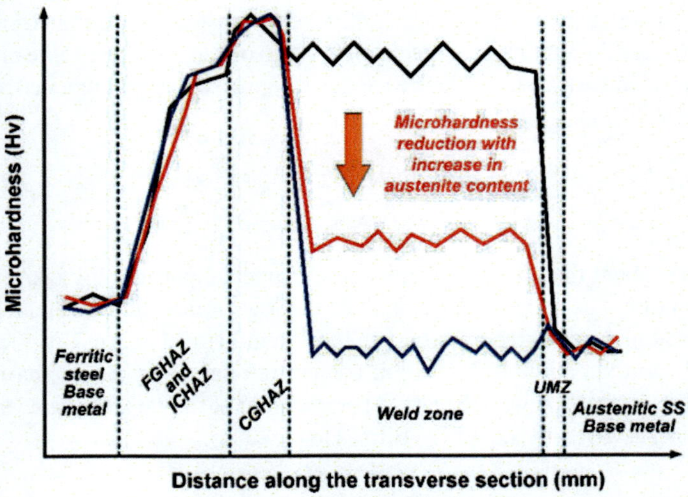

Fig. 7.6 Micro-hardness distribution in the A-TIG weldment developed using interlayers (blue/red lines showing influenced of interlayers and black one for without interlayer)

The interfacial failures can occur due to strain concentration caused by steep hardness gradient due to hard martensitic structure and soft austenitic structure on either side of the interface. The weld joints exhibiting dual-phase austenitic-martensitic structure in the weld zone result in overmatching strength and good ductility. In such weld joints, the austenite phase contributes to the ductility, while martensite contributes to the tensile strength (Kacar and Baylan 2004). However, a fully austenitic structure can lead to failures from the weld zones due to loss of solid solution strengthening and coarse-grain dendritic structure.

During impact toughness testing, the weld zone without interlayer fails in a brittle manner with low impact toughness value owing to its hard and brittle martensitic structure. The impact toughness is significantly improved if the weld joints are developed using interlayers. As a result, the specimens fail in a ductile manner. The improvement in impact toughness is attributed to the presence of austenite in the weld zone. The austenitic grains act as a crack blunter or shock absorber during impact testing, which results in deformation of the specimen rather than brittle fracture leading to a higher impact toughness value.

7.8 Stability of Retained Austenite

Austenite can transform into martensite under the influence of temperature change during cooling or mechanical strain such as cold deformation. Therefore, the estimation of metallurgical stability of retained austenite against transformation to martensite is important. The metallurgical stability of retained austenite under the action of mechanical strain can be assessed using Md_{30} (°C) temperature, which is given by Eq. (7.7). It is the temperature at which 50% of the austenite gets transformed into martensite under the action of the mechanical strain of a logarithmic value of 30%. A lower value of Md_{30} (°C) temperature suggests higher stability of retained austenite (Herrera et al. 2011).

$$Md_{30}(^{\circ}C) = 551 - 462(C + N) - 29(Ni + Cu) - 18.5Mo - 13.7Cr - 9.2Si - 8.1Mn \tag{7.7}$$

It is to be noted that during solidification, elemental partitioning takes place in different phases. The accurate estimation of the low atomic number elements such as carbon and nitrogen at the microscopic level in individual phases is difficult. The theoretical approximation of the carbon content in retained austenite phase can be made through its relation with the lattice parameter of austenite (calculated using the XRD pattern) as given by Eq. (7.8) (Bilmes te al. 2001).

$$(a_0)_{\gamma} = 3.572 + 0.033 \times wt\%C \tag{7.8}$$

If the estimated Md_{30} value of retained austenite phase is below the room temperature, it can be assumed that the austenite phase is stable against martensitic transformation under the action of mechanical strain.

The metallurgical stability can be validated by using the SEM micro-graph of the location adjacent to the crack in the fractured specimen during impact toughness testing. The strain rates during impact toughness testing are very high of order $10^4 s^{-1}$. If the micro-graph reveals two distinct phases (austenite and martensite) with deformed grain structure, then the very high metallurgical stability of austenite can be confirmed.

7.9 Effect on Carbon Migration

The phenomenon called carbon migration, basically involves the diffusion of carbon across the weld zone and parent metal interface during post weld heat treatments and/or high-temperature service, is an important issue in dissimilar ferritic steel-austenitic stainless steel-weld joints (Bhaduri et al. 1994; DuPont 2012). In such dissimilar metal joints, carbon diffusion takes place from the ferritic steel side HAZ to the weld zone. This, in turn, results in loss of carbon in the HAZ of ferritic steel, which weakens the HAZ side. On the other hand, the diffused carbon combines with alloying elements to form carbides in a narrow hard region of the weld zone called the carbon enriched zone (CEZ). Therefore, in a small region, narrow bands of hard and soft regions are created. The soft region, known as the carbon depleted zone (CDZ), exhibits lower creep strength than the rest of the zones. Additionally, the thermal strains due to the coefficient of thermal expansion mismatch and oxidation notches are located near the interface where CDZ is formed. All these factors contribute to the premature failures of the dissimilar steel joints from the CDZ. Therefore, efforts are required to reduce the tendency of carbon migration from the ferritic steel side HAZ to the weld zone.

The use of interlayers enriched in nickel increases nickel content in the weld zone and the phase fraction of retained austenite in the martensitic matrix. The austenite is a face-centred cubic structure that has a lower diffusivity coefficient (Mas et al. 2016). Therefore, the diffusion of carbon is slowed down across the interfaces.

7.10 A-TIG Welding Using Interlayers to Develop Functionally Graded Materials

Functionally graded (FG) materials are developed between two dissimilar materials having different physical, mechanical and chemical properties to enable a smooth

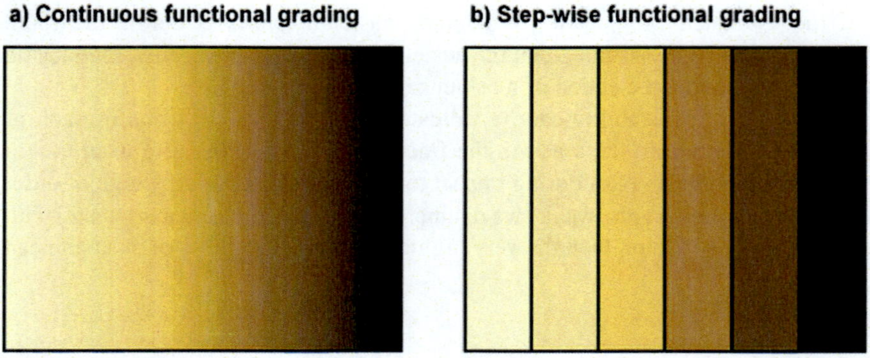

Fig. 7.7 Schematic representation of types of functional grading **a** continuous functional grading and **b** step-wise functional grading

changeover in chemical composition and mechanical properties across their interfaces. Such weld joints have wide applications in the nuclear and aerospace industries. The functional grading can be categorized as a) continuous functional grading in which spatial variation of the properties is very smooth (Fig. 7.7a, b) step-wise functional grading in which the changes occur over steps made at specific intervals (Fig. 7.7b). Although continuous FG joints are desirable, their development and fabrication using conventional manufacturing techniques require careful control and optimization of process parameters. The step-wise FG joints are comparatively easier to fabricate.

The FG materials (FGM) between metals-ceramics and metal-composites have been extensively researched (Reichardt et al. 2021). However, the FG between two dissimilar metals has received minimal attention so far. The recent emergence of additive manufacturing (AM) has made researchers work on functionally graded structures between metals. The AM processes such as direct energy deposition (DED) have been extensively employed to fabricate FG joints between dissimilar metals. However, AM techniques require the parent metals to be available in powder form. Moreover, benchmarking of properties requires extensive process parameter optimization to avoid defects such as lack of fusion, delamination and unmelted particles. On a few occasions, TIG welding using dual wire feeding has been used to produce FG joints between ferritic steels and austenitic stainless steels. It was found that the FG joint has a significant advantage over the conventional joint in terms of the reduced tendency of carbon migration and lower thermal stresses due to the reduced coefficient of thermal expansion (CTE) mismatch.

A-TIG welding is found to be beneficial in improving the productivity of the conventional TIG welding technique. One important observation in the A-TIG weld joint developed without interlayer is the intermediate chemical composition of the fusion zone with respect to the parent metals. This means that a gradual chemical compositional variation from one parent metal to the other is created in autogenous weldments of dissimilar metals. This phenomenon is known as the "self-grading" effect, which suggests that due to dilution from both parent metals, a step-wise chemical compositional variation is naturally achieved even if two metals are simply joined autogenously or deposited over one another (Pulugurtha et al. 2009).

In the earlier section, it was evident that the chemical composition and mechanical properties of the A-TIG weld joints can be modified using interlayers. Therefore, functional grading can be achieved by constructively employing/developing layer-by-layer A-TIG weldments between dissimilar parent metals and interlayers of parent metals only.

In this work, sandwiching of interlayers of the parent metals alternatively in between two dissimilar metals is proposed. The functionally graded material (FGM) was developed by making overlapping welds between parent metals and interlayers. The approach would result in multiple overlapping A-TIG welds of different compositions due to dilutions from metals/interlayers placed at different locations. The resulting FG joint would reduce the steep chemical composition gradients, and the microstructure is expected to vary in a step functional manner.

7.10.1 Development of FG Weld Joint

This approach uses multiple interlayers of two different widths during A-TIG welding. Figure 7.8 shows the interlayers of 5 mm and 1 mm width which are stacked alternately in between parent metals of ferritic steels and austenitic stainless steels. The substrates and interlayers were tacked together by TIG welding prior to A-TIG welding passes.

The welding sequence in this approach is schematically illustrated in Fig. 7.9a–c. The first weld (FGW-1) was made between austenitic stainless steel substrate, 1 mm ferritic steel interlayer and 5 mm wide austenitic stainless steel interlayer by ensuring a complete melting of ferritic steel interlayer. The second welding pass (FGW-2) was developed between FGW-1 and a 5 mm wide ferritic steel interlayer. The final welding pass (FGW-3) was made between FGW-2, 1 mm wide austenitic stainless steel interlayer, and ferritic steel substrate by warranting a complete melting of austenitic stainless steel interlayer.

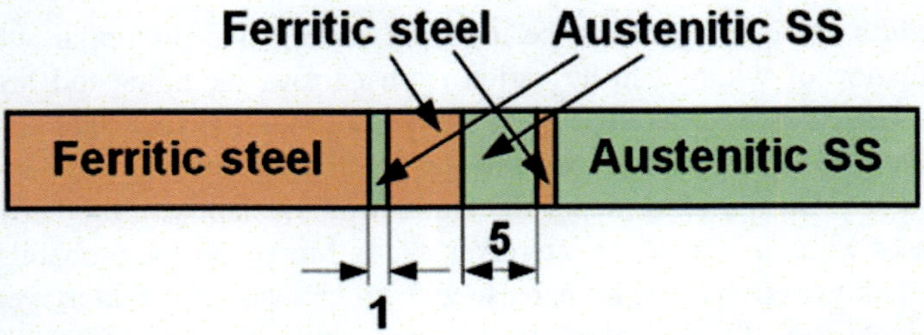

Fig. 7.8 Schematic representation arrangement of interlayers and substrates for FG weld joint development

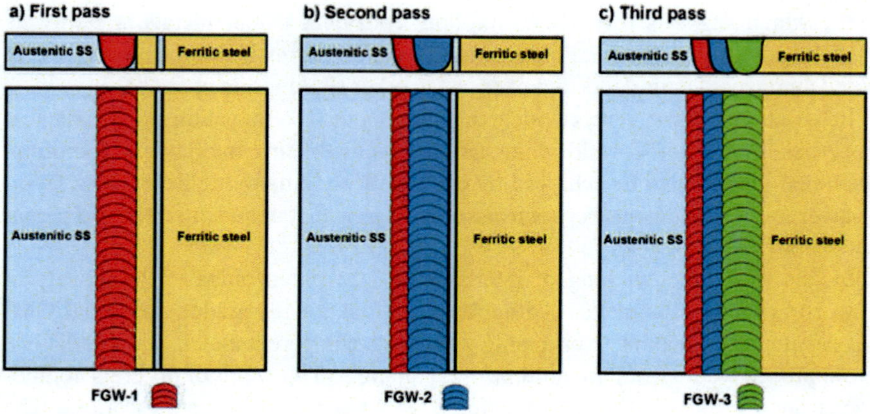

Fig. 7.9 Schematic representation of welding sequence in the development of FG weld joint **a** first welding pass, **b** second welding pass and **c** third welding pass

The activated fluxes are chosen such that each A-TIG welding pass results in a through-thickness penetration. The fluxes were chosen based on the exhaustive prior work. During the first two welding passes, most of the dilution would be from austenitic stainless steel. Therefore, a multi-component flux mixture comprising 25–35% SiO_2 + 30–40% TiO_2 + 20–30% MoO_3 + 5–15% CuO can be formulated. In the third welding pass (FGW-3), P91 steel parent metal would have maximum dilution. Therefore, a MoO_3 rich flux mixture comprising 50–70% MoO_3 + 10–30% TiO_2 + 10–30% Cr_2O_3 is suitable.

After every welding pass, the plates are allowed to be cooled to room temperature. Thereafter, the oxide layer on the weld surface needs to be cleaned. This is followed by the application of activated flux for the next welding pass. The A-TIG process parameters used for all three weldments are to be kept the same. The process parameters can be optimized based on the thickness of the workpieces.

7.10.2 Optimization of Interlayer Size

Different widths of interlayers can be used in between two dissimilar substrates. The combination of widths can be varied from narrow, medium- and broad-size interlayers. The effect of interlayer widths on the FG joint is schematically shown in Fig. 7.10a–c. With an increase in interlayer widths, the overall width of the FG region increases. On the other hand, the amount of overlap with the earlier weld pass decreases with an increase in interlayer width. Moreover, the probability of an unmelted interlayer is expected to increase with an increase in the interlayer width.

For FG weld joint development using narrow size interlayers, torch offsetting was required to avoid melting of the other interlayers, which are not expected to contribute to the welding pass. For example, for the FG joint using 0.5 and 3 mm

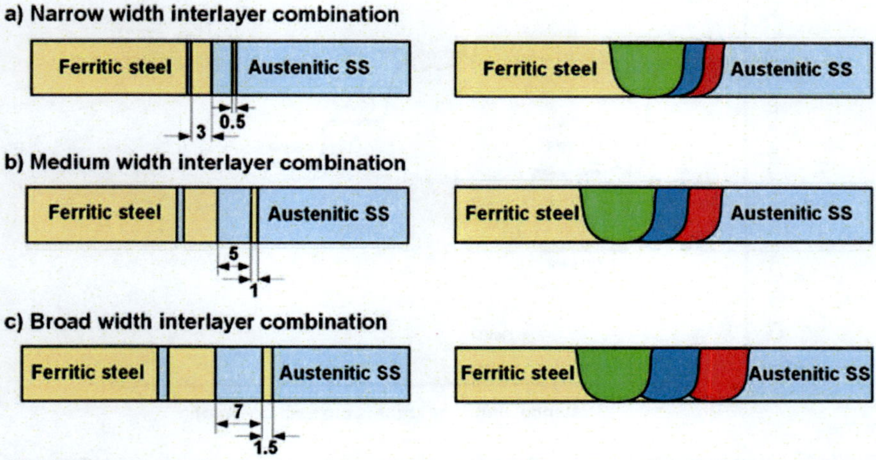

Fig. 7.10 Schematic of the FGWJ configuration with combinations of different size interlayers **a** narrow size (0.5 and 3 mm), **b** medium size (1 and 5 mm) and **c** broad size (1.5 and 7 mm)

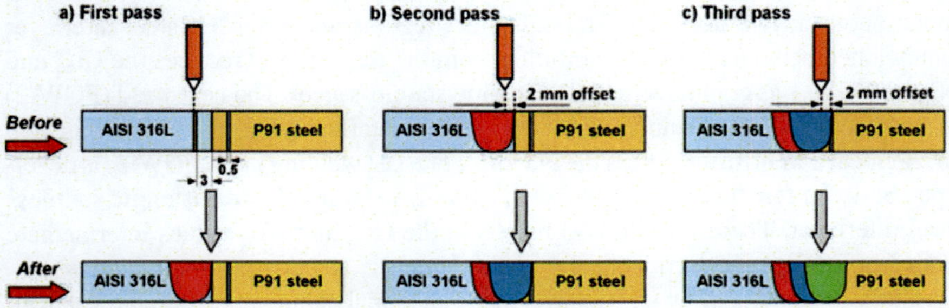

Fig. 7.11 Welding procedure showing torch offsetting technique for the development of FGWJ with small size (0.5 and 3) interlayers

wide interlayers, torch offsetting of 2 mm from the weld edge towards the centreline of the earlier welding pass was required. This is schematically shown in Fig. 7.11a–c. For medium-size and broad-size interlayers, torch offsetting is not required.

7.10.3 Chemical Composition of FG Joint

The chemical composition in the FG weld joints can be characterized by optical emission spectroscopy of the individual weld region or elemental line analysis using SEM–EDS or EPMA. The chemical composition in the FG weld joint exhibits a step-wise variation between two dissimilar steels as compared to the conventional A-TIG weld joint (Fig. 7.12). Additionally, a gradual change in the composition across all the steps is observed. This is attributed to the self-grading effect of parent

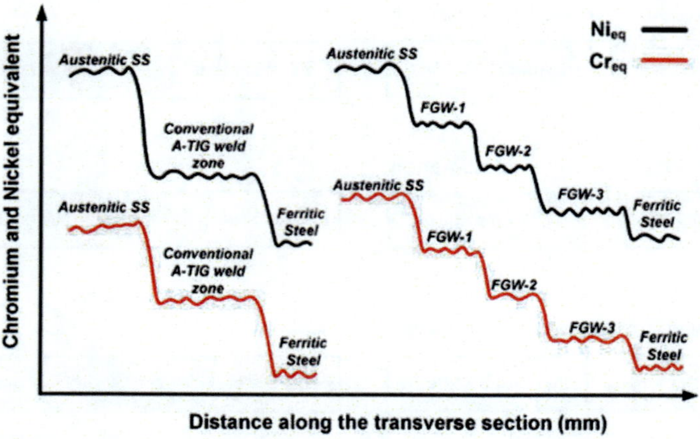

Fig. 7.12 Variation of Cr_{eq} and Ni_{eq} along the transverse section of the conventional A-TIG weld joint and FG weld joint

metals, their interlayers and overlapped weld passes. The first weld (FGW-1) is made between two austenitic stainless steel regions and a ferritic steel interlayer sandwiched between them. The dilution from the ferritic steel reduces the Cr_{eq} and Ni_{eq} of FGW-1 as compared to the austenitic stainless steel. The next weld (FGW-2) is made between FGW-1 and a wider ferritic steel interlayer that creates a further step in chemical composition across the FG joint. The last welding pass (FGW-3) is developed between ferritic steel parent metal, FGW-2 and a small size austenitic stainless steel interlayer. Therefore, FGW-3 results in the Cr_{eq} and Ni_{eq} values intermediate to that of FGW-2 and ferritic steel.

The functional grading with respect to the chemical composition is achieved in all the weld joints irrespective of the interlayer width combination, i.e. narrow, medium and broad. However, the overall fraction of alloying elements in the ferrous base of FG weld zones is lowered with an increase in the interlayer sizes. This was attributed to the higher overlapped area and the dilution from the earlier welding pass in case of narrow size interlayers. Additionally, in the FGW-1 pass, the lower dilution of small size P91 steel interlayer compared to AISI 316L results in the higher Cr_{eq} and Ni_{eq} in the fusion zone. Therefore, the subsequent passes are produced with higher Cr_{eq} and Ni_{eq} due to the self-grading effect. As a result, the narrow size interlayers result in higher Cr_{eq} and Ni_{eq} in the weld zones.

7.10.4 Microstructure of FG Joint

The Schaeffler diagram predictions and SEM micro-graphs of the various locations in the FG weld joint developed between ferritic steel and austenitic stainless steel are shown in Fig. 7.13a–d. The microstructure in the FG joint is heterogeneous across the

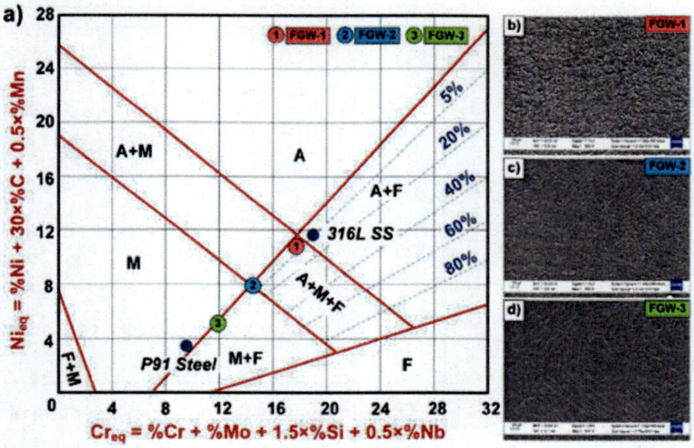

Fig. 7.13 **a** Schaeffler diagram representation of the various locations in the FG weld joint **b** micro-graph of FGW-1, **c** micro-graph of FGW-2 and **d** micro-graph of FGW-3

length of the joint. The microstructural heterogeneity in the joint is attributed to the difference in temperatures of martensitic transformation start (M_s) and martensitic transformation finish (M_f). The first weld (FGW-1) typically exhibits higher Cr_{eq} and Ni_{eq} which is close to austenitic stainless steel. Therefore, there is no driving force (Gibbs free energy) for the nucleation of martensite from austenite. However, delta ferrite is observed due to the high amount of Cr and Mo in the FGW-1 region. As a result, the micro-graph reveals a duplex austenitic-ferritic microstructure. The other two zones, FGW-2 and FGW-3, show a predominant martensitic structure as the M_s temperature is expected to be above room temperature. These zones also show some amount of austenite retained in the martensitic matrix, which is attributed to the M_f temperatures being lower than the room temperature.

7.10.5 Effect on Carbon Migration

As compared to the conventional A-TIG weld joint, FG weld joints alleviate the severity of carbon migration across the ferritic steel-weld zone interface. This can be validated through hardness measurements, optical or electron microscopy and simulated and experimental profiling of carbon content across the FG weld joint cross section. The reduction in the severity of carbon migration can be attributed primarily to the reduction in carbon potential. The amount of carbon and carbide forming elements (Cr, Mo, and Si) are the measure of carbon potential or activity in any zone. The carbide formers lower the free carbon content in the matrix by forming stable carbides. The ferritic steels typically contain a lower amount of chromium and higher carbon than austenitic stainless steel. Therefore, ferritic steels exhibit higher carbon potential. The FG weld joint, due to its smooth variation of alloying elements,

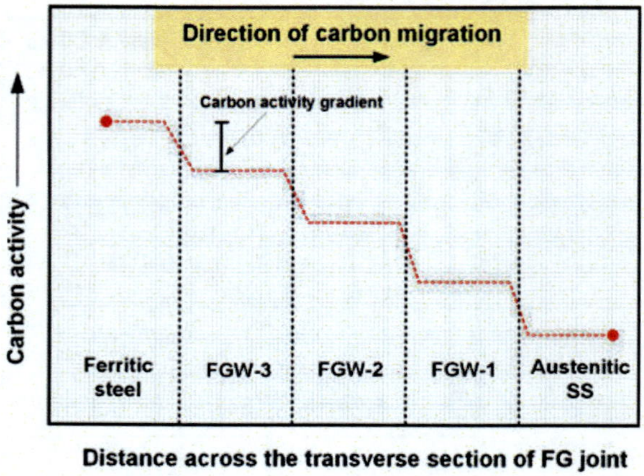

Fig. 7.14 Schematic representation of the variation of carbon activity in the FG weld joint

exhibits lower carbon potential gradients across the interfaces. Therefore, the loss of carbon and reduction of strength in ferritic steel is minimal.

It is expected that with an increase in interlayer size, the chemical compositional gradient at the ferritic steel side would be reduced. Therefore, the carbon migration tendency would be lower with the broader interlayers. However, as discussed in the earlier section, the issue of an unmelted interlayer region and poor overlapping may arise with broad-size interlayers (Fig. 7.14).

7.10.6 Mechanical Properties of FG Joint

The mechanical properties of the FG weld joint can be evaluated through hardness measurements, as shown in Fig. 7.15. In the as-welded and PWHT conditions, the FGW-1 exhibits slightly higher hardness than austenitic stainless steel. On the other hand, very high hardness values are observed in the FGW-2, FGW-3 and ferritic steel side HAZ in the as-welded conditions. In the PWHT condition, the hardness values get reduced in FGW-2, FGW-3 and HAZ of the ferritic steel side. Moreover, a steep hardness gradient is observed at the FGW-1 and FGW-2 interface.

A slight gain in hardness of FGW-1 as compared to austenitic stainless steel is due to an increase in delta ferrite content in FGW-1, which is a harder phase than austenite. The high hardness in the other regions is attributed to martensitic transformation due to weld thermal cycle experienced by the weld. The PWHT tempers the martensite and reduces its hardness.

During uniaxial tensile testing, failures occur from the interface of FGW-1 and FGW-2 due to a steep hardness gradient across this interface. The FG joint shows overmatching tensile strength in both as-welded and PWHT conditions. However, the joint exhibits poor ductility due to the existence of this steep hardness gradient facilitating localized yielding.

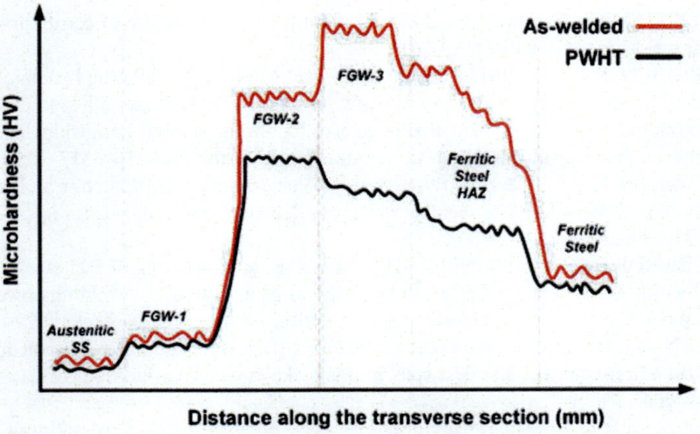

Fig. 7.15 Micro-hardness distribution in the FG weld joint in as-welded and PWHT condition

7.11 Summary

- Use of interlayers of Inconel 600 and Incoloy 800 in A-TIG welding of ferritic steel-austenitic stainless steel increases the chromium and nickel equivalents of the fusion zone. With an increase in the width of interlayers, chromium and nickel equivalents increase further.
- The use of interlayers modifies the weld zone microstructure. The nickel-based interlayer addition improves the impact toughness and ductility with a minimum loss of tensile strength.
- The use of interlayers with high Ni content reduces the severity of carbon migration across the ferritic steel-weld zone interface.
- A-TIG welding using multiple interlayers of parent metals and three overlapping passes results in functional grading of chemical composition and microstructure between the parent metals.
- During tensile testing, the FG weld joint exhibits an overmatching tensile strength.
- FG weld joint developed using A-TIG welding offers better resistance to carbon migration.

References

ASTM E975-13 (2013) Standard practice for X-Ray determination of retained austenite in steel with near random crystallographic orientation. ASTM International, West Conshohocken, PA

Bhaduri AK, Venkadesan S, Rodriguez P, Mukunda PG (1994) Transition metal joints for steam generators—an overview. Int J Press Vessels Pip 58(3):251–265

Bilmes PD, Solari M, Llorente CL (2001) Characteristics and effects of austenite resulting from tempering of 13Cr–NiMo martensitic steel weld metals. Mater Charact 46(4):285–296

DuPont JN (2012) Microstructural evolution and high temperature failure of ferritic to austenitic dissimilar welds. Int Mater Rev 57(4):208–234

Herrera C, Ponge D, Raabe D (2011) Design of a novel Mn-based 1 GPa duplex stainless TRIP steel with 60% ductility by a reduction of austenite stability. Acta Mater 59(11):4653–4664

Kacar R, Baylan O (2004) An investigation of microstructure/property relationships in dissimilar welds between martensitic and austenitic stainless steels. Mater Des 25(4):317–329

Kulkarni A, Dwivedi DK, Vasudevan M (2018) Study of mechanism, microstructure and mechanical properties of activated flux TIG welded P91 Steel-P22 steel dissimilar metal joint. Mater Sci Eng, A 731:309–323

Kulkarni A, Dwivedi DK, Vasudevan M (2019) Dissimilar metal welding of P91 steel-AISI 316L SS with Incoloy 800 and Inconel 600 interlayers by using activated TIG welding process and its effect on the microstructure and mechanical properties. J Mater Process Technol 274:116280

Kurt B, Orhan N, Somunkiran I, Kaya M (2009) The effect of austenitic interface layer on microstructure of AISI 420 martensitic stainless steel joined by keyhole PTA welding process. Mater Des 30(3):661–664

Mas F, Tassin C, Valle N, Robaut F, Charlot F, Yescas M, Bréchet Y (2016) Metallurgical characterization of coupled carbon diffusion and precipitation in dissimilar steel welds. J Mater Sci 51(10):4864–4879

Mills KC, Keene BJ, Brooks RF, Shirali A (1998) Marangoni effects in welding. Philosoph Trans Roy Soc London. Series A Math Phys Eng Sci 356(1739):911–925

Mortezaie A, Shamanian M (2014) An assessment of microstructure, mechanical properties and corrosion resistance of dissimilar welds between Inconel 718 and 310S austenitic stainless steel. Int J Press Vessels Pip 116:37–46

Pulugurtha SR, Newkirk J, Liou F, Chou HN (2009) Functionally graded materials by laser metal deposition. In: 2009 International Solid Freeform Fabrication Symposium. University of Texas at Austin

Reichardt A, Shapiro AA, Otis R, Dillon RP, Borgonia JP, McEnerney BW, Beese AM (2021) Advances in additive manufacturing of metal-based functionally graded materials. Int Mater Rev 66(1):1-29

Sudha C, Terrance ALE, Albert SK, Vijayalakshmi M (2002) Systematic study of formation of soft and hard zones in the dissimilar weldments of Cr–Mo steels. J Nucl Mater 302(2–3):193–205

Vidyarthy RS, Kulkarni A, Dwivedi DK (2017) Study of microstructure and mechanical property relationships of A-TIG welded P91–316L dissimilar steel joint. Mater Sci Eng, A 695:249–257

Wu W, Hwu LY, Lin DY, Lee JL (2000) The relationship between alloying elements and retained austenite in martensitic stainless steel welds. Scripta Mater 11(42):1071–1076

Chapter 8
Dissimilar Metal Joining of Steel-Aluminium Alloy by Spot Welding

This chapter presents the common issues encountered during resistance spot welding of dissimilar metals, namely steel and aluminium combination. Causes of various issues related to the steel-aluminium spot welding have been explained using heat and metallurgical interactions. Strategies to overcome these issues have also been described. Fundamentals of steel-aluminium resistance spot welding can be extended to the other approaches based on the brazing of such dissimilar metal combinations.

8.1 Background

The demand for energy efficient systems is growing continuously to cope with increasing challenges of depleting fossil fuels, improvement of fuel economy and protection of environment to avoid global warming and facilitate sustainable development. The development of such kind of energy efficient systems especially in automotive sector is conceptualized right from the initial design stage using multi-material system. This is leading to the application of wide range of materials in automobiles for manufacturing different components like rubber, plastics, cast, steel, aluminium, composite materials and bulk material with surface modification for enhanced the functionality, performance and service life. A typical car model shows use of steel and aluminium in design of different components (Fig. 8.1).

The application of different material combinations, however, imposes many issues related to manufacturing, service life and performance due to difference in physical, metallurgical and mechanical properties. Therefore, technologists are working hard globally to address such challenges in larger interest of economy and environment.

The development of energy efficient automotive invariably deals conflicting demand of economy and technology needs (material, design, manufacturing etc.). One of the common strategies is to develop energy efficient system of light dead weight using high specific strength materials, i.e. high-strength and low-density material while designing the components using newer approaches like use of fracture

© The Author(s), under exclusive license to Springer Nature Singapore Pte Ltd. 2023
D. K. Dwivedi, *Dissimilar Metal Joining*,
https://doi.org/10.1007/978-981-99-1897-3_8

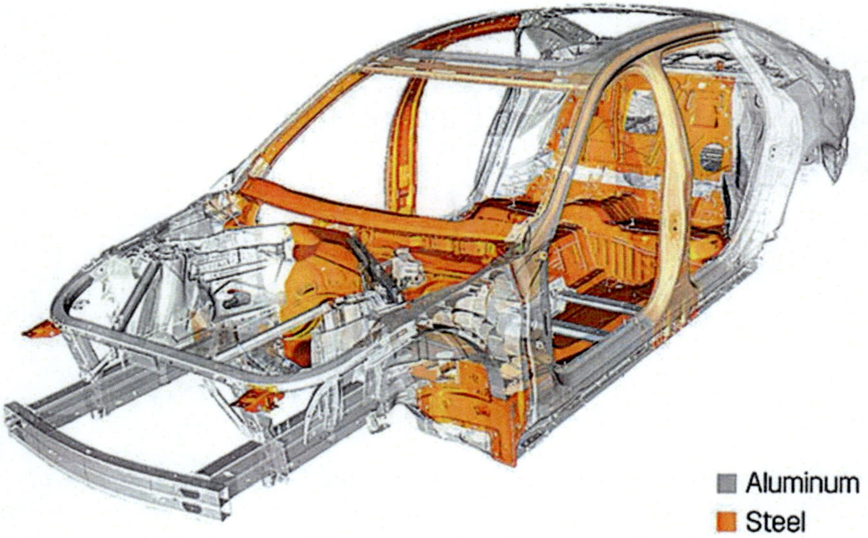

Fig. 8.1 Photograph showing typical use of steel and aluminium dissimilar in a car (New Techniques for Joining Steel and Aluminum)

mechanics, finite element methods, etc., so as to minimize the dimensions/section thickness and avoid unnecessary high factor of safety.

Considering these aspects, engineers/designers of automotive have started to use the different material combinations such as steel, aluminium and magnesium alloys. Therefore, joining of steel with other metals and alloys is inevitable in newer and futuristic automotive. The high heat input joining of ferrous and non-ferrous metals using fusion welding processes, however, suffers from various issues related to soundness and deteriorated joint properties in the form of low joint efficiency, intermetallic formation, cracking and porosity, softening of heat-affected zone and high residual tensile stress. Many of these issues originate due to differences in chemical, physical, mechanical, metallurgical properties and related incompatibilities of dissimilar metal combination to be joined.

The most common process used for joining of similar and dissimilar metals system in automotive is resistance spot welding. The spot welding process is performed in four steps, i.e. putting sheets together (in lap configuration) by apply electrode force (squeezing) followed by supply of welding current for generation of electrical resistance (joule) heat needed for thermal softening and melting to develop weld nugget (welding) and then consolidation (holding) and withdrawal of weld joint (off) as shown in Fig. 8.2. The weld cycle keeps on repeating.

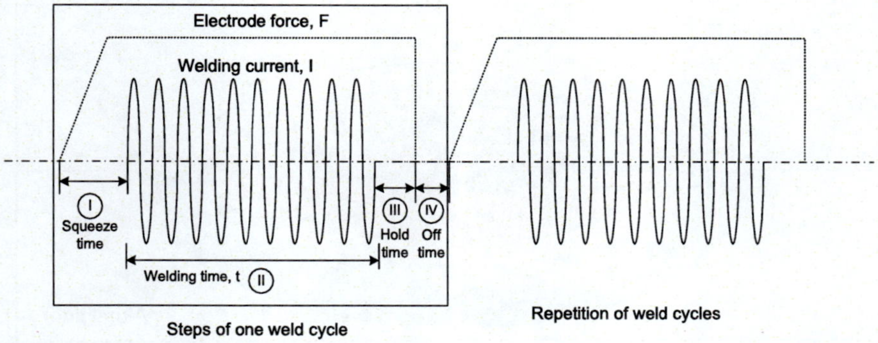

Fig. 8.2 Schematic shows the steps of resistance spot weld cycle

8.2 Challenges in Joining of Steel-Aluminium by Spot Welding

The joining of steel and aluminium dissimilar combination is considered difficult due to various issues hampering the development of sound joints. The difference in various characteristics (mechanical, metallurgical, physical and chemical) of steel and aluminium causes a range of joining problems (Table 8.1).

The difference in thermal properties, namely thermal conductivity and electrical resistivity, affects both the heat generation and dissipation to base metal, which in turn affects the weld thermal cycle, and so formation of weld nugget and heat-affected zone. Steel being of lower thermal conductivity results in narrow heat affected despite of more heat generation during spot welding than aluminium (Fig. 8.3). However, despite of higher heat generation/temperature on steel, no weld nugget is formed due to very higher melting temperature of steel (1538 °C) than aluminium (660 °C).

Table 8.1 Characteristics of aluminium alloy and steel

Characteristics	Aluminium alloy	Steel
Melting temperature, °C	463–671	1425–1540
Thermal conductivity, W/mK	140–240	14–54
Electrical resistivity, ohm.m	2.65×10^{-8}	1.43×10^{-7}
Thermal expansion coefficient, 10^{-6} m/m °C	21–24	10.8–12.5
Diffusion coefficient, $m^2\ s^{-1}$	1.8×10^{-4}, Al in Fe (1003–1673 K)	53×10^{-4}, Fe in Al (793–922 K)
Solid solubility, %	Nil, Al in Fe	10–12%, Fe in Al
Ductility, % elog	High	Moderate
Yield strength, MPa	200–400	350–1800

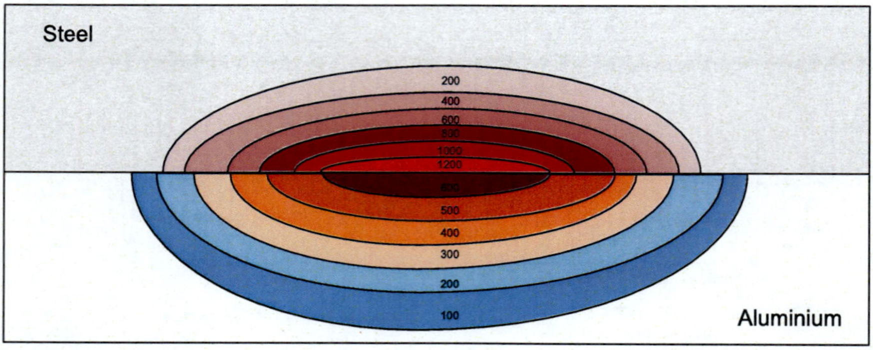

Fig. 8.3 Scheme of weld thermal profile in dissimilar steel-aluminium spot welding

The metallurgical incompatibility leads to the formation of intermetallic compounds and porosity formation due to zinc vaporization in case of galvanized steel. The differential thermal properties of dissimilar combination (steel and aluminium alloy), namely melting point, thermal conductivity and coefficient of thermal expansion, result in asymmetrical weld nugget formation, different widths and characteristics of heat-affected zones (HAZs) and high cracking tendency to due to residual tensile stress development in vicinity of weld nugget caused by differential expansion and contraction. The HAZ softening occurs in precipitation and work hardenable aluminium alloys while hardening and embrittlement on steel side. Poor wettability caused by alumina formation results in inclusions and poor bonding. The most of the above issues including the formation of unfavourable intermetallic compounds in turn reduce the mechanical performance of the spot weld joints. The severity of problems associated with steel-aluminium spot weld joints further increases with high heat input; therefore, attempts are always to optimize the heat input to minimize the related issues while realizing the sound spot weld joints (Chen et al. 2016; Gullino et al. 2019). Thus, the major challenges in joining of steel-aluminium dissimilar metal spot welding are following.

- The formation of hard and brittle intermetallic compound (IMC),
- Interfacial cracking due to residual tensile stresses and IMCs,
- Discontinuities in weld like porosity, indentation on aluminium sheet,
- Softening/hardening of heat-affected zone in aluminium/steel sheets,
- Alumina formation causing inclusion and poor metallurgical bonding.

8.2.1 *Formation of Intermetallic Compound*

The interaction of aluminium with iron in spot welding forms various type of intermetallic compounds (IMCs) due to limited mutual solid solubility. The formation of IMC is necessary for developing a sound and metallurgical weld joint between steel and aluminium. The morphology of IMC layer in form of a thin, fine grained

Table 8.2 Characteristics composition and hardness of Fe–Al IMCs

Al–Fe IMCs	Al in IMCs, wt%	Hardness, HV
$AlFe_3$	25	(344–368)
AlFe	50	(491–667)
Al_3Fe_2	63	–
Al_2Fe	66–67	(1058–1070)
Al_5Fe_2	68.7–73.2	(1000–1158)
Al_3Fe/ $Al_{13}Fe_4$	74–76	(772–1017)

and discrete/discontinuous IMCs results in a good joint strength. The IMCs formed at interface of spot weld nugget of steel-aluminium joint are of different types, e.g. Fe-rich Fe–Al IMC and Al-rich Fe–Al IMC (Table 8.2).

The solid solubility of Al in Fe is almost nil, while Fe has solid solubility in 10–12% in Al. Further, diffusion coefficient of Fe in Al is about 45–50 times higher than that of Al in Fe. Therefore, diffusion of Al in steel side is extremely limited/thin, while Fe diffuses in Al appreciably. A combination of high solid solubility and high diffusion coefficient of Fe in Al produces Al-rich Fe–Al IMCs ($FeAl_3$, Fe_2Al_5, Fe_2Al_3) on weld nugget side while high Fe-rich Fe–Al IMCs (Fe_3Al, FeAl, $FeAl_2$) on steel side as shown in Fig. 8.4. Due to the formation of Al-rich IMCs (Fe_2Al_5) at the weld nugget-steel interface deteriorates the mechanical properties and increases the embrittlement, which in turn increases the cracking tendency and reduces the load carrying capacity of the joint. Whereas the presence of Fe-rich IMCs (FeAl) in weld nugget of Al-steel spot weld is desirable considering their relatively high fracture toughness and positive effect on joint interface strength.

Thickness of IMC layer at interface of spot weld nugget of steel-aluminium joint is very crucial for mechanical performance. It is desirable to have a thin (0.5–2 μm) IMC layer for good mechanical performance of spot weld joint. Thick IMCs layer (> 10 μm) at weld interface deteriorates the mechanical performance and increases the cracking tendency. Thickness of IMCs layer is primarily determined by the heat input used for the developing a weld joint. Increase in heat input in general increases the thickness of IMCs layer. Moreover, thickness of IMC is maximum at the centre of the weld nugget, and it decreases gradually towards the periphery as shown in Fig. 8.5. Therefore, many innovative attempts have been made to spot weld joints using controlled and limited heat input for spot welding to reduce the related undesirable effects of high heat input using approaches like high-power density sources, using cover plate during spot welding to develop sound and good weld joints. Further, characteristics of the undesirable IMC layer can be modified using controlled alloying such as Zn in case of joining of galvanized steel sheet with aluminium to improve the joint performance.

The mechanism of IMC formation in spot welding of aluminium and steel dissimilar combination involves interaction of molten aluminium and solid hot steel surface, which in turn causes aluminizing of steel surface at elevated temperatures. One recently proposed hypothesis suggests that IMC is formed in four different stages

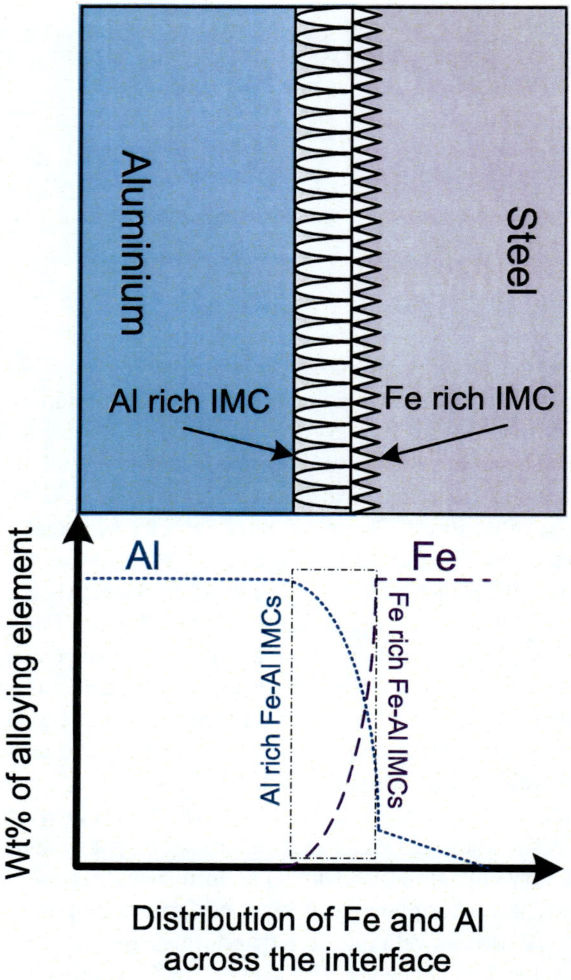

Fig. 8.4 Schematic showing distribution of Fe, Al in IMC formed at steel-weld nugget interface in spot weld joints developed between steel and aluminium

(Fig. 8.6). Initially, some Fe is dissolved with molten aluminium, and then, diffusion of aluminium in steel surface results in formation of Fe_2Al_5. This IMC (Fe_2Al_5) grows gradually from discontinuous and discrete thin layer to continuous and thickness layer. Thereafter, diffusion of Fe into Al rich Fe–Al IMC (Fe_2Al_5) develops the very thin (< 1 μm) serrated $FeAl_3$ on aluminium nugget side. The thickness of IMC layer increases with heat input during welding as high-temperature retention for longer time facilitates more interaction between Al and Fe at the interface, which in turn results in thicker IMC layer (Fig. 8.7a). Increase in thickness of hard and brittle IMC layer in turn reduces the load carrying capacity of the spot weld joint due to increased cracking tendency and interfacial embrittlement (Fig. 8.7b).

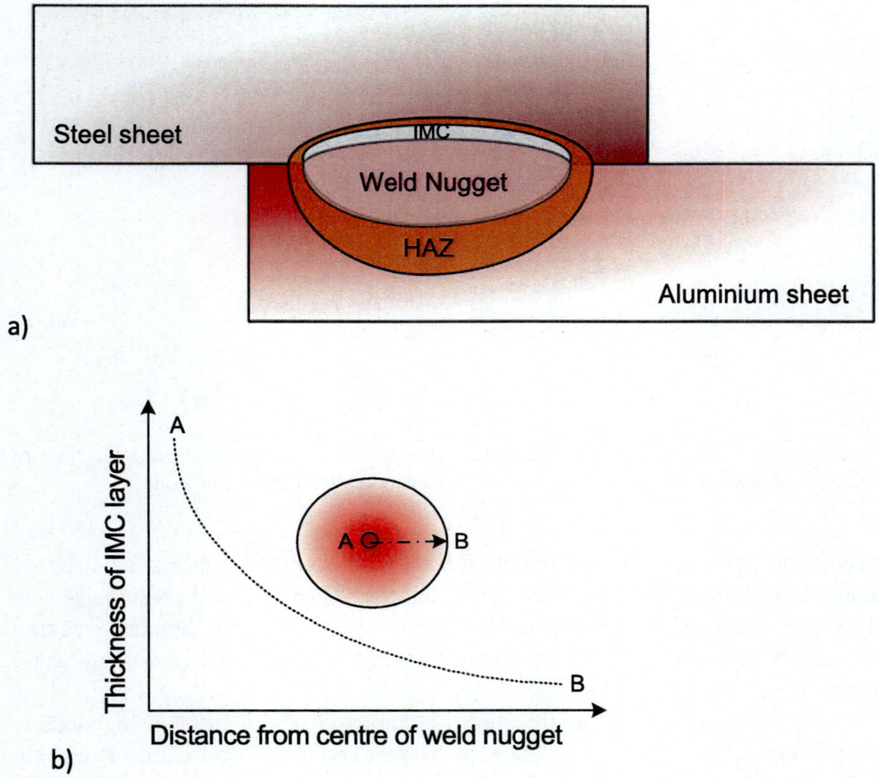

Fig. 8.5 Schematic both **a** and **b** showing variation in thickness of IMC layer formed at steel-weld nugget interface from the weld centre to periphery of the weld nugget in spot weld joints developed between steel and aluminium

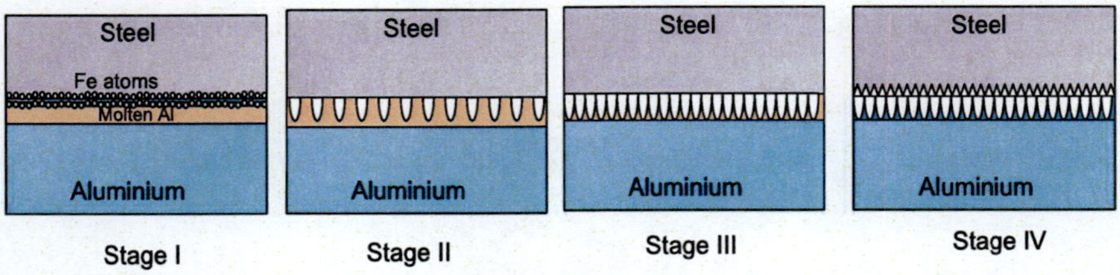

Fig. 8.6 Schematic showing stages in development of IMC layer formed at steel-weld nugget interface in spot weld joints developed between steel and aluminium

8.2.2 Interface Cracking Due to Residual Tensile Stress

The formation of hard and brittle Al-rich IMCs at weld nugget-steel base metal interface in presence of residual tensile stress at weld nugget due to differential thermal expansion and contraction imposed by weld thermal cycle during spot welding tends

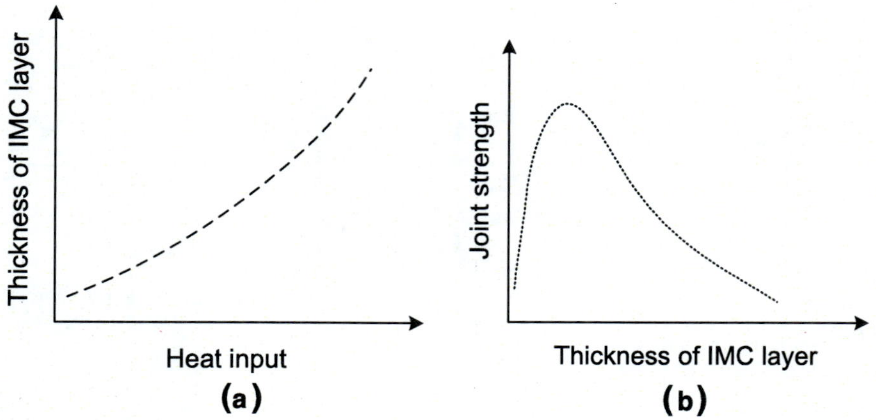

Fig. 8.7 Schematic showing effect of **a** heat input on thickness of IMC layer and **b** thickness of IMC layer on strength of spot weld joints developed between steel and aluminium

to cause the cracking in spot weld (Figs. 8.8 and 8.9). Therefore, any attempt to either (a) develop fracture tough IMCs layer at the interface or (b) reduce the residual tensile stress will eventually reduce the steel-aluminium spot weld joint interface cracking and embrittlement tendency. These factors improve the mechanical performance and load carrying capacity of the joint.

Controlled alloying and modification of characteristics of IMCs using suitable addition in the form of surface coating, interlayers and post weld treatment can help to develop fracture tough IMCs. Use of a suitable interlayer at the joint interface for more favourable gradient in property variation across the two dissimilar base metal sheets, while the post-weld stress relieving treatment can help in overcoming the interfacial cracking of the weld nugget by reducing tensile residual stress. Further, aluminium alloys of high solidification temperature range (usually > 50 °C) show solidification cracking in weld nugget zone and liquation cracking in partial melting

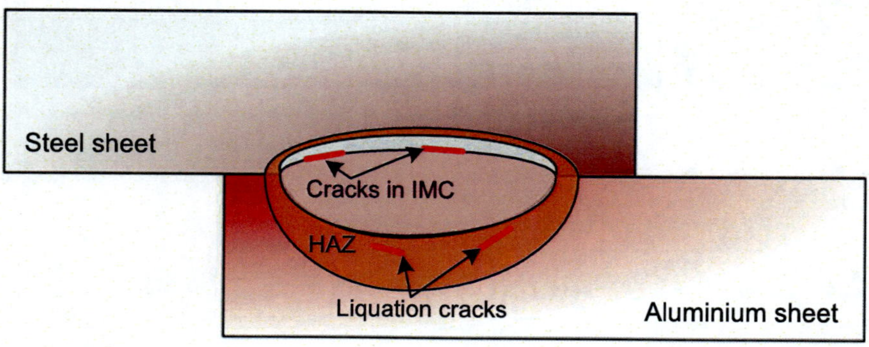

Fig. 8.8 Schematic showing cracks at IMC and HAZ in Al side in spot weld joint developed between steel-Al

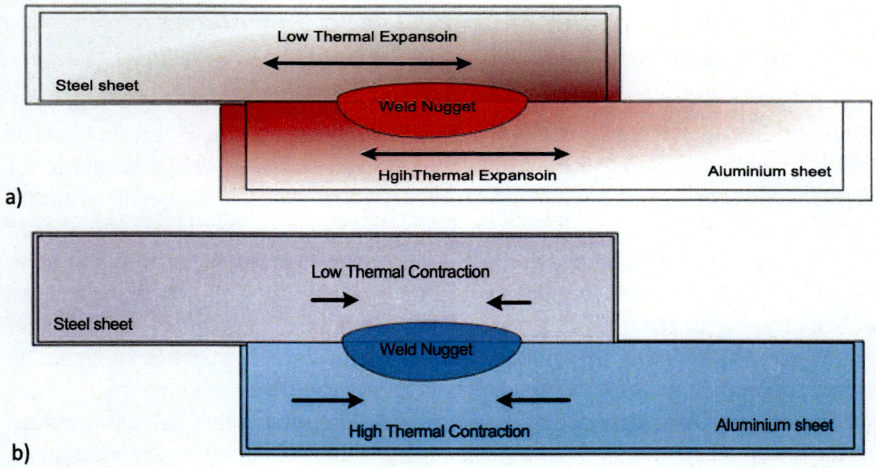

Fig. 8.9 Schematic showing dimensional variation during **a** heating due to thermal expansion and **b** cooling due to thermal contraction while developing spot weld joint between steel and aluminium

zone in vicinity of weld nugget. These cracks can be controlled by decreasing residual tensile stress through choice of suitable preheat and post-heat treatment as a part of weld cycle itself (Fig. 8.10). Preheating reduces the heating and cooling rate, which in turn lowers the residual stress in the spot weld joint. Similarly, post-weld heating relaxes the locked-in strain to reduce the residual tensile stress (Fig. 8.9). The pre-weld and post-weld heating current and their respective times need to be established while developing welding procedure specifications.

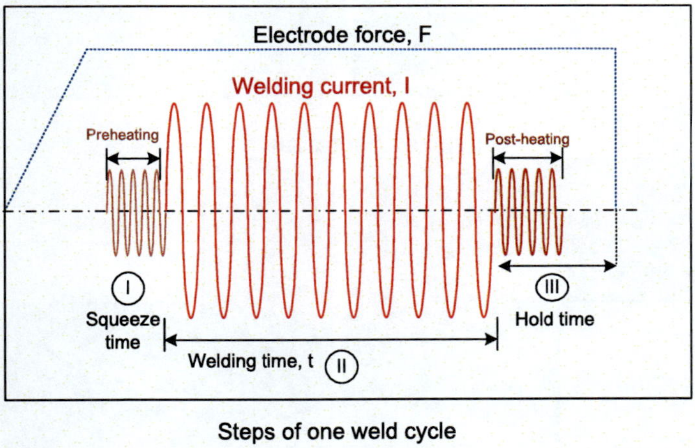

Fig. 8.10 Schematic of resistance spot welding showing additional steps of pre and post heat treatment cycles

8.2.3 Discontinuities like Porosity in Weld, Indentation on Aluminium Sheet

Inappropriate spot welding procedure results in discontinuities in spot weld joints of steel-aluminium dissimilar metal combinations. The common weld discontinuities such as cracks, porosity and indentation are observed in spot weld nugget of dissimilar metal joint (Fig. 8.11). Porosity can be in form of large gas pocket, hydrogen induced pin-hole porosity and shrinkage pores. Further, gaseous porosity can also be due to vapours generated due to evaporation of the zinc during the welding. Improved surface cleaning to remove oxides, impurities, organized compounds and moisture from the surface reduces the gaseous porosity, while suitable selection of electrode pressure and heat input for welding helps to control the shrinkage porosity.

High temperature generated at the electrode-aluminium sheet contact interface due to electrical resistance heating causes more thermal softening of aluminium than steel sheet. The aluminium sheet therefore under influence of inappropriately high electrode force gets penetration to a greater depth. This in turn appears in the form indentation on the aluminium sheet. The indentation reduces the load resisting sectional area of the aluminium sheet, which in turn increases the stress localization. This leads to the formation of a weak zone prone to the fracture. Thermal softening

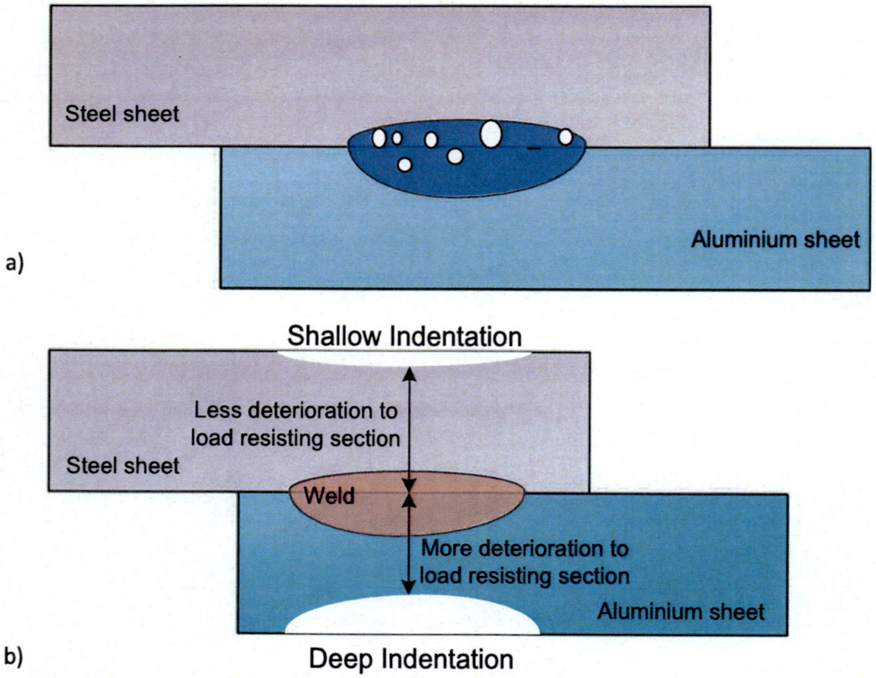

Fig. 8.11 Schematic showing various spot weld discontinuities in dissimilar steel-Al combination **a** pores in weld nugget and **b** indentation at the surface Al and steel due to electrode force

of steel is very limited and so it remains harder and stronger than aluminium and therefore indentation by electrode under identical welding condition is very shallow on the steel surface.

8.2.4 Softening/Hardening of Heat-Affected Zone in Aluminium/Steel Sheets

The steel-aluminium dissimilar spot weld joint can experience both HAZ softening and hardening depending upon the base metal sheets and their heat treatment conditions (Fig. 8.12). In general, both work hardened, precipitation hardened aluminium alloy sheet and quenched-tempered (Q and T) steel sheet experience HAZ softening near spot weld. HAZ softening of work hardened aluminium alloy sheet occurs due to recovery, recrystallization and grain growth, while that in case of precipitation, hardened aluminium alloy takes place due to reversion (dissolution of precipitates) and grain growth. The softening of HAZ in Q and T steel sheet is primarily attributed the over tempering of already tempered martensite. However, HAZ of hardenable steel sheets in hot/cold rolled, annealed, normalized and quenched condition is hardened due to typical martensitic transformation caused by weld thermal cycle during welding. HAZ softening of Q and T steel leads to weaken of the joint, while hardening of steel and their heat treatment condition promotes the cracking and embrittlement tendency.

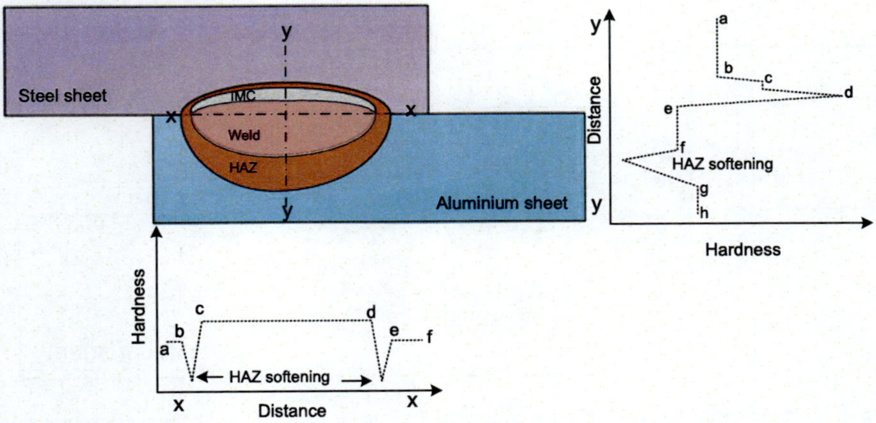

Fig. 8.12 Schematic of hardness profile of spot weld joint developed between steel-Al combination showing hardening in HAZ of steel and softening in HAZ of Al in two xx and yy directions

8.2.5 *Inclusion and Poor Bonding Due to Alumina Formation*

The aluminium during the spot welding in steel-aluminium dissimilar metal combination shows high affinity to atmospheric oxygen at the high temperature generated during the welding and forms refractory and adherent alumina. A thin layer of alumina affects the spot welding in many ways (a) deteriorates the wettability of molten aluminium and (b) forms inclusions (Fig. 8.13). Reduction in wettability of molten aluminium in turn decreases the metallurgical bonding in spot welding between the aluminium and steel sheet through the weld nugget. The presence of inclusion acts as stress raiser besides decreasing the load resistance cross-sectional area. Therefore, poor bonding and inclusion both in turn adversely affect the mechanical performance of steel-aluminium spot weld joints.

8.3 Characteristics of Resistance Spot Weld Affecting Tensile/Shear Strength

The mechanical performance of steel-aluminium spot weld joints depends on many factors governing the load resisting cross-sectional of the joint, metallurgical properties, soundness and stress raisers, if any. Accordingly, the tensile shear strength of steel-aluminium spot weld joints depends on weld nugget diameter, metallurgical characteristics, soundness and indentation.

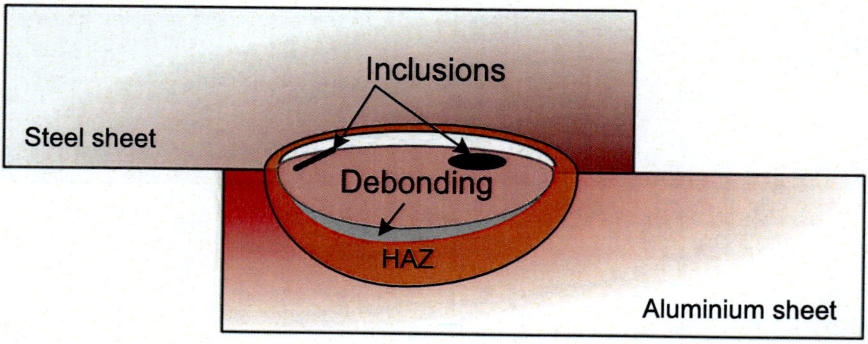

Fig. 8.13 Schematic showing weld discontinuities like inclusions and poor bonding in spot weld joint developed between steel-Al combination

8.3.1 Weld Nugget Diameter

The load resisting cross-sectional area across the weld joint mainly depends on the weld nugget diameter. An increase in weld nugget diameter increases the load resisting cross-sectional area of the weld joint. Increase of nugget diameter (within limits) therefore reduces external stress for a given load, which in turn results in increased load carrying capacity and mechanical performance of the joint due to enhanced metallurgical continuity between two sheets. In general, increase in spot weld nugget diameter for a given thickness (of metallic sheets) increases the load carrying capacity, and after reaching maxima, it starts decreasing gradually due to expulsion, thinning of weak base metal in the combination and stress concentration. Therefore, an optimum weld nugget diameter is identified during the development of welding procedure specification for establishing welding parameters, namely welding current, time and electrode force. Weld nugget diameter primarily depends on the heat input. Very low heat input may lead to no metallurgical bonding or very limited bonding with small nugget diameter. Initially, weld nugget diameter increases rapidly with increase in heat input (current, time); then, rate of increase in nugget diameter with welding current/time decreases due to shunting effect and expulsion tendency (Fig. 8.14). However, weld nugget diameter in case of dissimilar metal spot welding mainly develops on the weak and low melting point base metal sheet side due to both differential heat generation and heat required for melting during welding. In this case, therefore, weld nugget is mainly formed on aluminium side while steel sheet remain unmelted/unaffected. Still the metallurgical bond is formed at the steel-aluminium interface.

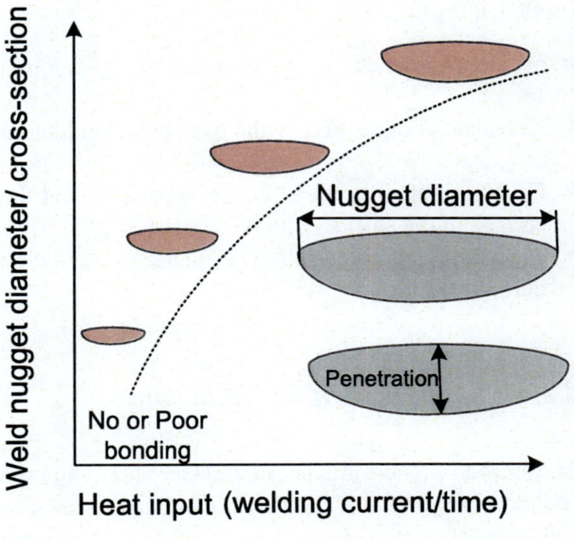

Fig. 8.14 Schematic showing effect of heat input on weld nugget size in spot welding

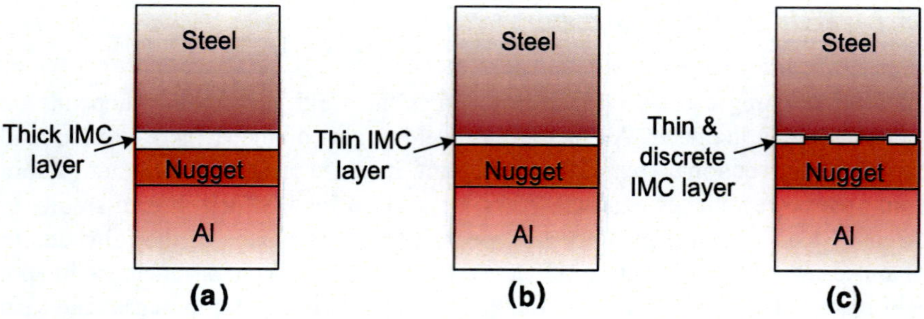

Fig. 8.15 Schematic showing **a** thick and continuous, **b** thin and **c** thin and discreet IMC layer at steel-weld nugget interface in spot weld joints developed between steel and aluminium

8.3.2 Metallurgical Characteristics of Spot Weld Joint

The mechanical performance of the steel-aluminium spot weld joint from the metallurgical characteristics point of view depends on (a) the type and thickness of IMC layer formed at weld nugget-steel sheet interface, (b) microstructure of the weld nugget and (c) softening/hardening of HAZ of steel/aluminium sheet (Fig. 8.15). The failure location of the steel-aluminium spot weld joint subjected to tensile shear load will indicate the weak zone of the joint, namely weld nugget, IMC, softened HAZ and base metal itself. The formation of thick (> 10 μm) hard and brittle continuous Al-rich IMC layer reduces the ductility and tensile shear load carrying capacity of the joint while thin (< 10 μm) fracture tough discrete Fe-rich Fe–Al IMC increases tensile shear load carrying capacity of the joint. The characteristics of IMCs layer formed at the interface depend on.

- Welding parameters, namely heat input (welding current, time and electrode force),
- Chemical composition of the base metals (aluminium alloy and steel) sheets to be joined,
- Quality of interlayer/coating (composition and thickness of zinc layer) on the surface of the steel sheet,
- Thickness and composition of interlayer sandwiched between steel-aluminium during welding.

8.3.3 Soundness of the Weld Joints

The mechanical performance of the steel-aluminium spot weld joint with regard to the soundness depends on (a) presence of stress raisers, if any, in weld joint, (b) reduction in load resisting cross-sectional area occurring due to presence of the weld discontinuities (Fig. 8.16). Further, the effect of weld discontinuities on the stress concentration and load resisting cross-sectional area dictated by size, type, orientation and location

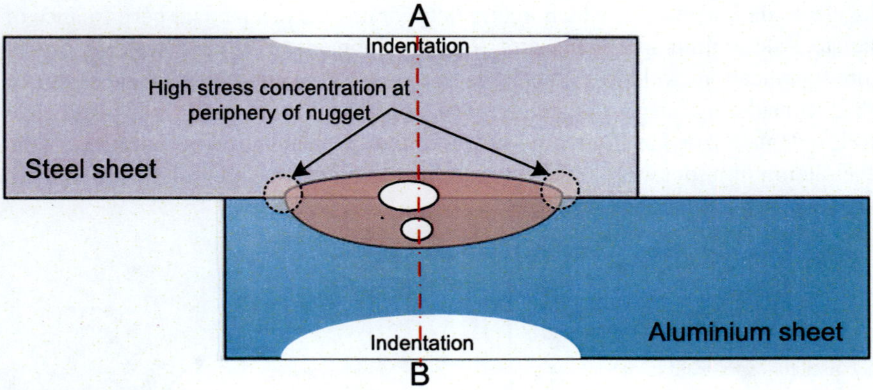

Fig. 8.16 Schematic showing weld discontinuities like stress raiser, pores and indentation in spot weld joint developed between steel-Al combination reducing load resisting cross-sectional area across the section A–B

of weld defects/discontinuities. Large and sharp edge crack like discontinuities-oriented parallel to the direction of loading at periphery/circumference spot weld decreases the shear load carrying capacity appreciably than those small, round-tipped (pores, inclusions), oriented normal to the external load and those discontinuities located internally within the weld nugget.

The common weld defects/discontinuities determining the mechanical performance of the steel-aluminium spot weld joint include the weld expulsion at the circumference of the weld nugget, gas and shrinkage pores and voids, poorly bonded zone due to limited wetting in presence of alumina. The development of proper welding procedure specifications including cleaning methods and welding parameters, namely welding current, time and electrode force helps to avoid weld discontinuities and improve the mechanical performance of the steel-aluminium spot weld joint.

8.3.4 Electrode Indentation on the Aluminium Sheet

The mechanical performance of the steel-aluminium spot weld joint is compromised due to indentation of the electrode on steel and aluminium leading to the significant localized reduction in cross-section of mainly on aluminium sheet near the weld nugget. The indentation on the aluminium is primarily a penetration of the electrode into the thermally softened aluminium due to electrical resistance heating at electrode-aluminium sheet contact interface (Fig. 8.17). The resistance to thermal softening reduces the indentation on sheet of base metal due to electrode force. Therefore, excessive welding current for longer time coupled with high electrode increases the electrode indentation, which in turn deteriorates the mechanical performance of the steel-aluminium spot weld joint. The development of suitable welding procedure specifications welding parameters, namely welding current, time and electrode force, considering the geometry of electrode tip and aluminium alloys to avoid electrode indentation on the aluminium so as improve the mechanical performance of the steel-aluminium spot weld joint. Electrode geometry is selected/modified suitably to reduce such indentations (Fig. 8.18).

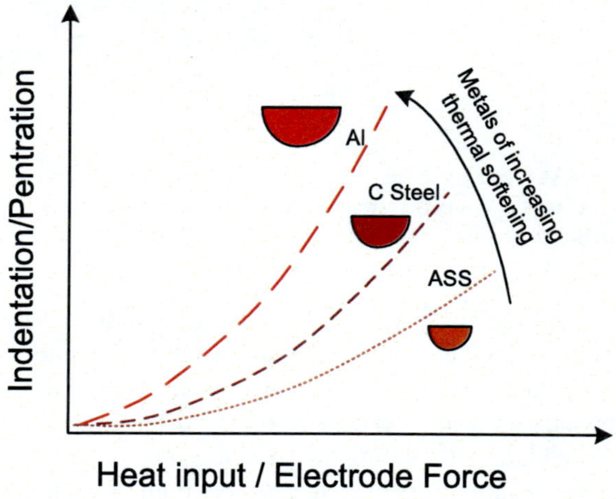

Fig. 8.17 Schematic showing effect of heat input/electrode on indentation in spot weld joint developed between steel-Al combination

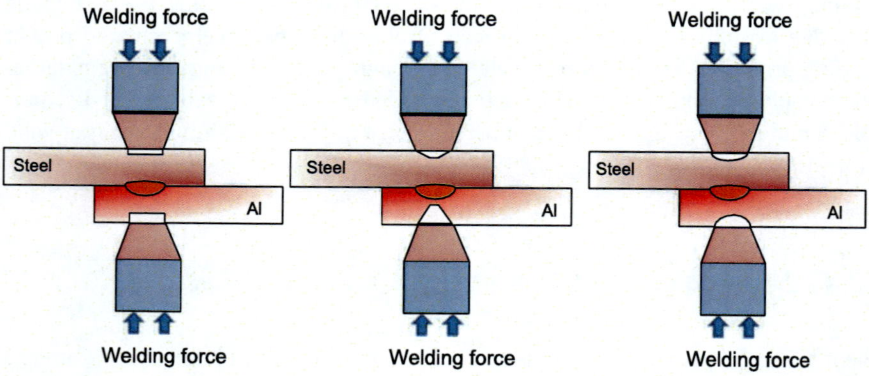

Fig. 8.18 Schematic showing effect of electrode tip geometry on indentation in spot weld joint developed between steel-Al combinations

8.4 Resistance Spot Welding and Its Parameters

The resistance spot welding for developing a welding joint between metallic sheets uses heat generated by electrical resistance (joule) heating at contact interface (Fig. 8.19). The heat causes the thermal softening and partial melting of base metal being joined followed by consolidation under pressure to produce a weld nugget. Therefore, parameters related to the spot welding determining the heat generation (H: I^2Rt) include welding current (I, A) and time (t, ms) for which welding current applied and interfacial contact resistance (R, ohm) depending on the surface cleaning, roughness and electrode force. The joule heating equation suggests that heat generation increases with square of welding current and linearly with welding time and contact resistance. The contact resistance depends on the metallic intimacy between the sheets being joined, which in turn depends on the surface finish, cleanliness and electrode force applied during welding.

Smooth, perfectly clean surfaces of base metal sheets subjected to high electrode force result in very good metallic intimacy and electrical contact across the interface for the flow of current. Good metallic contact in turn decreases the interfacial contact resistance so the joule heating for the given set of welding parameters (I, t). Reduction in heating in turn decreases the weld nugget diameter even for a given welding current. An optimum combination of welding current (I) and welding time (t) produces sound weld with desired size of weld nugget (Fig. 8.20). Excessive current causes undesirable features in the joint like expulsion, shrinkage, porosity, cracks, stress concentration and thickness reduction (due to electrode indentation) and electrode degradation.

Increase in the weld nugget diameter or increase in number of weld nuggets during joining of large sheets causes the shunting of welding current leading to the flow of the most of the welding current through low contact resistance area of already welded joints. Therefore, shunting of the welding current decreases the weld nugget size.

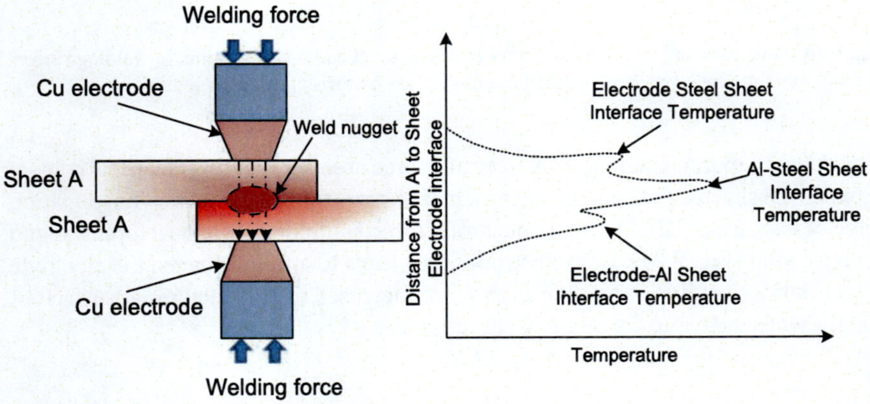

Fig. 8.19 Schematic showing resistance spot welding and temperature variation at interfaces across the electrode during joining of steel-Al

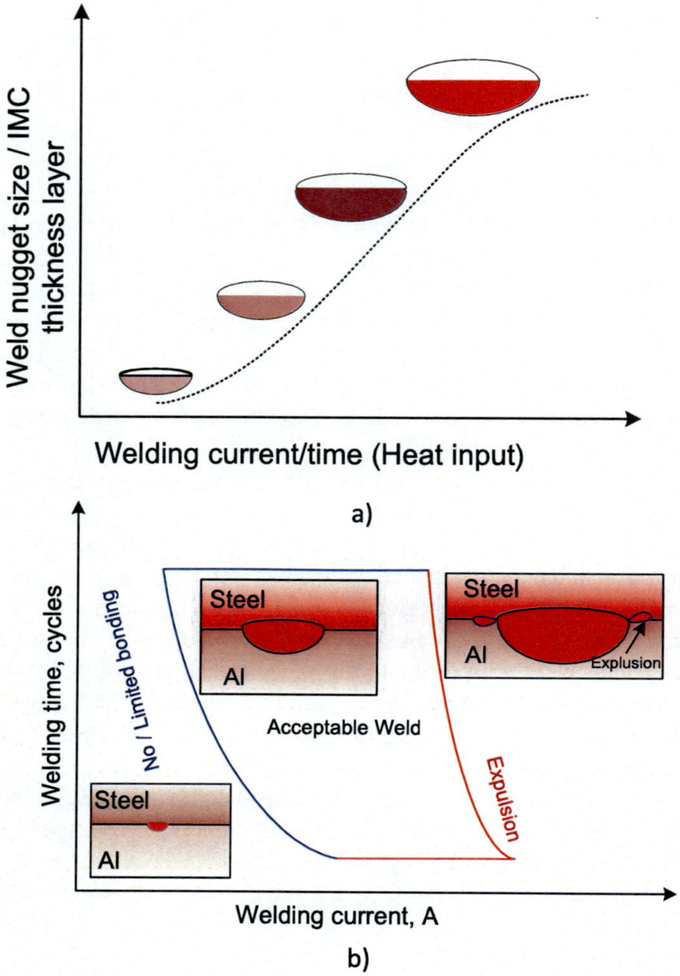

Fig. 8.20 Schematic indicating need of proper **a** heat input and **b** combination of welding current and welding time for developing sound joint of steel-Al by resistance spot welding

Another important parameter is electrode force used for ensuring electrical contact between the sheets being joined during squeeze stage (I) and thereafter consolidation stage of hold time (III). Optimum electrode force is important for developing sound welding joint (Fig. 8.21). Low electrode force leads to arcing between the electrode sheets during welding, while too high force increases the indentation at surface of metal sheets and expulsion at the weld interface.

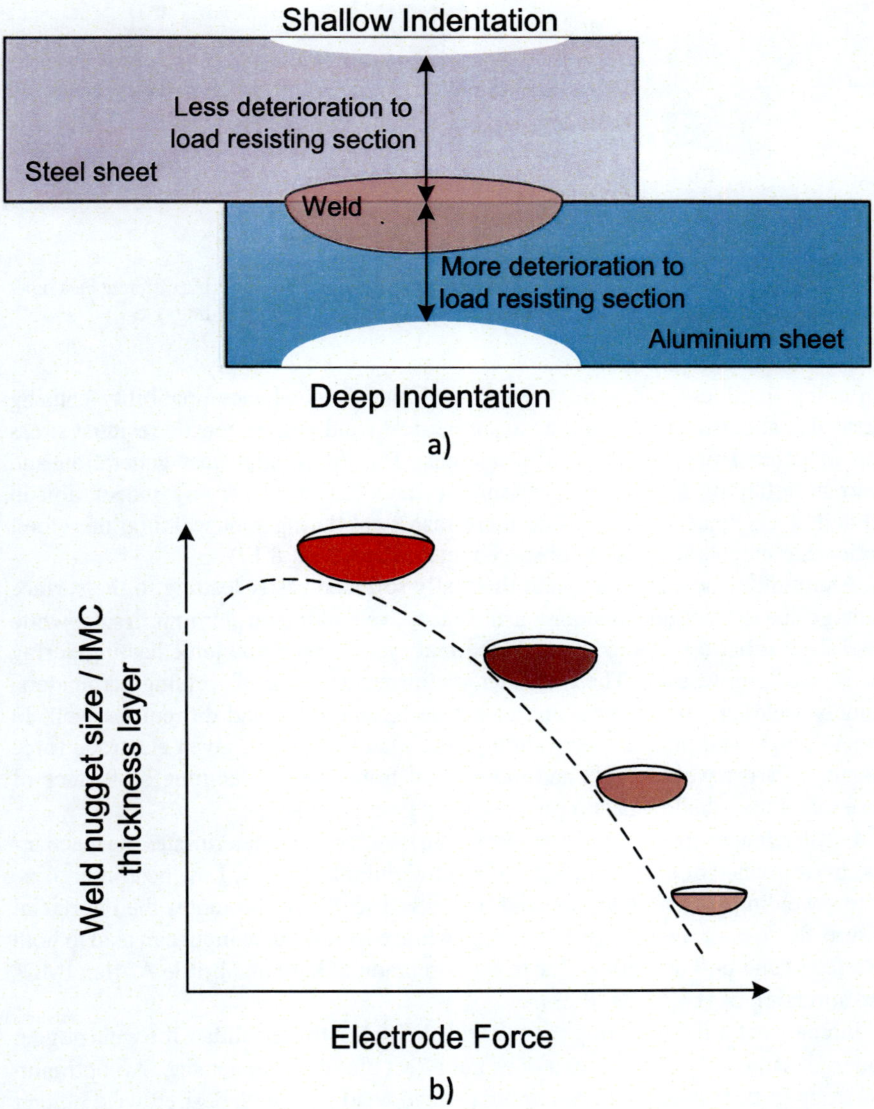

Fig. 8.21 Schematic indicating **a** formation of indentation on steel and aluminium sheets due to electrode force and **b** effect of electrode force on weld nugget size during steel-Al by resistance spot welding

8.4.1 Steel-Aluminium Spot Welding

The resistance spot welding steel-aluminium dissimilar metal combination imposes altogether different kind of challenges due to large difference in their electrical resistance (conductivity), thermal expansion coefficient, melting temperature, thermal

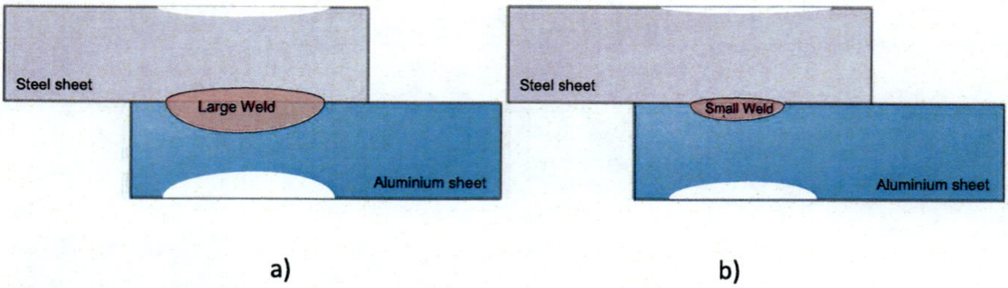

Fig. 8.22 Schematic showing indentation on steel and aluminium sheets during different heat input condition **a** high heat input and **b** low heat input during resistance spot welding

softening, hardness and yield strength and metallurgical incompatibility causing skewed heat generation leading to asymmetrical weld nugget, tensile residual stress and in general poor mechanical properties. The differential heat generation and thermal softening behaviour of aluminium and steel lead varying indentation in Al and steel. Since Al is softened more than steel during spot welding, therefore, noticeable indentation mainly observed on Al sheet (Fig. 8.17).

Aluminium sheet is melted and thermally softened easily leading to the surface damage due to electrode indentation and excessive melting on aluminium side, while steel sheet is hardly affected by weld thermal cycle imposed by joule heating during resistance spot welding. These characteristics make choice of welding parameters, namely welding current, time and electrode force crucial and difficult as well. In general, high heat input as per welding current and time for a given electrode force results in larger weld nugget and indentation from copper electrode on surface of base metal than the low heat input spot weld joint (Fig. 8.22).

Further, inappropriate heat input (low/high) used for joint deteriorates the mechanical performance of spot weld joint of steel-aluminium sheets. Low heat input (due to low welding current, low time and high electrode force) promotes the interfacial failure due to poor metallurgical bonding, while excessive heat input can lead to both interfacial and pull out failure due to the formation of hard and brittle Al-rich IMCs and softening of HAZ (Fig. 8.23).

Increase in hold time during spot welding helps just to consolidate the weld nugget under pressure only; therefore, it does not affect the weld nugget size. An optimum electrode force is needed for developing sound weld joint with desired weld nugget size. Very low electrode force can cause arcing between the sheets during welding, while too high electrode force increases indentation on the surface of aluminium sheet and reduces the weld nugget diameter. Increase in welding current and time increases the heat input. Like electrode force, an optimum combination of welding current/time are needed for developing sound weld joint with desired weld nugget size. Very low heat input (due to low welding current/time) for given electrode force results in poor metallurgical bonding and so interfacial failure of the weld joint. Excessive heat input (due to high welding current/time) forms hard and brittle thick IMCs layer and electrode indentation on the surface of aluminium sheet beside excessively large the weld nugget diameter on aluminium sheet.

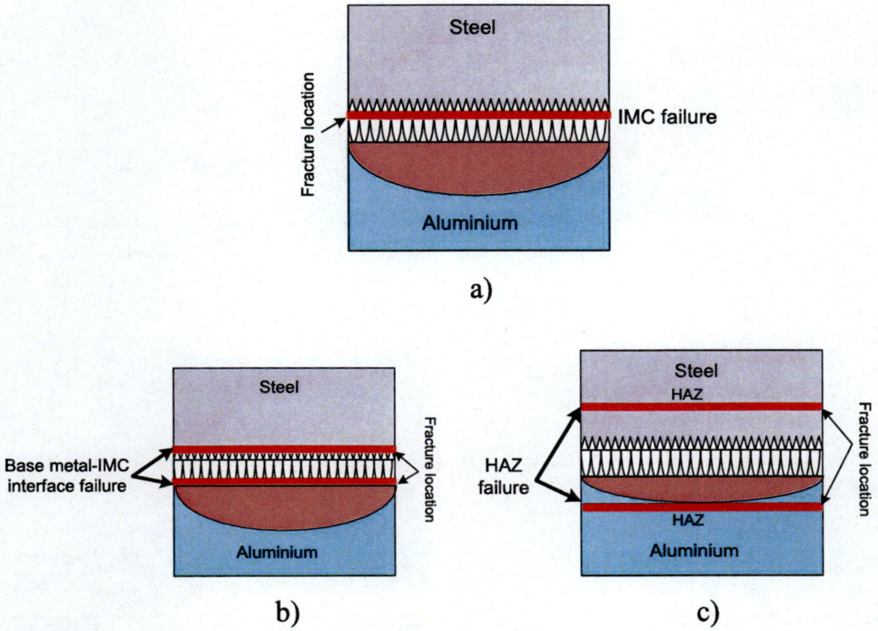

Fig. 8.23 Schematic showing failure of steel-aluminium dissimilar metal spot weld joint from different locations **a** IMC layer, **b** base metal-IMC layer interface and **c** HAZ of weak base metal, i.e. Al

8.5 Failure Modes in Resistance Spot Weld Joints

The spot weld joints of steel-aluminium dissimilar metal combination are known as a joint with very large variation in mechanical properties not just in respect of base metal properties but also in many other ways:

- Hardened and softened HAZs of steel and aluminium alloys,
- Hard and brittle IMCs at weld nugget/steel sheet interface,
- Moderate properties of cast weld nugget.

Depending upon the hardenability of steel and its heat treatment, coating on the steel surface, strengthening mechanism of aluminium and its temper condition and welding procedure (surface cleaning, welding current, time and electrode force), the weak zone (softened HAZ/hard and brittle IMC) can occur at any of above three locations, i.e. HAZ, weld nugget and interface IMCs (Fig. 8.24). The failure location, accordingly, can be at weld nugget, interfacial IMCs, HAZs (Zhang et al. 2014).

The fracture under tensile shear load condition can take place through any of the following modes (a) weld nugget pull out, (b) fracture of weak member of dissimilar metal combination, i.e. Al, (c–d) interfacial fracture (Fig. 8.25). Even combined interface and nugget pull out failure are also observed.

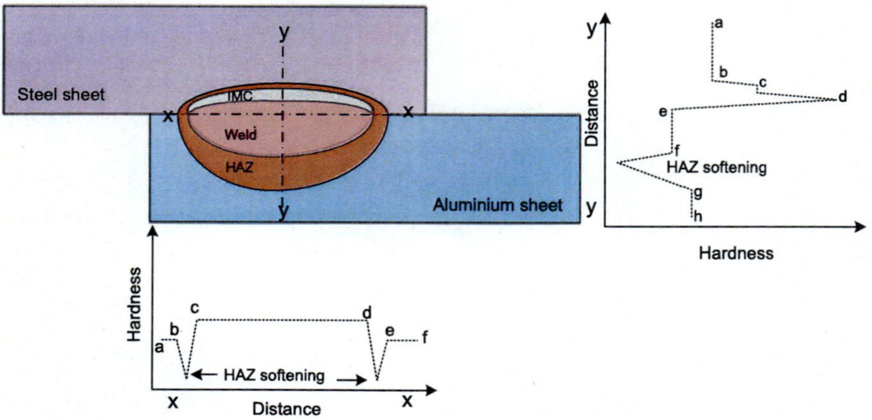

Fig. 8.24 Schematic hardness profile across the weld nugget of steel-aluminium dissimilar metal spot weld joint

Fig. 8.25 Schematic of failure mode in steel-aluminium dissimilar metal spot weld joint **a** nugget pull out, **b** HAZ failure and **c–d** interfacial failure

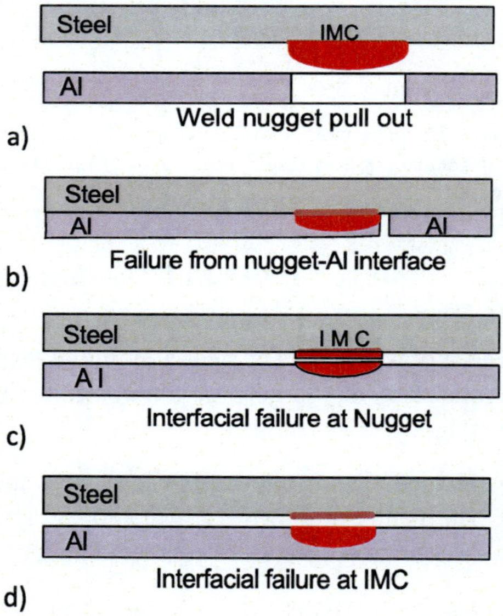

The interfacial fracture of steel-aluminium spot weld joint is very common due to the formation of hard and brittle IMCs, poor metallurgical bonding due to low heat input or presence of alumina, high stress concentration at circumference of the weld nugget, large and harmful defects like pores, cracks, inclusion, etc., in weld nugget due to inappropriate welding procedure specifications. Increase of heat input through control of welding current and time for spot welding mode of fracture changes from interfacial fracture to pull out provided very harmful IMC which is not formed. The load carrying capacity, energy absorption (until fracture) and ductility offered by

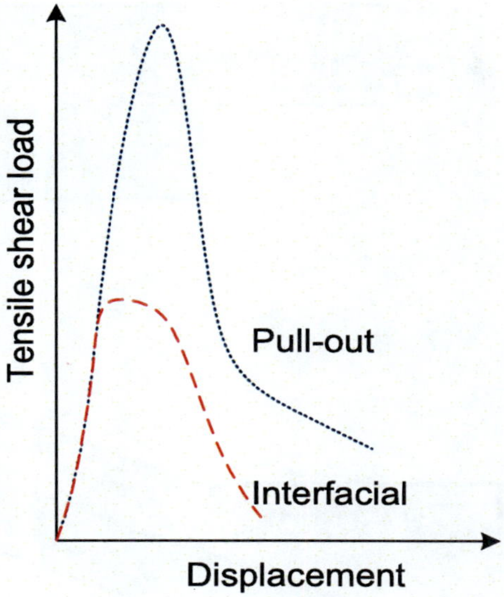

Fig. 8.26 Schematic of load–displacement curves for different failure modes during tensile shear testing steel-aluminium dissimilar metal spot weld joint

steel-aluminium spot weld joint in case of interfacial fracture are very low. This fracture is mostly triggered by presence of crack in brittle IMC layer at the interface and mostly produces flat fracture surface (Wan et al. 2017).

The weld nugget pull out mode fracture of steel-aluminium spot weld joint is observed when weld nugget and interfacial IMCs layer are strong and fracture tough. Under such conditions, tensile shear stress mainly acts at (a) periphery of spot weld and (b) weak softened HAZ region mainly on aluminium side. Moreover, bending stress generated due to deformation of the sheets under the influence of tensile shear loading (due to misalignment of sheets) further promotes tensile fracture at periphery of the weld nugget. The weld joint and more specifically HAZ allow greater load carrying capacity along with deformation. Thus, weld nugget pull out fracture offers higher load carrying capacity, energy absorption and ductility than interfacial fracture (Fig. 8.26). Further, the interfacial-pull out mixed mode fracture of spot weld joints of steel-aluminium sheets offers the intermediate mechanical performance as compared to above these two independent modes of fracture.

8.6 Mechanism of Fracture of Spot Weld Joints

The fracture mechanism of spot weld joint needs careful consideration of deformation and stress localization occurring during tensile shear loading (Fig. 8.27a). The spot weld joint developed in lap joint configuration inherently imposes bending on weld nugget zone as applied tensile load on the two sheets is not purely tensile in nature but acts eccentrically leading to the bending and shear loading conditions (Fig. 8.27b). The weld nugget zone under the externally applied tensile loading deforms and tends

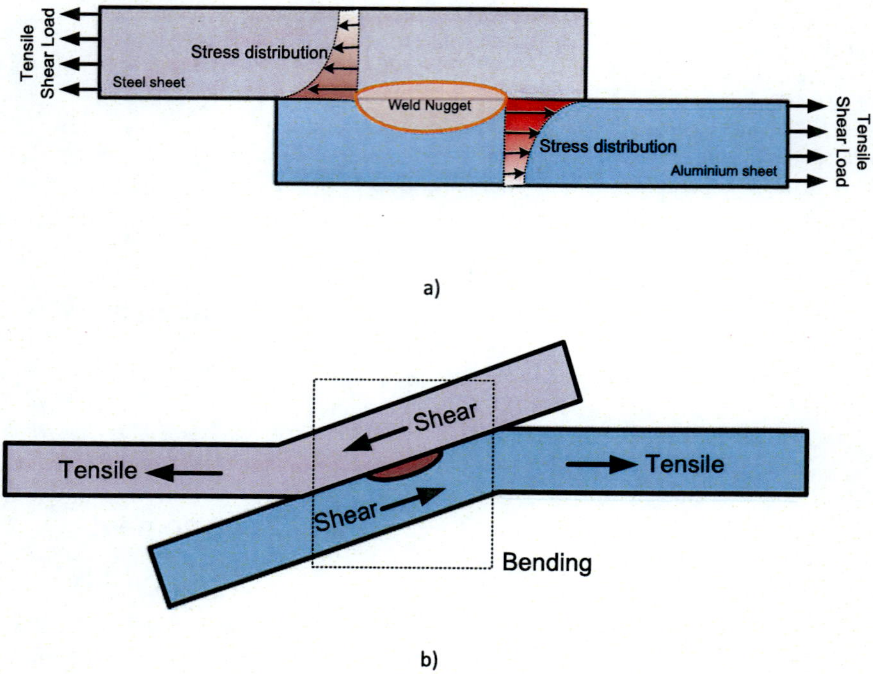

Fig. 8.27 Schematic showing **a** stress profile across the spot weld joint due to inherent stress concentration, **b** typical changes experienced during tensile shear testing steel-aluminium dissimilar metal spot weld joint

to align with loading direction. Such deformation of weld nugget zone imposes the (a) tensile stress near the weld nugget and (b) shear stress on the weld nugget itself. Depending on the weak zone of either HAZ or weld nugget and interface, the failure of the spot weld joint can occur due to (a) tensile stress acting from periphery of the weld zone, i.e. HAZ or (b) shear stress at weld nugget/interface zone (Fig. 8.27).

The fracture of spot weld joint under tensile stress acting in vicinity of HAZ at periphery of the weld zone is encouraged by presence of crack like weld defect and softening of HAZ due to reversion, recovery, recrystallization and grain growth in aluminium alloys while over temperature in case of Q and T steels. The HAZ experiences significant deformation prior to fracture and allows high load carrying capacity.

The fracture of spot weld joint under shear stress acting at weld nugget is promoted by presence of weld defects and formation thick, continuous layer of hard and brittle Al-rich IMC layer. Severity of shear stress acting on the weld is somewhat less as compared to the tensile stress. Therefore, interfacial fracture is discouraged unless the interface is extremely weak. The interfacial fracture takes place as soon as shear stress due to external loading exceeds the shear strength of the weld. Generally, the interfacial fractures occur at very low load; therefore, the deformation/ductility experienced by base metal and HAZs is very limited (Fig. 8.28). The typical stereoscope images

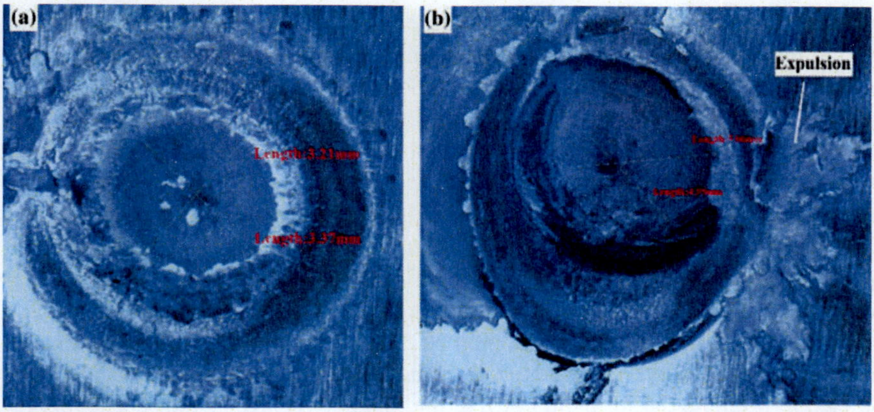

Fig. 8.28 Stereoscopic photographs of tensile shear fracture surfaces of steel-aluminium dissimilar metal spot weld joint developed using **a** very low heat input led to interface failure and **b** high heat stress profile causes expulsion and nugget pull (Singh 2019)

of interfacial fracture surfaces of weld joint show the flat fracture surface; while overlapping elongated dimples are observed under a scanning electron microscope.

8.7 Approaches to Enhance Joint Efficiency

The failure location of the steel-aluminium spot weld joint during testing or service indicates the weak zone of the joint. The failure in such a dissimilar metal joint can occur from weld nugget, weld nugget-steel interface, HAZs and base metal itself. The failure from the base metal suggests that steel-aluminium spot weld joint is perfect under given service/test condition and shall give 100% joint efficiency (ratio of joint strength to weak base metal strength between two steel/aluminium sheets under consideration). However, failure from other three locations, namely weld nugget, weld nugget-steel interface and HAZs, certainly indicates the weakening of base metals due to welding procedure applied for developing a spot weld joint. Therefore, it needs proper care/intervention to develop suitable welding procedure specification to avoid said weakening of the base metals (Singh 2019).

Further, nugget pull out failure from the heat-affected zone of either aluminium or steel sheet shows the thermal damage in mechanical and metallurgical properties of the base metals due to weld thermal cycle imposed during the welding for given base metals. Therefore, it needs either more effective control over the heat input for welding or post weld treatment to restore the HAZ properties similar to the respective base metals. The heat input can be controlled using suitable set of welding parameters, namely welding current, time (squeeze, welding, hold) and electrode pressure. While the post weld treatment can have applied either using electrical resistance heating as a part of welding sequence or later using heat treatment furnace. The type of heat treatment will depend on whether the failure is taking place from HAZ of steel or aluminium side accordingly suitable heat treatment is designed for stress reliving or structure and properties modification using normalizing, Q and T, T6, T4, etc.

The interfacial failure of the weld joint from weld nugget or weld nugget-steel sheet interface occurs due to high stress concertation (an inherent feature of spot weld joint), defect in weld nugget and formation of hard, brittle and weak IMCs layer. Failure due to weld defects like pores, inclusion, cracks, etc., can be avoided by eliminating weld defects from the weld nugget using suitable welding procedure specifications. The failure due to formation of thick, continuous hard, brittle and weak IMC layer can be reduced by modifying the welding procedure to encourage the formation of thin, discrete, fine-grained fracture tough IMC layer. This can be done using suitable control/reduction of the heat input, alloying of elements like Zn, Si, etc., to promote the formation of fracture tough IMC layer. Zinc acts as fluxing element (Faseeulla Khan et al. 2010, 2012a, b, Faseeulla Khan et al.; Mittal and Dwivedi 2012, 2013).

The mechanical performance and efficiency of steel-aluminium spot weld joints can be enhanced through many approaches. In the following section, few approaches are presented (Singh 2019).

- Cover sheet approach,
- Zn alloying,
- Weld defect prevention through optimization of welding process parameters.

8.7.1 Cover Sheet Approach

Aluminium and steel dissimilar weld combination differs significantly in terms of physical properties, melting point, thermal and electrical conductivity. For a given welding current, joule heat generation is very less on aluminium side due to low electrical resistivity than steel. Further, the heat generated on aluminium side is also very dissipated rapidly to copper electrode due to high thermal conductivity. Therefore, net heat available for the weld nugget formation is reduced.

The usage of low thermal conductivity and high electrical resistivity metal sheet as a cover sheet on aluminium alloy sheet side during spot welding of dissimilar steel-aluminium alloy combination helps to.

- Increase the supply of heat (through joule heating) to aluminium sheet (in dissimilar metal combination) which is generating lesser heat due to low electrical resistivity than steel sheet during welding,
- Reduce the heat dissipation to the copper electrode from the aluminium sheet side, thus, the most of the heat generated used in development of weld nugget (Fig. 8.29).

The cover sheet of steel sandwiched between copper electrode and aluminium sheet and welding current is passed from the electrode to the sheets to be welded through cover steel sheet. Cover steel sheet generates more joule heat under the same

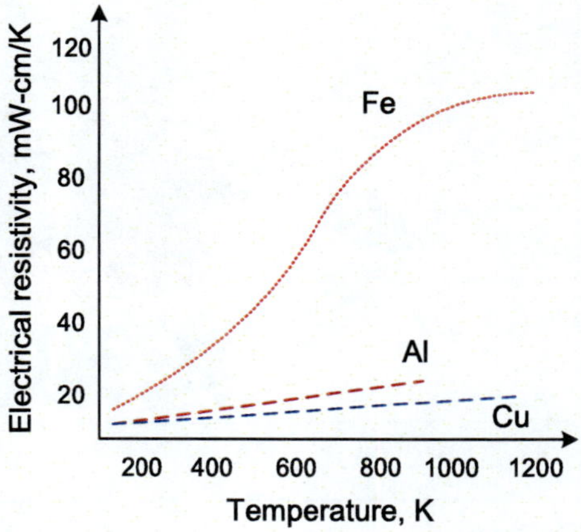

Fig. 8.29 Variation in electrical resistivity as function of temperature for Fe, Al and Cu

welding parameters (current, time) than aluminium itself. This heat is transferred to aluminium sheet to be welded for weld nugget formation. This in turn leads to little higher temperature on aluminium side; which in turn increases weld nugget size and HAZ as well than steel sheet (Figs. 8.30 and 8.31).

Another important purpose of using steel cover sheet is to avoid the formation of electrode indentation on aluminium as steel sheets being hard and strong distribute the electrode force over a larger area, which in turn allows application of high heat input (welding current/time) even at higher electrode force without forming indentation on aluminium sheet. In absence of cover sheet, the indentation on aluminium side reduces the aluminium sheet thickness at the weld nugget area and acts as stress raiser (due to stress concentration) which in turn lowers the mechanical performance of spot weld joint in terms of reduced the load carrying capacity, fatigue life and joint efficiency (Fig. 8.31).

Ability to apply more heat and electrode force without indentation on aluminium sheet side during spot welding of steel-aluminium dissimilar metal combination in turn results in increased (a) the weld nugget diameters and (b) soundness/cleanliness of the weld as impurities/low melting point/oxides, etc., constituents forced out from interface to the circumference/periphery of the weld nugget. Both these factors increase the load carrying capacity, fatigue life and joint efficiency. The low melting point eutectic and other phases formed due to interactions of aluminium/iron with zinc, silicon and other element present at the interface are removed. This might affect the IMC layer thickness and continuity and so weld joint performance.

8.7.2 Role of Interlayers in Steel-Aluminium Welding

The presence of interlayers in steel-aluminium spot welding affects the weld nugget and mechanical performance due to many aspects such as interfacial contact resistance (so joule heating), metallurgical interactions with Al and iron, diffusion of

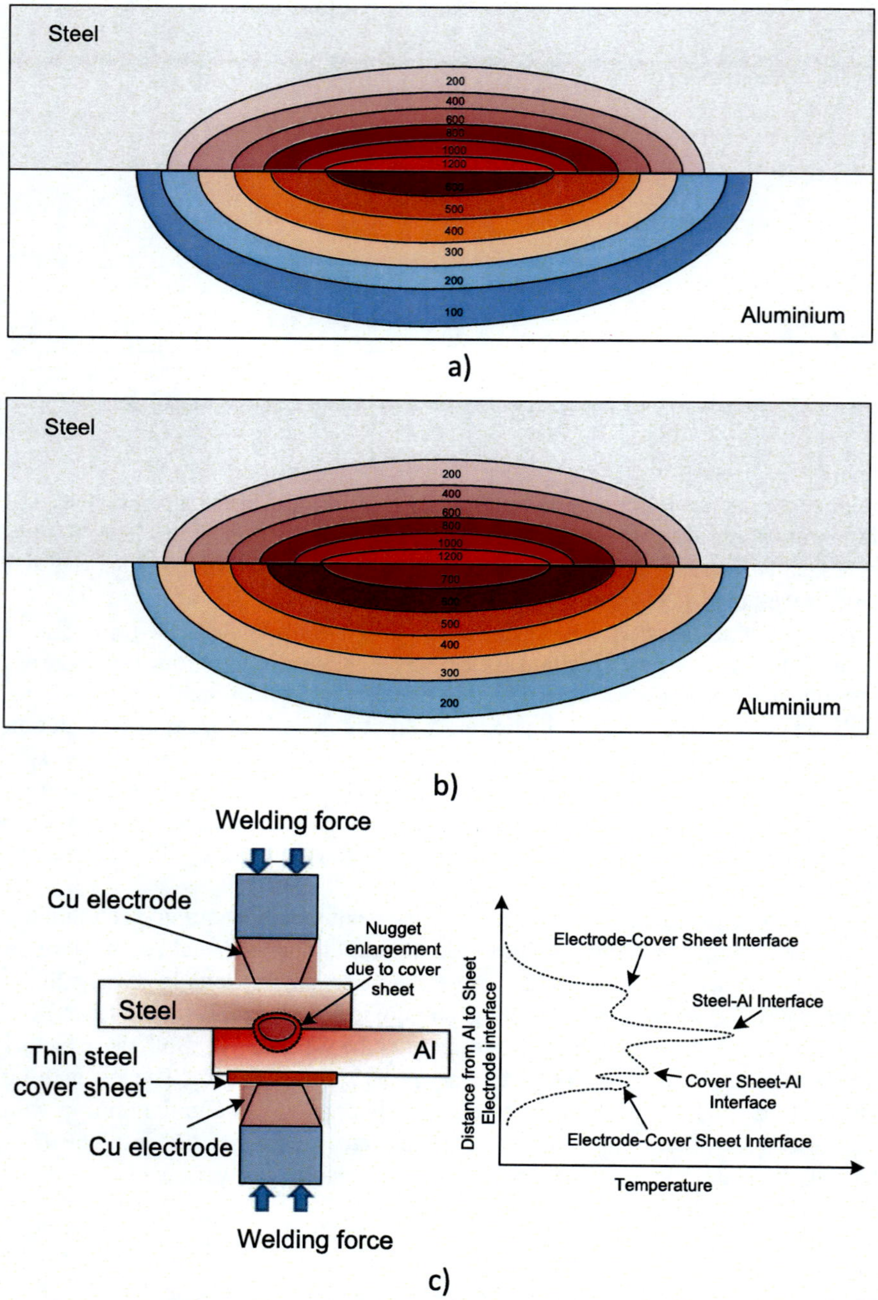

Fig. 8.30 Schematic showing temperature profile across the spot weld joint in steel and Al sheets during spot welding **a** without cover sheet, **b–c** with cover sheet steel-aluminium dissimilar metal welding

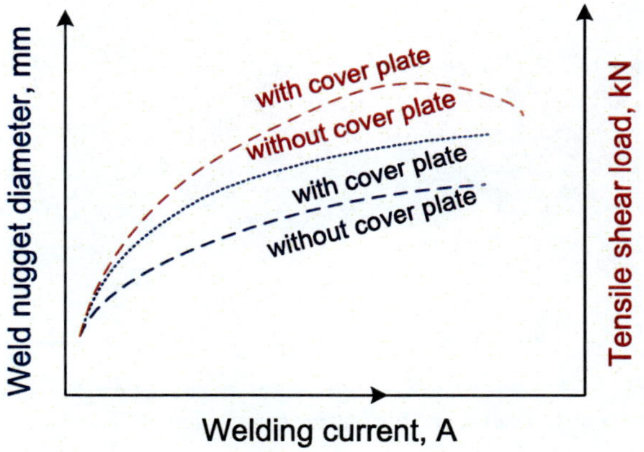

Fig. 8.31 Schematic showing variation in weld nugget size and tensile shear load carrying capacity of spot weld joint with and without cover sheet during steel-aluminium dissimilar metal welding

elements across the interface leading intermixing and formation of variety of IMCs. The alloying at the interface can done by either sandwiching the thin film of desired interlayer (Zn, Al–Si, etc.) between steel and aluminium before spot welding or coating of the desired element on the one of the base metal sheets like zinc coating on the steel surface. The choice of the interlayer depends on the multiple factors including metallurgical and thermo-physical compatibility between base metal sheets (being joined), need of forming fracture toughness IMCs, improving the fluidity and wettability for metallurgical bonding, possibility of formation of low melting constituents, cleaning action required for removing impurities from the weld interface/nugget and expulsion tendency.

The electrical conductivity of the interlayer affecting the interfacial contact resistance can change the weld nugget diameter for a given welding parameters accordingly load carrying capacity may increase or decrease. The low electrical conductivity of the interlayers will generate more joule heat, which in turn can increase the weld nugget diameter while that of high electrical conductivity interlayer will deteriorate the nugget size and joint strength.

Another aspect related to joule heating with use of the interlayers in steel-aluminium spot welding is that a part of heat generated (due to joule heating I^2Rt) used for melting (and even sometimes evaporation) of elements in the interlayers. This makes somewhat less effective heating available for the formation of the weld nugget, which is expected to reduce the weld nugget diameter and so strength of the joint as well. The element having low boiling point (like Zn) evaporates during welding so formed metal vapour if entrapped in the weld nugget results in porosity which in deteriorates the joint performance (Fig. 8.32).

The zinc offers low contact resistance, low melting (419 °C) and boiling point (907 °C) so joule heat generated is reduced under identical welding condition, and at the same time, a part of heat generated is used for melting (sensible and latent heat) and evaporation of Zn layer. Thus, effective heat available for melting of base metal (Al) is reduced which in turn reduces the weld nugget size. Therefore, optimum welding current needed to development given size of weld nugget/joint strength can be higher than spot weld uncoated same steel-aluminium combination

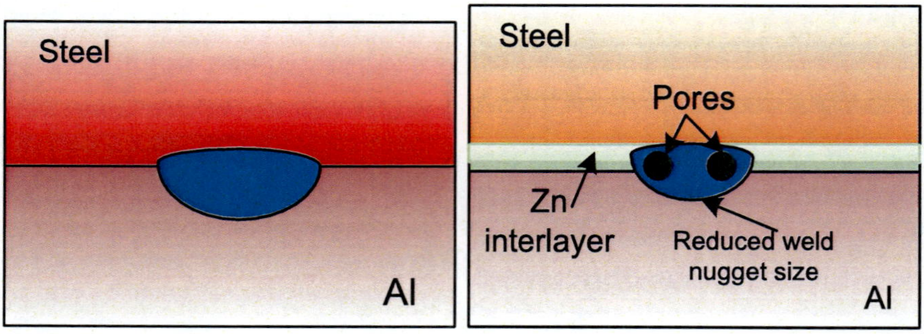

Fig. 8.32 Schematic showing effect of Zn coating on steel sheet on welding nugget formation **a** without Zn coating and **b** with zinc coating during steel-aluminium dissimilar metal welding

or steel-aluminium combination without interlayer. Al-Si alloy interlayer having silicon improves the fluidity and wettability, which can result in better the metallurgical bonding and the joint performance. Further, the presence of Cr in steel forming thin, non-porous, coherent and refractory protective oxide layer may interfere with diffusion and metallurgical interactions of iron with aluminium during spot welding. Therefore, elements like Cr can reduce the IMCs formation and improve the properties of sound spot weld joint; however, any reduction in wettability may compromise with metallurgical properties and encourage weld discontinuities.

8.8 Resistance Spot Welding of Galvanized Steel-Aluminium Alloy Sheets

A typical case study on spot welding of galvanized steel-aluminium alloy with and without cover plate is presented in following section. Effect of welding parameters, namely welding current, time and electrode force, on weld nugget diameter, metallurgical and mechanical performance of spot welding joint developed with and without cover plate has been presented.

Increase of welding current and time for spot welding increased the weld nugget diameters and tensile shear load carrying capacity of the weld joint, while IMCs layer has been found to decrease with increase of heat input (welding current and time). Increase in tensile shear load carrying capacity of the weld joint with heat input (welding current/time) has been found to be consistent with increase of weld nugget diameter. Further, initially increase in tensile shear load carrying capacity and nugget diameter of the weld joint increases with welding current rapidly but at higher current, rate of increase in both these weld size and strength decreases and the same tends to get stabilize due to two factors (a) weld nugget diameter approaches to electrode tip diameter determining the path of flow of welding current for joule heating and (b) increased shunting effect and reduction in effective welding current available for joule heating. Further, increase in heat input with increase of welding current and time increases the formation of low melting constituents in presence of zinc at the interface in vicinity of steel sheet. These low melting point constituents along with some IMC are pushed to the periphery of the weld nugget under effect of electrode force during weld consolidation stage (hold time) which in turn decreases the thickness of IMC layer (Fig. 8.33a–c). Therefore, increase in the mechanical performance of galvanized steel-aluminium spot weld joint can be attributed to both increase of weld nugget diameter coupled with decrease in thickness of IMCs layer.

Further, the application of steel sheet as a cover plate on aluminium sheet side under identical welding parameters (welding current, time and electrode force) increases the weld nugget diameters, decreases both indentation on aluminium sheet and the thickness of IMCs layer because steel sheet as cover plate on the aluminium side during welding plays many roles (a) generates heat during the welding, (b) reduces dissipation of aluminium-steel interfacial heat to copper electrode and (c) reduces the electrode indentation on aluminium sheet due to distribution of electrode force over a larger area of steel cover plate. Therefore, mechanical performance (tensile shear load carrying capacity, fatigue strength) of galvanized steel-aluminium spot weld joint developed using cover plate becomes higher than those developed without a cover plate.

8.9 Summary

The joining of steel-aluminium by spot welding imposes many difficulties due to a large difference in thermal and metallurgical properties. A systematic understanding of physics of resistance welding process its effect on weld nugget formation and metallurgical interaction can help in developing sound weld joints with desired combination of mechanical properties. Therefore, it needs attention to developed fracture tough IMC at weld nugget-steel interface, avoid stress raiser in form of weld discontinuities and indentation while taking care of undesirable HAZ softening. These factors in turn can encourage the failure either from the weak base metal or weld nugget pull out failure leading to improved mechanical performance of dissimilar steel-aluminium steel-weld joint.

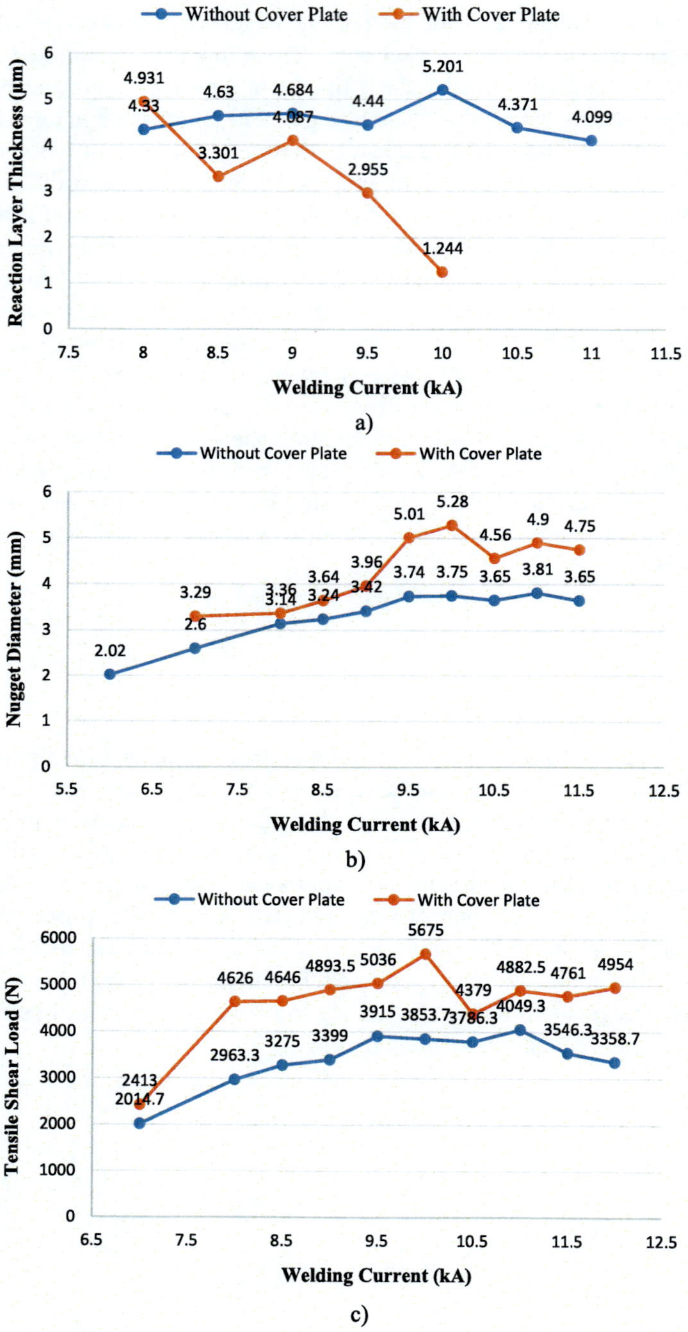

Fig. 8.33 Effect of welding current on **a** thickness of IMC layer formation, **b** weld nugget diameter and **c** tensile shear load of Al 5052-galvalized steel sheets spot weld joints developed with and without sheet cover sheet (Singh 2019)

References

Chen J, Yuan X, Hu Z, Sun C, Zhang Y, Zhang Y (2016) Microstructure and mechanical properties of resistance-spot-welded joints for A5052 aluminum alloy and DP 600 steel. Mater Charact

Faseeulla Khan MD, Dwivedi DK, Ghosh PK (2010) Studies on the effect of process parameters on the shear performance of weld- bonds of aluminium alloy. In: Proceedings of 36th international MATADOR conference held in manchester 14th–16th July 2010

Faseeulla Khan Md, Dwivedi DK (2012a) Development of response surface model for tensile shear strength of weld-bonds of aluminium alloy 6061 T651. Mater Des 34:673–678

Faseeulla Khan MD, Dwivedi DK (2012b) Mechanical and metallurgical behaviour of weld-bonds of 6061 aluminium alloy. Mater Manuf Process 27(6):670–675

Faseeulla Khan MD, Sharma G, Dwivedi DK, Weld-bonding of 6062 aluminium alloy. Int J Adv Manuf Technol 78(5–8):863–873

Gullino A, Matteis P, Aiuto FD (2019) Review of aluminum-to-steel welding technologies for car-body applications. Metals (basel) 9:1–28

Mittal M, Dwivedi DK (2012) Statistical analysis of influence of input process parameters on characteristics of weld-bonds of Al 5052 H32 alloy using Box-Behnken Design (BBD). Pro Instit Mech Eng Part B, J Eng Manuf 226(6):1001–1017

Mittal M, Dwivedi DK (2013) Studies on fatigue behavior of weld-bonds of Al–Mn–Mg alloys. In: Proceedings of international conference on manufacturing research (ICMR 2013) held at Cranfield University, Bedford, UK during 19–20 Sep 2013

New Techniques for Joining Steel and Aluminum, Assembly, 93777, (assemblymag.com)

Singh GK (2019) Studies on structure and properties of steel-aluminium alloy weld joints by spot welding, M. Tech. Dissertation (welding engineering), IIT Roorkee, pp 53–56

Wan Z, Wang HP, Chen N, Wang M, Carlson BE (2017) Characterization of intermetallic compound at the interfaces of Al-steel resistance spot welds. J Mater Process Technol 242:12–23

Zhang H, Qiu X, Xing F, Bai J, Chen J (2014) Failure analysis of dissimilar thickness resistance spot welded joints in dual-phase steels during tensile shear test. Mater Des 55:366–372

Chapter 9
Joining of Dissimilar Metals by Diffusion Bonding

This chapter presents the fundamentals of diffusion bonding including mechanism and process parameters like bonding temperature, pressure, time and vacuum with regard to dissimilar metal joining using suitable schematics. Selection of interlayer for diffusion bonding and diffusion brazing has been elaborated. The principle, mechanism, process parameters and metallurgy of diffusion brazing of dissimilar metals have also been described.

9.1 Introduction

The diffusion bonding is a solid-state joining process, which primarily depends on diffusion of atoms of elements across the joint interface. The pre-requisite (s) for diffusion to take place is metal-to-metal contact between parent metals to be joined besides compositional gradient, favourable temperature conditions and complete good metallic intimacy of faying surfaces free from all adsorbed gases and reaction products like oxides, oil, paint, etc. Therefore, surface preparation (surface finish, cleanliness and flatness) of the parent metals to be joined by diffusion bonding becomes very crucial (Fig. 9.1) (Sharma and Dwivedi 2017a, b; Sharma et al. 2018). Depending upon the metal system being joined under diffusion bonding (temperature and pressure) conditions, surface oxides (Ti, Ta, Zr) may get dissolved/decomposed, oxide films and layers like (Al_2O_3) agglomerate and spheroidize. The time for dissolution increases with square of oxide film thickness. Therefore, diffusion bonding is performed in a controlled vacuum environment so that surfaces of parent metal do not chemically interact with atmospheric gases to form reaction products and reduce the metallic intimacy. Additionally, good interfacial metallic contact for diffusion bonding is facilitated using suitable compressive load (5–20 MPa) depending on the parent metals, vacuum conditions, bonding temperature, type and thickness of interlayer, if any.

© The Author(s), under exclusive license to Springer Nature Singapore Pte Ltd. 2023

D. K. Dwivedi, *Dissimilar Metal Joining*,
https://doi.org/10.1007/978-981-99-1897-3_9

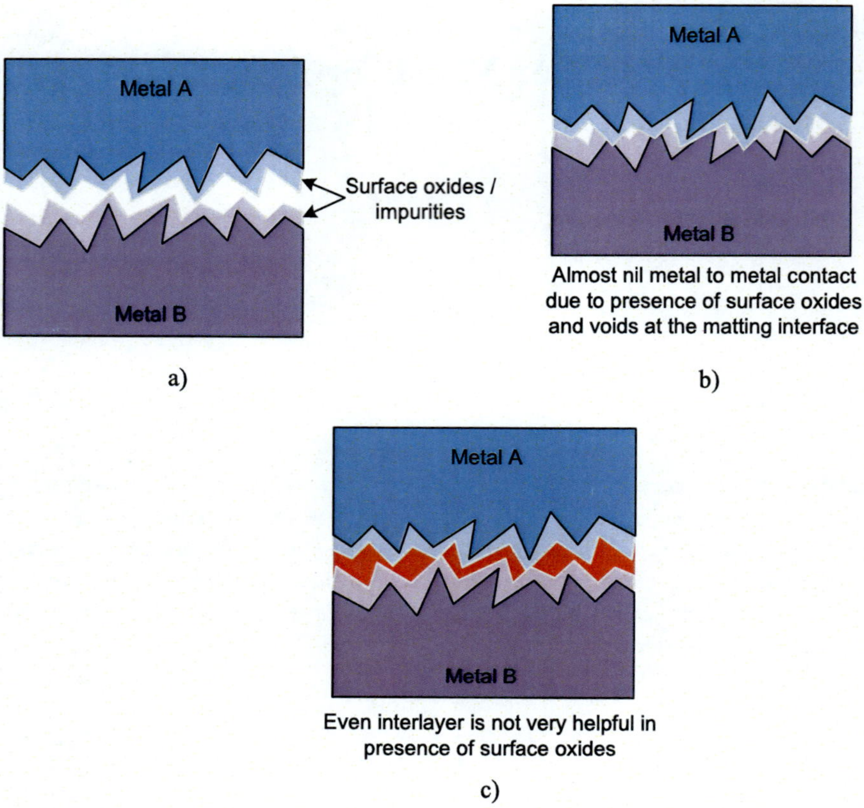

Fig. 9.1 Schematic showing effect of surface impurities (in case of limited or no vacuum) on diffusion bonding **a** oxidized surfaces, **b** limited or no metallic intimacy and **c** even an interlayer does not bring desired metal-to-metal contact in presence of surface impurities

9.2 Mechanism of Bonding

Depending upon the interaction of elements (due to diffusion) from parent metals of dissimilar combination at the interface, either completely recrystallized grains are formed or new individual phases and/or compounds are developed. The formation of new grains at the joint interface free from any intermetallic and unfavourable compounds results in a good joint efficiency and service performance; however, development of hard and brittle intermetallic compounds (in case of poor metallurgical compatibility between parent metal being joined by diffusion bonding) can lead to reduced joint toughness, ductility and efficiency (Fig. 9.2). Therefore, suitable interlayer compatible with both the parent metals can be used to avoid/reduce the formation of undesirable intermetallic compounds and phases at the interface to

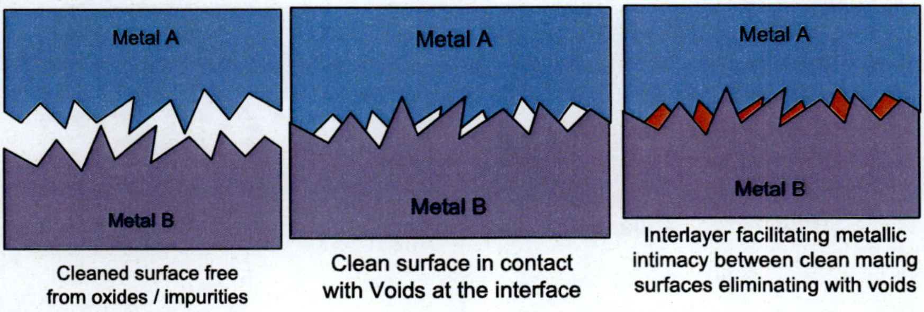

Fig. 9.2 Schematic showing how the clean the surfaces affect diffusion bonding **a** clean surfaces, **b** reasonably good metallic intimacy and **c** interlayer results in perfect metal-to-metal contact

improve the mechanical and corrosion performance for the diffusion bonds. Application of an interlayer in similar metal joining can help to (a) increase the metallic intimacy by filling surface irregularities under pressure at elevated temperature and (b) reduce the bonding time by acting as diffusion accelerator. The interlayers apart from above two functions can play a more crucial role in diffusion bonding of dissimilar metals by developing favourable intermetallic compounds and phases at the interface. An interlayer of suitable metal is placed at the interface which either causes solid-state metallurgical interactions for diffusion bonding or diffuses, melts and solidifies for diffusion brazing (Sharma and Dwivedi 2017a; Sharma et al. 2018). In view of above, it can be noted that surface preparation, surface roughness, vacuum, pressure and temperature, bonding time and interlayer are few important aspects, which should be chosen and applied carefully.

9.3 Stages of Diffusion Bonding

The diffusion bonds are developed in the following stages (a) establishing metal-to-metal contact through localized interface micro-level yielding of peaks and valleys through plastic deformation and creep at elevated temperature, (b) diffusion across the interface resulting in dynamic recrystallization, development of new grains across the interface and gradual elimination of initial stage voids present at the interface between parent metals and (c) volume diffusion towards the voids at the interface, and grain boundary movement eventually leading to the elimination of all voids from the interface and producing sound diffusion bonds. These stages are exhibited schematically in detail using the following six Figs. 9.3a–f.

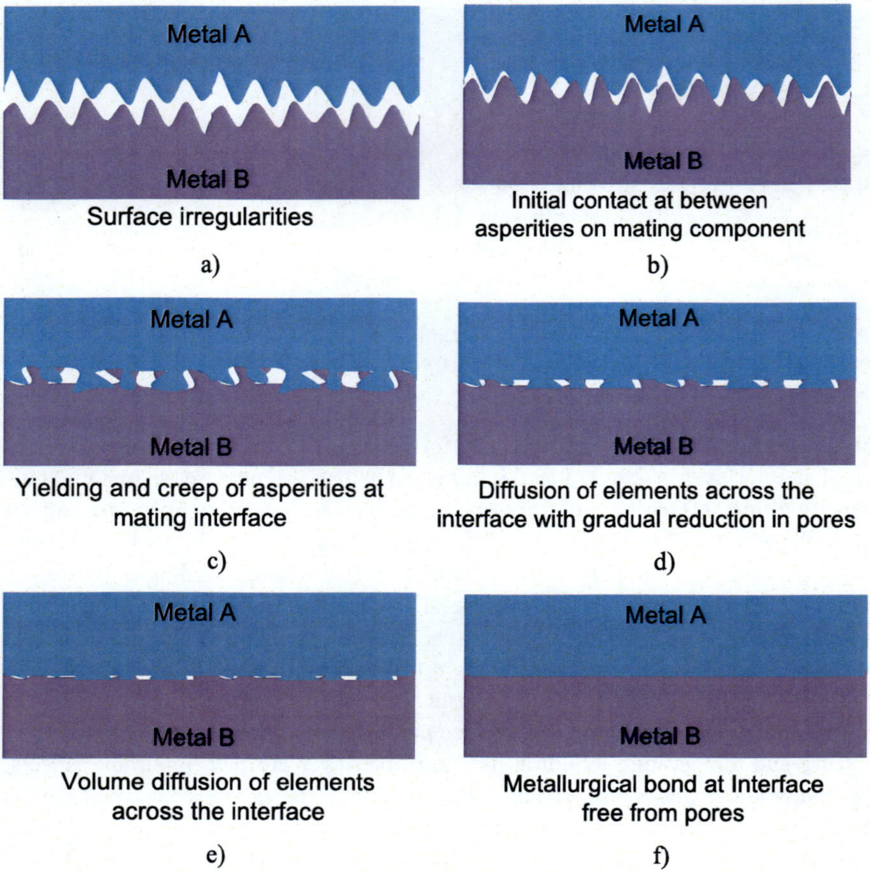

Fig. 9.3 Schematic showing sequential steps of diffusion bonding **a** clean surface with irregularities, **b** initial stage contact between asperities on the mating surfaces, **c** localized yielding through plastic deformation and creep of the surface layers, **d** commencement of bonding through diffusion of elements across the interface, **e** volume diffusion of elements across the interface leading to gradual elimination inartificial voids and **f** metallurgical bond at the interface free from voids

9.4 Diffusion Bonding Conditions

The diffusion of elements across the interface of mating parent metals is the key and the factors like surface cleanliness, surface roughness, bonding temperature, pressure, vacuum and interlayers, if any, all affect the diffusion in one or other ways, which in turn determines the time for developing diffusion bonds and so productivity. Faying surfaces of parent metal free from surface impurities, having good surface finish ($R_a < 0.1\ \mu m$), reasonably high bonding temperature (0.5–0.7 T_m), moderate pressure (5–20 MPa) and reasonably high vacuum (10^{-3} to 10^{-5} Torr) to avoid atmospheric gas

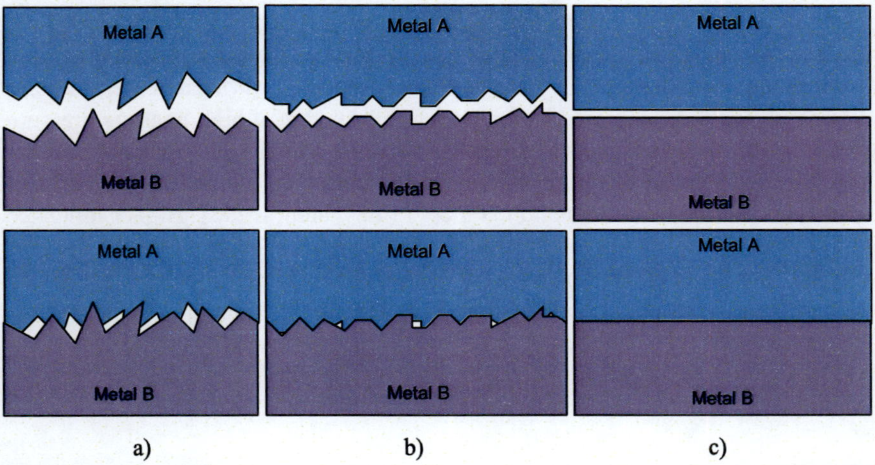

Fig. 9.4 Schematic showing effect of surface roughness on soundness of diffusion bond interface **a** very rough surface leaving many voids and unbonded zones, **b** relative less rough surface resulting in leaving few voids and small unbonded region and **c** very smooth surface and finished surface leading perfect metallic bond without any voids

metal interactions all improve metal-to-metal contact to facilitate diffusion across the bonding interface.

9.4.1 Surface Roughness

Surface roughness directly affects the metal-to-metal contact at the initial stage of the diffusion bonding. An increase in surface roughness increases tendency of interfacial void formation which in turn adversely affects the diffusion across the interface (Fig. 9.4a). Conversely, bonding time needed to produce a sound and void-free diffusion bond also increases with surface roughness. A rough surface needs high bonding pressure, longer diffusion bonding time to develop sound/void-free metallurgical interface. On the other hand, good surface finishing (low surface roughness) results in good metal to metal contact, which in turn encourages the diffusion of elements and reduces the interfacial void formation tendency (Fig. 9.4b, c) (Sharma and Dwivedi 2018, 2019, 2021).

9.4.2 Bonding Pressure

Pressure determines interfacial surface layer deformation and creep under diffusion bonding conditions, which in turn affect metallic intimacy between the parent metals at the stage I of diffusion bonding. An increase of bond pressure increases the metallic intimacy at the interface that in turn (a) reduces the diffusion bonding time for developing sound, void-free joint interface of bonds, and (b) increases the joint efficiency (Fig. 9.5). Increased surface layer deformation at high bonding pressure forms more diffusion channels leading to accelerated diffusion and recrystallization. However, the extent of effect of bonding pressure on joint interface characteristics depends on surface roughness and hardness of the parent metals. Higher influence of the bonding pressure on joint interface characteristics is observed on the rougher surfaces. The choice of bonding pressure depends on bonding temperature, time and metal systems of dissimilar combination. The maximum bonding pressure is generally limited by ability of component to take up the load and capacity of press. Bonding pressure requirement decreases with reduction in yield strength of weak parent metal of dissimilar combination and interlayer metal at the bonding temperature.

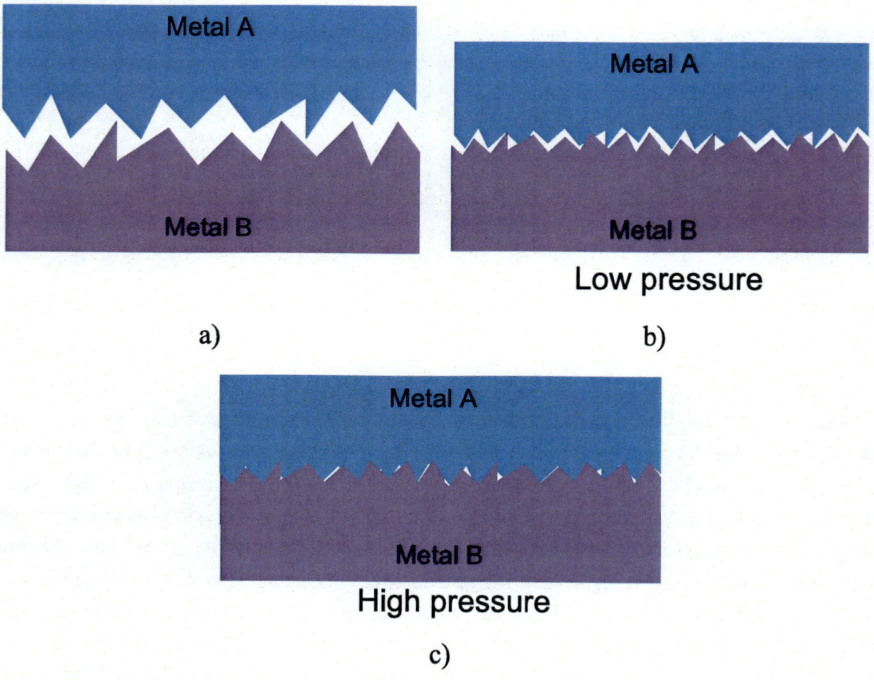

Fig. 9.5 Schematic showing effect of bonding pressure on diffusion bonding **a** surface irregularities, **b** low bonding pressure leading to many interfacial voids and **c** high bonding pressure resulting in good metallic intimacy and fewer interfacial voids

9.4.3 Bonding Temperature

Temperature is the most important parameter in diffusion bonding as it affects both mechanical properties like thermal softening, yield strength and ductility of parent metals during diffusion bonding. These changes in mechanical properties in turn determine the surface layer deformation, creep and so metallic intimacy at the stage I and diffusion coefficient affecting the bonding time during the stage II and III. An increase of bonding temperature in general increases the metallic intimacy between parent metals owing to the increased surface layer deformation due to thermal softening, reduction in yield strength and increase of ductility and reduces the bonding time due to increased diffusion coefficient (Fig. 9.6). The diffusion coefficient (D) increases exponentially with increase of temperature (T) and is expressed by following relationship.

$$\text{Diffusion coefficient at temperature}(T) D = D_o e^{-Q/KT}$$

where Q is the activation energy for diffusion, T is the absolute bonding temperature, K is the Boltzmann coefficient and D_o is the proportionality constant. Above relationship suggests that small change in temperature results in significant change in diffusion coefficient and thus the process kinetics.

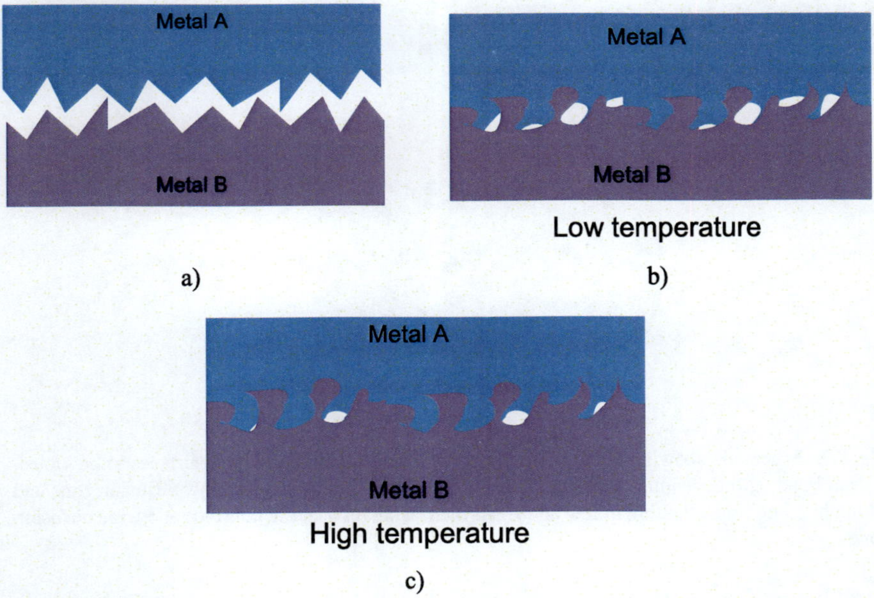

Fig. 9.6 Schematic showing effect of bonding temperature on diffusion bonding **a** surface irregularities, **b** low bonding temperature leading to many interfacial voids and **c** high bonding temperature resulting in good metallic intimacy and fewer interfacial voids

9.4.4 Bonding Time

The diffusion bonding time for developing sound bonds free from interfacial void is needed to (a) establish the metal-to-metal contact during the stage I involving surface layer plastic deformation and creep of mating components and (b) complete stages II and III primarily involving diffusion of elements across the interface, movement of grain boundaries and gradual elimination of pores from the interface through the volume diffusion (Fig. 9.7). The bonding time is strongly influenced by bonding temperature and parent metals of dissimilar combinations to be joined, composition and thickness of interlayer, if any. The diffusion length (x) across the interface approximately varies with the square root of bonding time (t) for the given diffusion coefficient (D) at bonding temperature (T). Increasing bonding time and pressure increases the joint strength up to a limit then it stabilizes.

$$\text{Diffusion length}(x) = C(Dt)^{1/2}$$

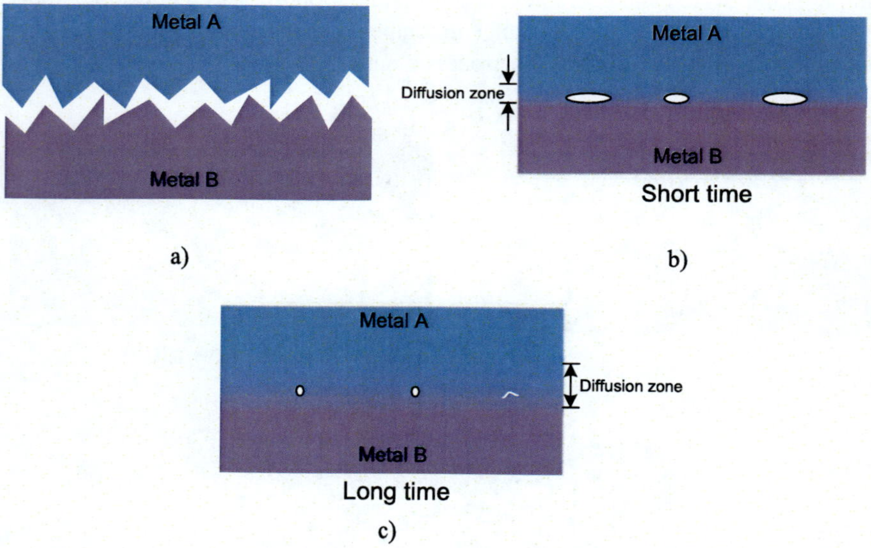

Fig. 9.7 Schematic showing effect of bonding time on diffusion bonding **a** surfaces to be joined, **b** short bonding time leading large interfacial voids at the interface and narrow diffusion zone and **c** long bonding time resulting in few and small interfacial voids at the interface and wide diffusion zone

9.4.5 Vacuum

Diffusion bonding is carried out in vacuum to avoid the formation of the reaction products like oxides and nitrides on the mating surfaces of parent metals at diffusion bonding temperature, which reduce the metallic intimacy, compromise the diffusion across the bonding interface and increase the time required for developing void-free interface of diffusion bonds. The vacuum depending upon the reactivity and affinity of metals to atmospheric gases may vary from 10^{-3} to 10^{-7} Torr. Higher the reactivity and affinity of metals to atmospheric gases, greater the vacuum level desired for diffusion bonding. Vacuum (of the desired level) results in surfaces free from surface oxidation and good metal-to-metal contact for diffusion bonding (Fig. 9.8).

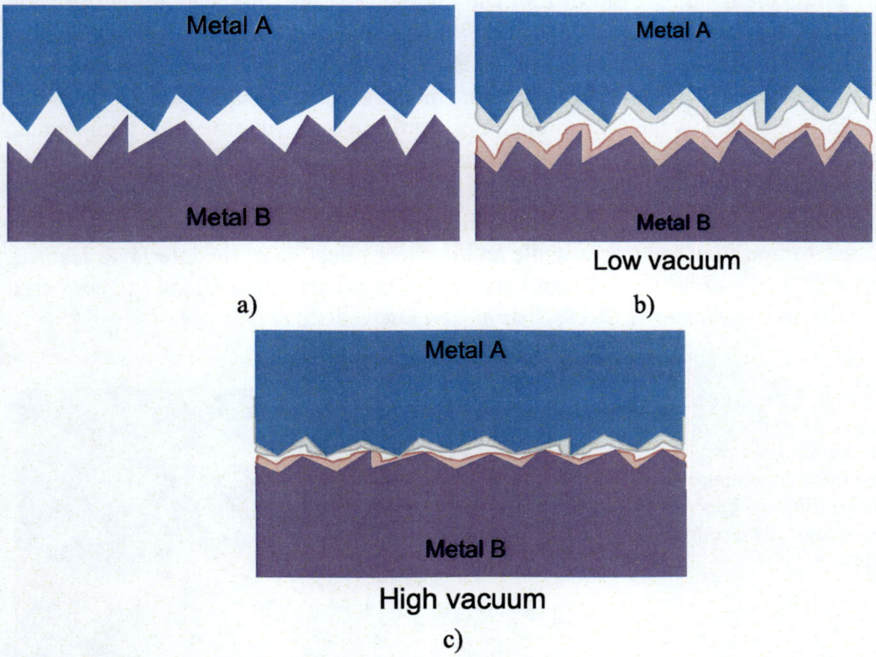

Fig. 9.8 Schematic showing effect of vacuum on diffusion bonding **a** surfaces to joined, **b** low vacuum causing surface oxidation and less metallic intimacy and **c** high vacuum leading to less surface oxidation and good metallic intimacy

9.4.6 Metallurgical Aspects

The metallurgical interactions and changes taking place during diffusion bonding affect the diffusion coefficient and productivity of process. There are two metallurgical aspects related to diffusion bonding which needs attention namely metallurgical transformation like phase transformation and recrystallization occurring during diffusion bonding and chemical compositional modification due to the addition of a few elements intentionally to accelerate the diffusion in form of films, foils and interlayers at the interface. The metallurgical transformations in metals during diffusion bonding cause more plastic behaviour and higher diffusion coefficient, which in turn increases the metallurgical intimacy due to improved interfacial surface deformation and reduces the diffusion bonding time, respectively. Similarly, the presence of certain elements in form of film and interlayer at the interface during diffusion bonding acts as diffusion accelerators which reduce the time required for volume diffusion to eliminate interfacial voids. However, these diffusion-accelerating interlayers should have high solubility in parent metals else these may have side effects like the formation of low melting point phases, hard and brittle intermetallic compounds, and other metallurgically unstable phases (Fig. 9.9). An appropriately selected interlayer (of thickness less than 25 μm) for diffusion bonding helps to reduce bonding temperature, pressure, time and undesired element from the joint interface in the form of impurities and traces while an improper selection of the interlayer leads to reduced ability of diffusion bonds withstand at high temperature, load and corrosion conditions (Sharma et al. 2018; Sharma and Dwivedi 2021).

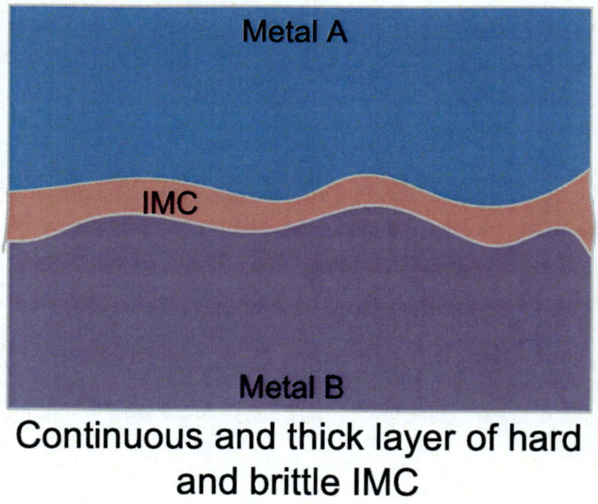

Fig. 9.9 Schematic showing formation of thick and continuous layer of intermetallic compound during diffusion bonding of metallurgically incompatible metals

9.5 Dissimilar Metal Bonding

The dissimilar metal joining using diffusion bonding is more attractive than fusion-based joining processes especially for the metal combinations having (a) large difference in thermo-physical properties (melting point, solidification temperature range, thermal expansion coefficient, thermal conductivity and thermal diffusivity), (b) incompatible chemical composition, (c) metallurgical incompatibility and (d) increased tendency of deterioration in toughness, and other mechanical and metallurgical characteristics of one of the parent metals on heating to high temperature during fusion welding. Issues related to dissimilar metal diffusion bonding are elaborated in the following section. Since heating (temperature) of dissimilar metal combination during diffusion bonding is initially limited (0.5–0.8 T_m) therefore issues related to differences in thermo-physical, mechanical and metallurgical properties are found to be lesser. Still, these need attentions and must be addressed properly and are presented in the following section.

9.5.1 Thermo-Physical Properties

Among the thermo-physical properties, thermal expansion coefficient and melting temperature are the most important properties of parent metals of dissimilar combination affecting ease of joining by diffusion bonding. A large difference in thermal expansion coefficient leads to development of opposite types of residual stress across the joint interface (Fig. 9.10). The presence of tensile residual stress in turn increases the cracking tendency of the diffusion bonds, especially in presence of hard and brittle phases and IMCs at the joint interface. Issues of residual stress and IMC formation in dissimilar metal bonding can be somewhat reduced using a suitable interlayer.

Similarly, a large difference in melting temperature makes the selection of bonding temperature more difficult as bonding temperature is usually 0.5–0.8 times of melting temperature (T_m) of low melting point metal of dissimilar combination, for example, the diffusion bonding of Ti (1668 °C) and Al (662.5 °C). In general, increase in diffusion bonding temperature in dissimilar metal joining due to increased differential expansion and contract increases the residual stresses developed at bonding interface (Fig. 9.11).

9.5.2 Mechanical Properties

Mechanical properties of dissimilar metal combination to be joined by diffusion bonding need consideration (at a given bonding temperature), particularly from the selection of bonding pressure point of view. A large difference in yield strength, thermal softening and ductility may cause bulging and undesired deformation of

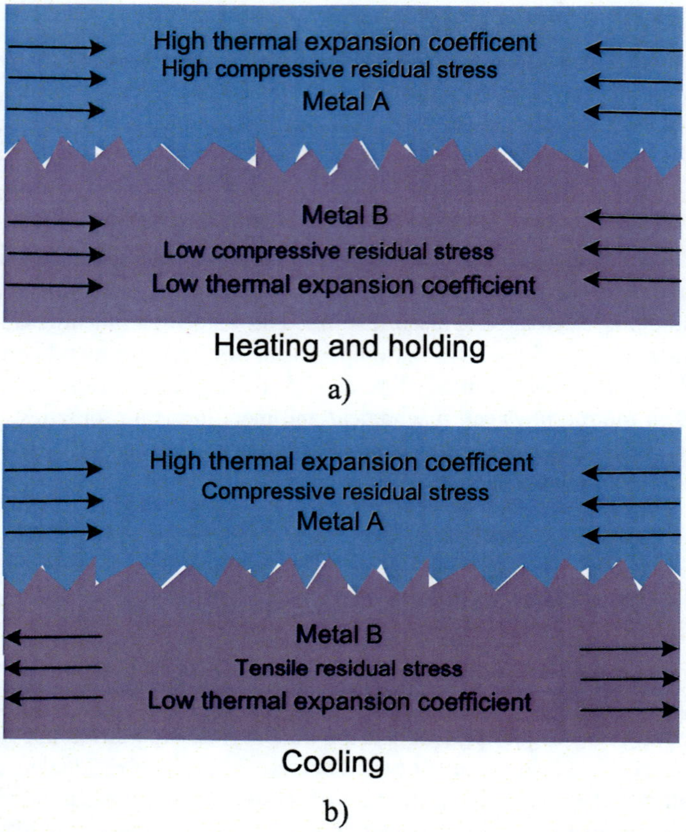

Fig. 9.10 Schematic showing residual stress developing in diffusion bonding on **a** heating and holding period at bonding temperature and **b** cooling to room temperature after bonding

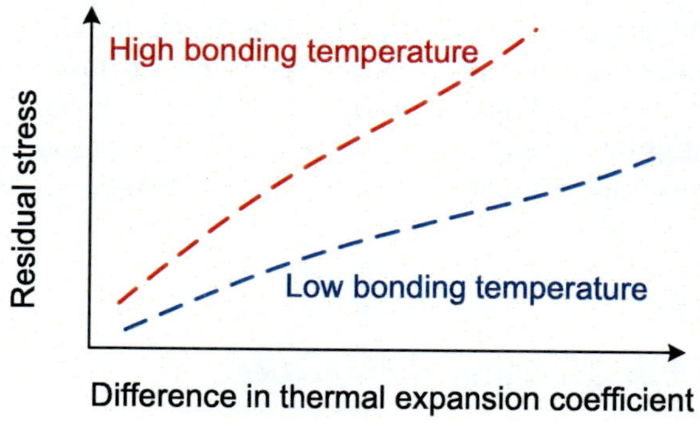

Fig. 9.11 Schematic showing effect of difference in thermal expansion coefficient and bonding temperature and on residual stress in diffusion bonding of dissimilar metals

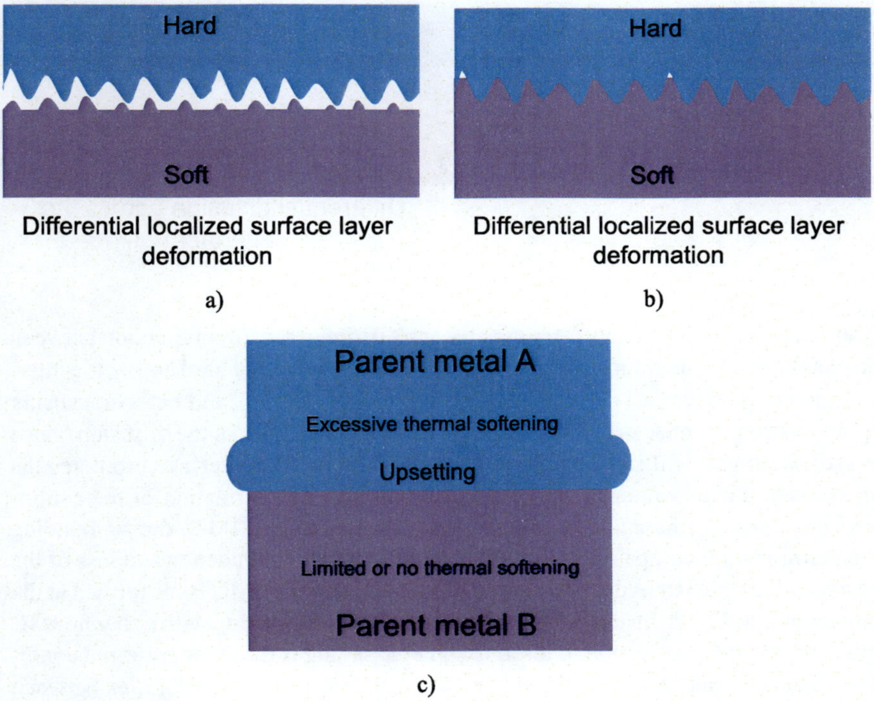

Fig. 9.12 Schematic showing effect of difference in mechanical properties on diffusion bonding of dissimilar metals **a** components to be joined, **b** deformation soft metal filling all cavities at the interface establishing good metallic intimacy and **c** bulging of soft metal subjected to excessive thermal softening

weak parent metal/component of dissimilar combination in case of inappropriate selection of bonding pressure for a given bonding temperature (Fig. 9.12).

9.5.3 Metallurgical Aspects

The metallurgical aspects of diffusion bonding should be looked into separately for similar and dissimilar metal joining because significant difference in chemical composition, microstructure at diffusion bonding temperature complicates the metallurgical transformations and changes occurring at the joint interface. Diffusion of elements is the key in diffusion bonding and in dissimilar metal joining, the diffusion of elements across the interface is not equal and uniform which in turn leads to segregation of elements selectively in vicinity of bond interface (Fig. 9.13).

The metallurgical changes may occur in the different ways such as (a) segregation of few elements at the interface and form undesirable bands including phases and compounds, (b) differential diffusion coefficient of certain elements in two dissimilar

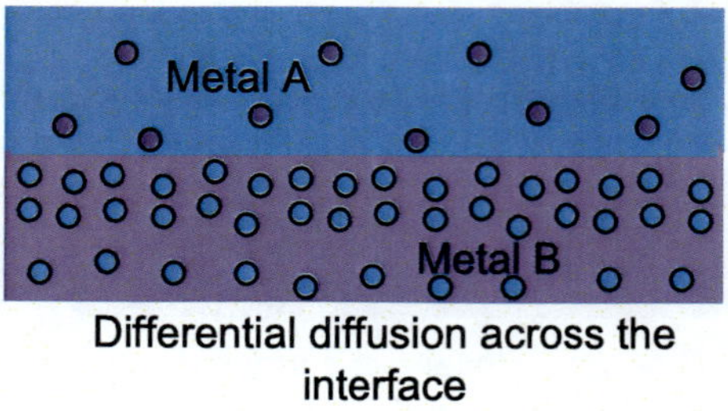

Fig. 9.13 Schematic showing differential diffusion of elements across the bonding interface

parent metals causing the porosity in parent metals/interface, (c) interaction between the metallurgically incompatible metals leading to formation of hard and brittle intermetallic compounds, (d) development of low melting phases and eutectic causing significant compromise in mechanical and corrosion properties of the diffusion bonds and (e) sharp gradient in distribution of element across the interface promotes the corrosion and deteriorates the mechanical properties. The formation of the continuous thick layer of hard and brittle intermetallic compound (IMC) during bonding of metallurgical incompatible metals causes more cracking tendency and loss of the tensile and fatigue strength as compared to thin and discrete IMC layer formed at the joint interface. The diffusion bonds having metallurgical bonding with either no IMC or favourable IMC formation in dissimilar metal joining is the most preferred conditions. This can happen when there is either metallurgical compatibilities between parent metals or suitable metallurgically compatible interlayer is used (Fig. 9.14).

Few issues to some extent can be addressed using carefully chosen interlayer of suitable metal and thickness considering the dissimilar metal combination, bonding temperature and other relevant bonding conditions. Depending upon the metallurgical compatibility and diffusion of elements across the two interfaces formed by interlayer with two parent metals, the diffusion zone and IMC formation can be only on one side or both the sides (Fig. 9.15). An improperly selected interlayer can cause formation of low melting phases, hard and brittle intermetallic compounds, porosity in parent metals and joint interface, deterioration in mechanical properties and corrosion resistance of the diffusion bonds due to unfavourable chemical and metallurgical heterogeneity. The degree of chemical heterogeneity when using interlayer to some extent is reduced by giving high temperature exposure for long duration during diffusion bonding.

Diffusion bonding parameters such as bonding temperature, pressure and time for diffusion bonding of dissimilar metals need consideration of possibility of formation of low melting phases, partial melting, melting of interlayer, yield strength and ductility of weak metal in dissimilar combination, diffusion coefficient and metallurgical compatibility of elements present in two different metals and interlayer if any. Thickness and hardness of IMC layer significantly affect the tensile strength of the diffusion bonds. A thick and hard IMCs layer results in much lower tensile strength than thin and soft IMC layer primarily due to reduced tendency of the interfacial void formation during external loading and increased tolerance to the discontinuities (Fig. 9.16).

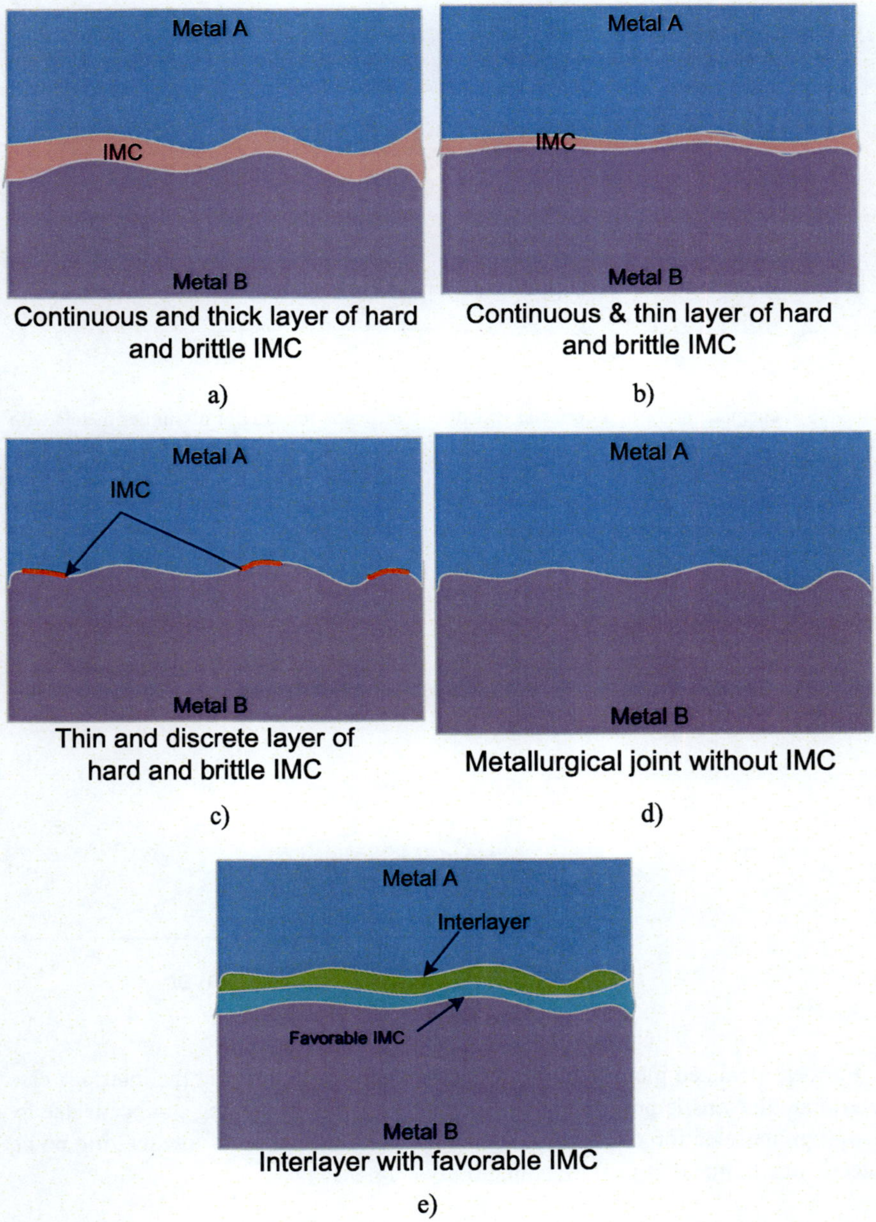

Fig. 9.14 Schematic showing various types of IMC formation in diffusion bonding of dissimilar metals **a** continuous thick layer of IMC, **b** continuous think layer of IMC, **c** a discrete thin layer of IMC deformation, **d** no IMC and **e** more favourable IMC formation in presence of interlayer

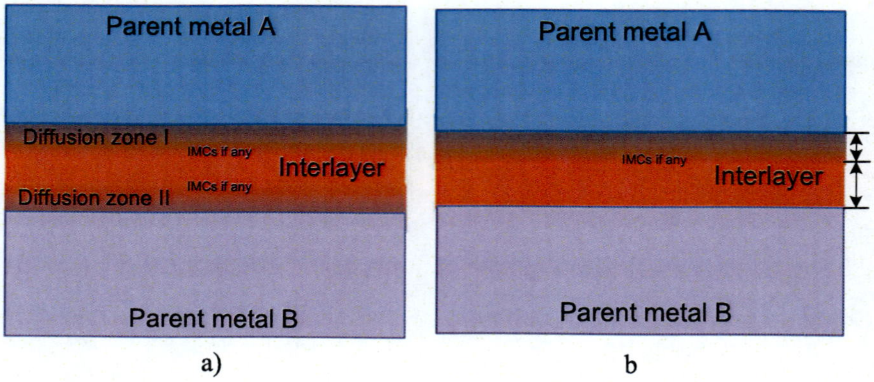

Fig. 9.15 Schematic showing interfacial structure in diffusion bonding of dissimilar metals with interlayer **a** diffusion zone (and IMC if any) formation both sides of interlayer and **b** diffusion zone (and IMC if any) on one side of interlayer only

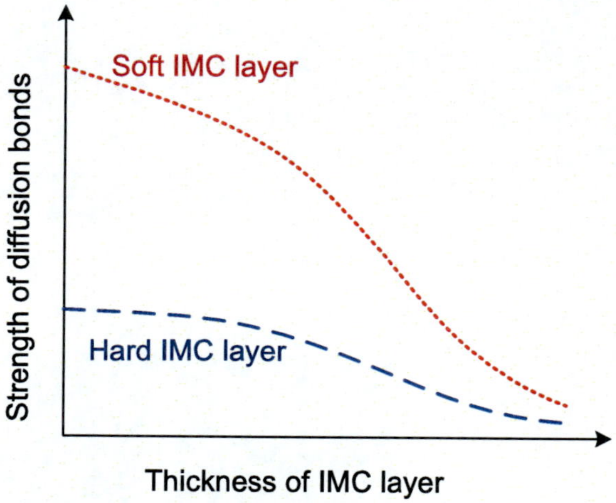

Fig. 9.16 Schematic showing effect of IMC layer thickness and its characteristic

Further, localized melting of phases and compounds formed at the interface can deteriorate the tensile properties. The localized interfacial melting can occur due to inappropriate selection of bonding temperature and formation of low melting point eutectic due to metallurgical incompatibility (Fig. 9.17).

9.6 Diffusion Brazing

The diffusion brazing (similar to the conventional brazing) uses low melting point film/interlayer as brazing metal which is melted for developing diffusion-brazed

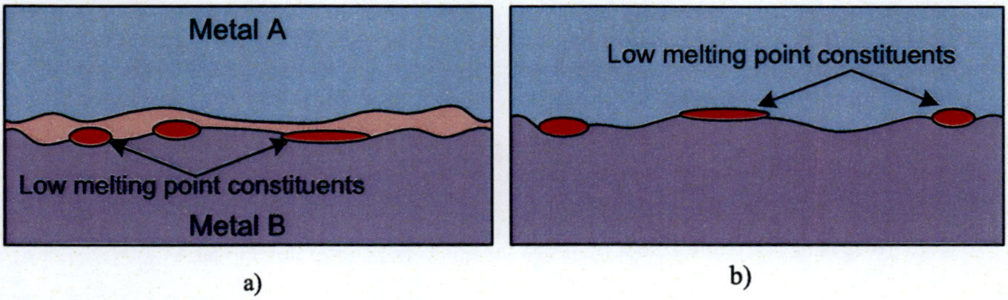

Fig. 9.17 Schematic showing localized melting at the bonding interface **a** with interlayer and **b** without interlayer

joint. The diffusion brazing takes longer than conventional brazing as diffusion brazing is based on diffusion, which heavily depends on time and temperature. Moreover, diffusion-brazed joints offer joint strength matching with parent metal or even higher. Diffusion-brazed joints are usually stronger than the conventional brazed joints (Fig. 9.18).

Filler may be similar to one of the (low melting point weak) parent metal of dissimilar combination but with melting point reducing elements, which form either eutectic or low melting temperature constituents such that filler metal has little lower melting temperature than weak parent metal in dissimilar metal joining. The filler metal for diffusion brazing placed between the mating surfaces of components to be joined melts and wets the surface of the parent metals. Filler then gradually diffuses into the parent and eventually filler is vanished (Fig. 9.19).

Fig. 9.18 Schematic showing comparison of thermal cycle of conventional and diffusion brazing

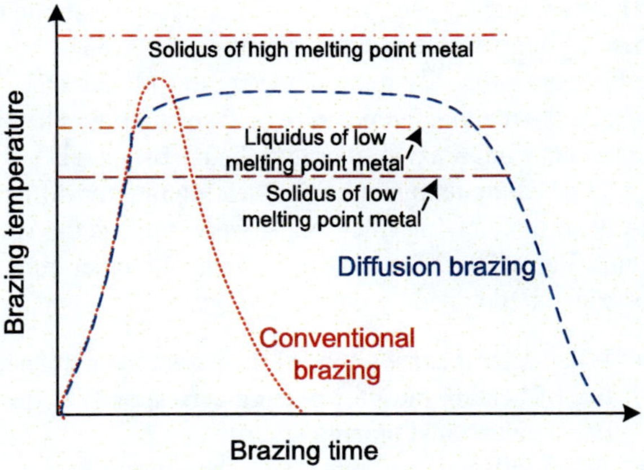

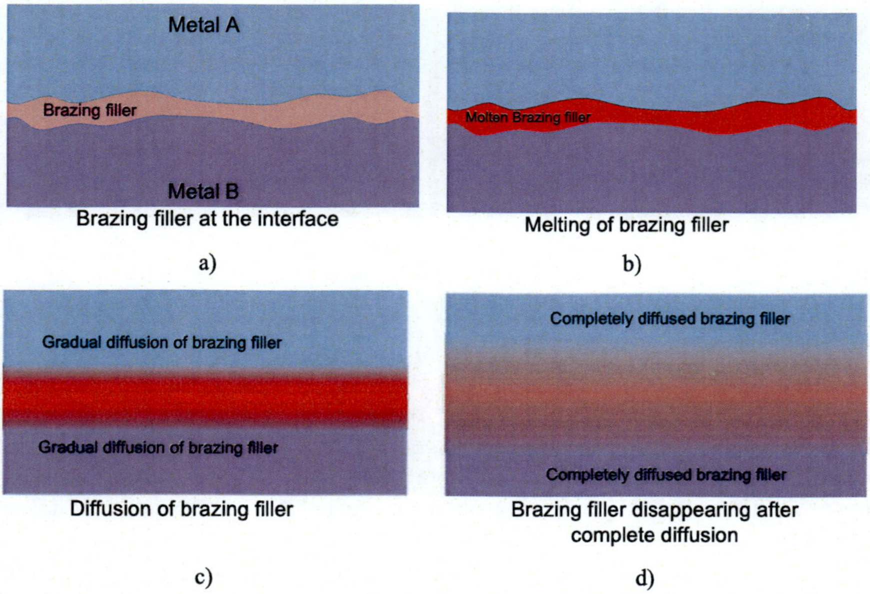

Fig. 9.19 Schematic showing sequential steps of diffusion brazing **a** placement of brazing filler, **b** melting of filler, **c** gradual diffusion of elements across the interface and **d** brazing filler eventually vanished

9.6.1 Thermal Cycle for Diffusion Brazing

There are three aspects related to thermal cycle applied for diffusion brazing namely heating temperature and rate of heating followed by holding time at high temperature, which are primarily determined by the filler metal. The filler metal composition affects the melting temperature (and so maximum heating temperature), diffusion of elements from brazing filler into the parent metals.

The heating temperature and heating rate for diffusion brazing are selected carefully so that brazing filler melts, wets and fills the voids and forms a metallurgical joint (Fig. 9.20a). The maximum heating temperature and heating rate for diffusion brazing are dictated by the following:

- When filler is same as one of the parent metal then heating temperature (irrespective of heating rate) for brazing corresponds to the melting temperature of filler like conventional brazing.
- When filler is different from the parent metals and is expected to produce eutectic/peritectic reactions at the interface, then both heating temperature and heating rate for brazing become important. The heating rate determines the formation of molten eutectic. Heating temperature should be little higher than the eutectic temperature while heating rate should be high. Low heating rate may result in excessive diffusion from the filler to the parent metal, which in turn may

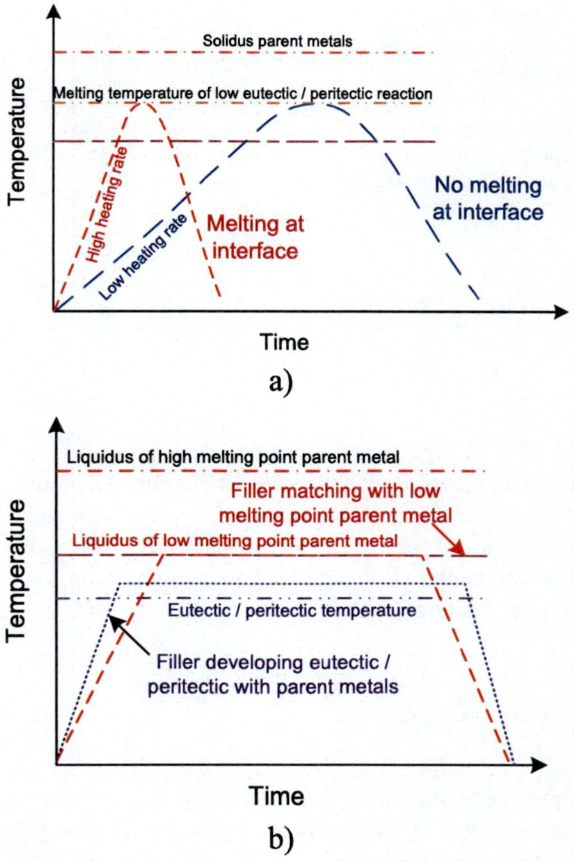

Fig. 9.20 Schematic showing **a** effect of heating rate on diffusion brazing and **b** thermal cycle for diffusion brazing using matching filler and filler forming eutectic

avoid/limit the desired eutectic/peritectic reactions to produce molten filler for brazing (Fig. 9.20b).

Apart from filler metal, the metallurgical/mechanical property alterations occurring in parent metals affect the selection of maximum heating temperature for diffusion brazing else it may require post-brazing heat treatment of diffusion-brazed joints to restore the mechanical properties of the parent metals.

9.6.2 Brazing Time

Time for hold the parent metals of dissimilar combination at brazing temperature is determined by the heating temperature which in turn affects diffusion coefficient of elements in filler to parent metals and maximum acceptable concentration of elements in filler in vicinity of joint interface. After solidification of brazing filler, joint must

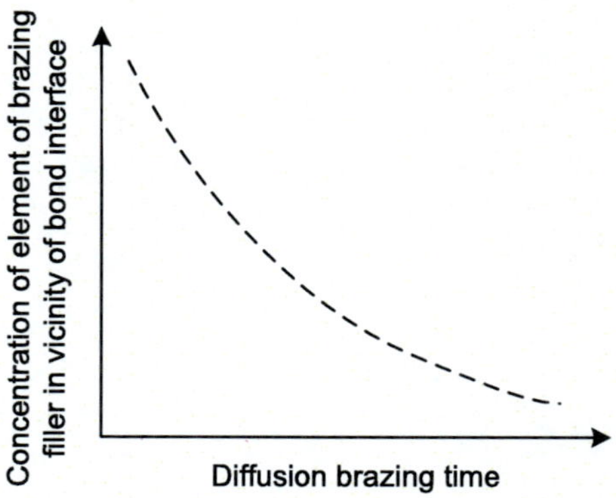

Fig. 9.21 Schematic showing effect of diffusion brazing time on the concentration gradient of alloying elements as a function of increasing distance from the interface

be kept at high temperature for some time so that the concentration gradients of element across joint interface/filler can be reduced and more chemical homogeneity is established (Fig. 9.21). Otherwise, excessive chemical heterogeneity across the joint interface can lead to (a) the formation of undesirable compounds, microstructural transformation at the interface and (b) phase transformation coupled with response to post-brazing heat treatment.

9.6.3 Brazing Pressure

Pressure applied for the diffusion brazing is either nil or very little as pressure can squeeze out the molten brazing metal from the interface leading to the failure of primary purpose of filler application at the interface. Sometimes even specially designed attachment as spacers are used to maintain specific gap/clearance between components to be diffusion brazed.

9.6.4 Metallurgy of Diffusion Brazing

Metallurgical aspects of diffusion brazing are very important for mechanical and corrosion performance of diffusion-brazed joints. The brazing filler metal is expected to completely diffuse out into both the parent metals and establish the chemical homogeneity across the interface conversely brazing filler should be vanished from the interface and does not remain as an individual alloy/filler at the interface. However, despite of absence of brazing filler at the interface as an individual component, diffusion-brazed joints always exhibit the compositional gradient across the interface. This compositional gradient can be gradual or sharp depending upon the diffusion coefficient under diffusion brazing conditions (Fig. 9.22). The low diffusion coefficient results in sharp gradient while high diffusion coefficient results in gradual

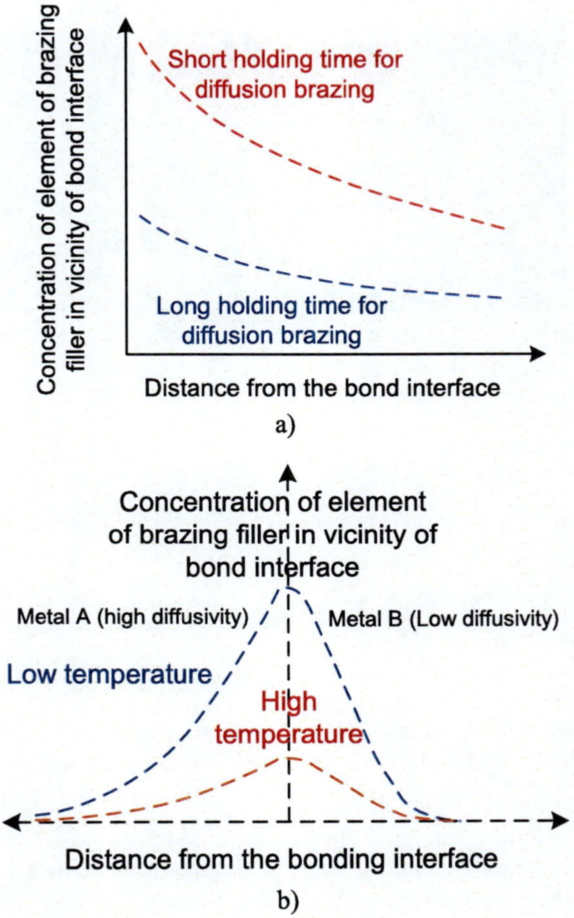

Fig. 9.22 Schematic showing effect of **a** diffusion brazing time and **b** diffusion brazing temperature on concentration gradient of alloying elements as a function of increasing distance from the interface

compositional gradient. The chemical composition significantly affects the metallurgical characteristics of interface including phase transformation, compounds and response to the post-brazing heat treatment. Therefore, diffusion-brazed joints of dissimilar metals can produce varying microstructure and mechanical properties on both the parent metals in vicinity of joint interface.

9.6.5 Filler Metal for Diffusion Brazing

Filler metal should form a low melting point constituent like eutectic after interactions with one or both the parent metals during diffusion brazing and these low melting point constituents should be compatible with both the parent metals. The filler in the form of pure metal can be applied using coating, sputtering and electroplating.

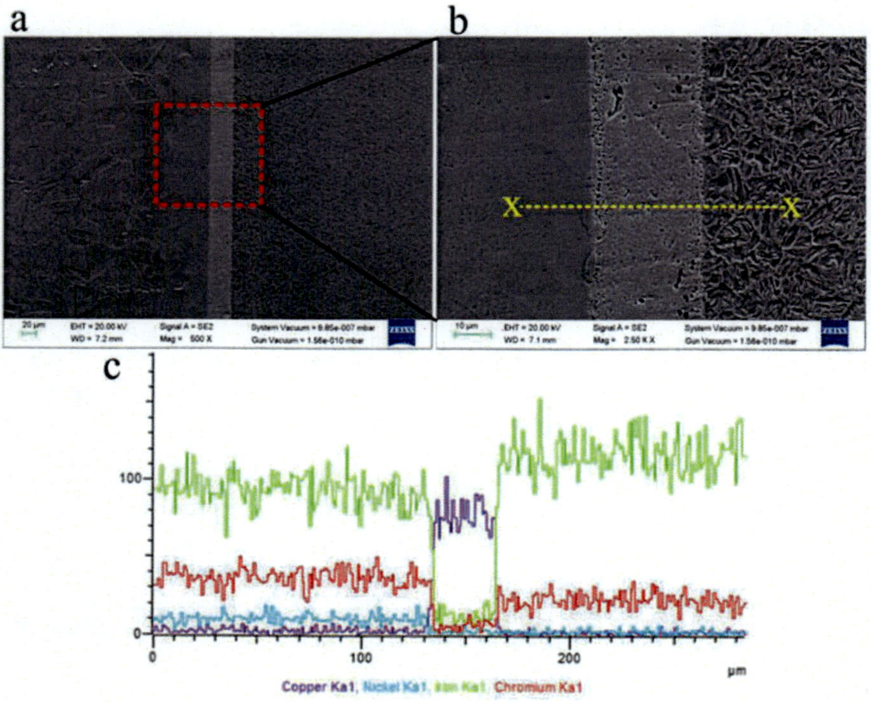

Fig. 9.23 EDAX analysis showing line scanning across the diffusion bond of ferritic-martensitic steel (P91) and austenitic stainless steel (AISI 316) developed using copper interlayer (Sharma and Dwivedi 2022)

These may be added with melting point reducing elements, e.g. Ni, Co, B, Si, to lower the melting point of filler. However, addition of such elements can result in the formation of hard and brittle micro-constituents. Addition of B (2–3 wt%) in Ni base filler reduces the melting temperature significantly. Diffusion brazing of ferritic-martensitic steel (P91) and austenitic stainless steel (AISI 316) is done using copper interlayer (Fig. 9.23) (Sharma and Dwivedi 2022).

References

Sharma G, Dwivedi DK (2017a) Diffusion bonding of pre-friction treated structural steel with reversion of deformation induced grains. Mater Sci Eng, A 696:393–399

Sharma G, Dwivedi DK (2017b) Effect of pressure pulsation on bond interface characteristics of 409 ferritic stainless steel diffusion bonds. Vacuum 146:152–158

Sharma S, Dwivedi DK (2018) Impulse pressure assisted diffusion bonding of AISI 304 austenitic stainless steel at different bonding temperatures. In: International conference on advances in mechanical engineering, Istanbul, 19–21 Dec 2018

Sharma G, Dwivedi DK (2019) Study of metallurgical and mechanical properties of CSEF P92 steel diffusion bonds developed using pressure pulsation. J Manuf Process 38:196–203

Sharma G, Dwivedi DK (2021) Diffusion bonding of 304 austenitic stainless-steel using pressure pulses. Mater Today: Proc 44:2135–2141

Sharma G, Dwivedi DK (2022) Role of surface roughness in impulse pressure diffusion bonding of dissimilar steels using copper interlayer. Mater Today: Proc 64:1384–1391

Sharma G, Tiwari L, Dwivedi DK (2018) Impulse pressure assisted diffusion bonding of low carbon steel using silver interlayer. Trans Indian Instit Metals 71:11–21

Chapter 10
Dissimilar Metal Joining of Steel-Aluminium Alloy by Friction Stir Welding

This chapter presents various applications of steel-aluminium joints in industries as well as the modern techniques employed in the fields to join this combination. There are multiple issues in joining a high melting point material, such as steel with a comparatively low melting point material, such as aluminium. These issues have been discussed, and a non-fusion process, namely friction stir welding, is presented as an alternative process for steel-aluminium joining. Fundamentals related to heat generation and deformation are discussed thoroughly. Characteristics of FSW joints in butt and lap configurations are presented to show the viability of FSW for dissimilar welding.

10.1 Background

Emerging trends of multi-material components for desired safety, strength and weight-saving have brought dissimilar metal joining to the forefront. The recent shifting of the world towards minimum carbon emission standards aided the usage of lightweight materials in automotive industries. Lightweight materials such as aluminium, polymers and composites are considered to be the most suitable for automotive industries from a technological and economic point. Despite the continuous reduction in the usage of steel, it is still the most used material in industrial applications and automobiles. The joining of lightweight aluminium and rigid high-strength steel by applying sustainable joining processes is highly sought. Integration of aluminium and steel typically gives around 25% weight saving in a car subframe (Kusuda 2013). The reduction of the weight of a component, in turn, reduces power consumption and carbon emission.

Various processes have been tried by the researchers to join these materials (Kaushik and Dwivedi 2021). Although the joining of Al-steel is carried out using processes such as tungsten inert gas (TIG) welding, resistance spot welding (RSW), laser welding and electron beam welding, issues with the fusion welding processes

© The Author(s), under exclusive license to Springer Nature Singapore Pte Ltd. 2023
D. K. Dwivedi, *Dissimilar Metal Joining*,
https://doi.org/10.1007/978-981-99-1897-3_10

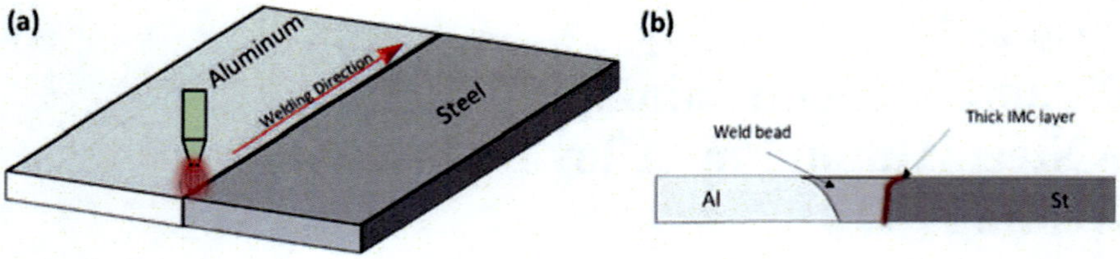

Fig. 10.1 Schematic diagram showing **a** butt weld developed using fusion welding process and **b** weld cross section

render the joint ineffective in critical service conditions. The formation of a thick intermetallic compound (IMC) layer along the weld faying surface is the prominent cause of the unsatisfactory performance of the welding joints made by the fusion joining processes, as shown in Fig. 10.1. The IMC layer formed in the Al-steel joint is inherently hard and brittle, which deteriorates the mechanical performance of weld joint.

The formation of the IMC layer depends on the interdiffusion distance of both the materials (Al and steel). The diffusion coefficients increase with the increase of temperature. Therefore, as the temperature of the two incompatible metals in contact increases, interdiffusion across the contact interface leads to the formation of intermetallic compounds. The temperature attained in fusion-based welding processes of Al-steel is generally in the range of 700–1000 °C or even higher. The high temperature causes the growth of the IMC layer beyond the safe limits of 10 μm (Kaushik and Dwivedi 2022a). Therefore, solid-state joining processes like friction stir welding (FSW) help to reduce IMC growth. The peak temperature reached in FSW of Al-steel is well below the melting point of Al alloy. The interdiffusion of Al and steel up to this temperature is considerably low, and the thickness of the IMC layer formed in the FSW process remains under 10 μm. Thus, avoiding melting and solidification of Al alloy during the joining process helps in controlling the thickness of the IMC layer.

The friction stir welding process, a relatively new process, developed by The Welding Institute (TWI) in 1991, has also found its application in dissimilar metal joining of Al-steel, Al-copper, Al–Mg and many more dissimilar metal combinations.

10.2 FSW Fundamentals

In FSW, a hard tool is stirred against a soft metal to cause its plastic deformation with the assistance of friction. The tool is a non-consumable type, and its material is harder than the parent metal plates. The tool rotated at high speed and then plunged at a location along the weld line to start the welding. After some dwell time, the tool is moved in the traverse direction along the weld line over the entire length to

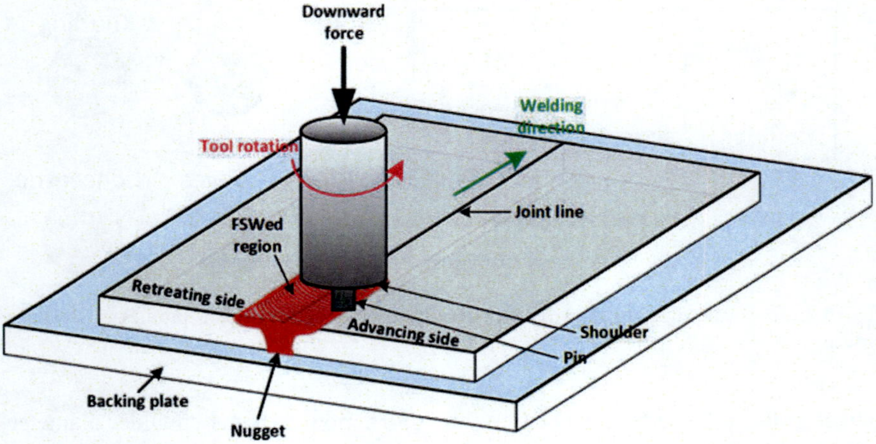

Fig. 10.2 Schematic diagram of friction stir welding process

develop a weld joint. The FSW tool generally consists of two major components: shoulder and pin, as shown in Fig. 10.2.

The area covered under the stirring action of the tool shoulder and pin is termed as stirred zone or nugget zone. There are two sides (with respect to weld line) namely, advancing and retreating side depending upon the effective stirring velocity experienced by the base material. The side where tool rotational velocity vector and welding direction are in the same direction is termed as advancing side, and where tool rotational velocity and welding direction are in the opposite direction is termed as the retreating side of the weld, as shown in Fig. 10.3.

The dimensions of the tool's pin are chosen according to the thickness of the plates to be joined. The length of the pin should be approximately in the range of 85–95% of the plate's thickness to ensure complete penetration and avoid the bottom flowing out of the stirred material. The shorter length of the tool pin may cause an incomplete depth of penetration, and a larger length may cause the material to flow downwards and stick to the backing plate.

The shoulder of the tool is stirred against the top surface of both the parent metals help in containing the material from expulsion. The size of the shoulder (i.e. shoulder

Fig. 10.3 Schematic of advancing and retreating side of a joint

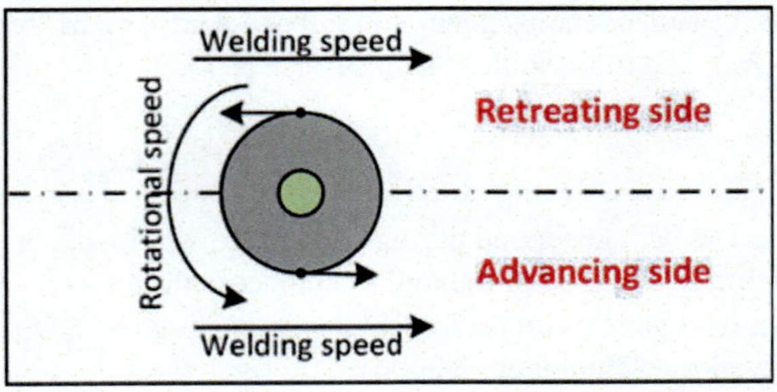

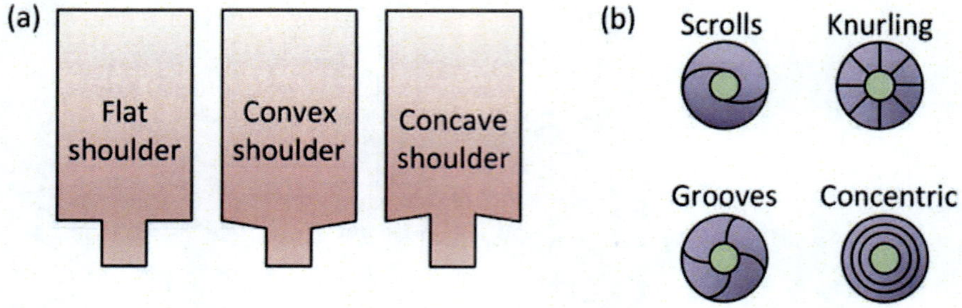

Fig. 10.4 Schematic diagram showing various **a** types of shoulders and **b** profiles of shoulder surface

diameter) dictates the heat generation through friction. A larger shoulder diameter will stir a bigger area and cause a higher heat generation and vice versa (Kaushik and Dwivedi 2020). Multiple types of tool shoulder profiles are in trend. A flat shoulder surface is mainly used because of its easy and economical machining. A concave or convex-shaped shoulder tool is also used in special requirements, as shown in Fig. 10.4a. Tool shoulder can be of many types of surface features for better stirring action, as shown in Fig. 10.4b.

Similar to tool shoulder, tool pin profiles can also vary for better stirring action. A flat cylindrical tool pin is the most common shape of the pin. Further tool pin cross section can be tapered, conical, threaded, triflute and A-skewed.

10.2.1 Heat Generation in FSW

Heat generated in FSW is primarily due to friction between tool and workpiece surface and deformation of parent metal around tool pin and shoulder. The heat generation is directly related to the rotational speed of the tool and the area available for friction between tool and workpiece metal. Two basic tribological processes act in FSW, i.e. pure sliding (friction) and pure sticking (deformation). A significant amount of heat is generated by the FSW tool rotational speed, and the heat generated from the traverse movement of the tool is marginal. Considering a pure sliding case, a perfect coulomb friction condition is assumed, and heat generated can be calculated as the sum of energy generated at the tool pin tip surface (Q_{pt}), tool pin curved surface (Q_{ps}), and tool shoulder's tip surface (Q_{st}).

$$Q = Q_{pt} + Q_{ps} + Q_{st} \tag{10.1}$$

The heat generated during friction stir welding is directly related to the rotational velocity and area of the tool in contact with the metal during welding. The formula for heat generation (in joules) due to friction in FSW (if pure sliding is assumed) is given as (Schmidt et al. 2004):

$$\text{Heat generation from shoulder tip}(Q_{st}) = \frac{2}{3}\pi\mu\omega p\left(R_s^3 - R_p^3\right) \tag{10.2}$$

$$\text{Heat generation from pin tip surface}\left(Q_{pt}\right) = \frac{2}{3}\pi\mu\omega p\left(R_p^3\right) \tag{10.3}$$

$$\text{Heat generation from pin curved surface}\left(Q_{ps}\right) = \pi\mu\omega p H\left(R_p^2\right) \tag{10.4}$$

Here μ is the friction coefficient between tool and workpiece, whereas ω is tool angular speed in radians per sec. R_s is the tool shoulder radius in mm, R_p is the tool pin radius in mm, p is the pressure exerted in MPa, and H is the height of the tool pin in mm. The tool shoulder is assumed to be flat, and the tool pin is of cylindrical profile for simplification.

Total heat can be calculated as

$$Q_{\text{total sliding}} = \frac{2}{3}\pi\mu\omega p\left(R_s^3 + 3R_p^2 H\right) \tag{10.5}$$

The same formula for pure sticking conditions becomes

$$Q_{\text{total sticking}} = \frac{2}{3}\pi\omega\tau_{\text{yield}}\left(R_s^3 + 3R_p^2 H\right) \tag{10.6}$$

But, generally, the heat generation in the FSW process is a very complex phenomenon, and the contact conditions are a combination of both pure sliding and pure sticking. Therefore, a dimensionless contact condition variable δ is introduced. When there is pure sliding condition $\delta = 0$ and for pure sticking condition $\delta = 1$ and in case of partial sliding/sticking, it can range from 0 to 1.

$$Q = \delta Q_{\text{total sticking}} + (1 - \delta)Q_{\text{total sliding}} \tag{10.7}$$

10.3 Aluminium Steel Metal System

Al-steel metal combination is very hard to weld because of the issues related to a significant difference in their physical, mechanical and metallurgical properties, as discussed earlier. The problems of thick IMC layer, non-solubility in each other and cracking due to differential expansion and contraction are prevalent during welding of this dissimilar metal combination. Therefore, the joining processes, reducing these problems are highly sought after for dissimilar welding of Al-steel. Assembling methods such as nuts and bolts and rivets are generally used to prevent the metallurgical intimacy of these two metals. However, using an additional weight element in the form of nuts and bolts in these methods is a major concern. Thus, the processes using minimum heat input for welding are found to be suitable for Al-steel joining.

Friction stir welding process, generating solid-state weld and avoiding the fusion and solidification-related issues is potential alternative to conventional welding processes for joining Al-steel and other metallurgically incompatible systems.

The most common types of welded joints used in industries are butt joints and lap joints. Joining Al-steel metals in butt and lap joint configuration poses different challenges and issues. Therefore, Al-steel welding depending upon the type of joint is presented in the following section.

10.4 Al-Steel Butt Joint

The square butt joint of the Al-steel is formed by placing Al and steel sheets in a butting position with square edge preparation. Both plates are placed such that there is no gap between them. Properly cleaned edges of the same thickness are fixed together, and the rotating tool after plunging is traversed over the joint line longitudinally. Generally, more heat is generated on the advancing side of the tool than on the retreating side because of the higher relative velocity between the tool and the workpiece. Therefore, the material having a higher melting point and hardness is placed on the advancing side to facilitate softening of parent metals on both sides. In Al-steel dissimilar metal butt joining, steel is invariably placed on the advancing side to have a better softening effect and avoid excessive heating of the Al plate, as shown in Fig. 10.5.

10.4.1 Zones in Friction Stir Welded Joints of Al-Steel

There are multiple zones in friction stir welded butt joint of Al-steel because of exposure of both parent metals to varying amounts of heat and deformation. The

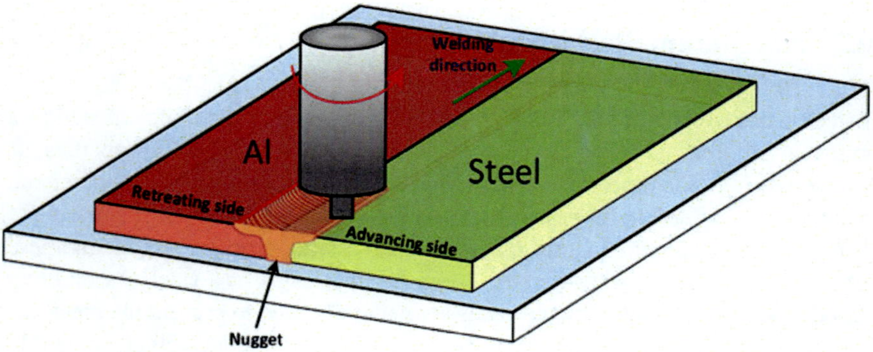

Fig. 10.5 Schematic of Al-steel friction stir welding joint

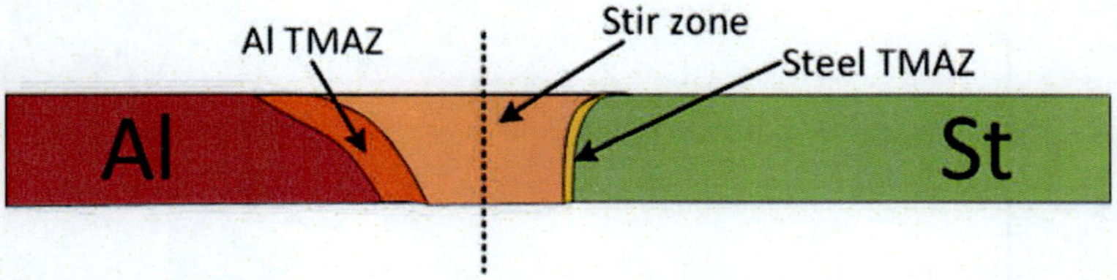

Fig. 10.6 Schematic diagram showing various zones present in Al-steel friction stir welded joint

central part, which is under the stirring action of the tool pin, is known as stirred zone or the weld nugget. Mechanical and metallurgical changes occur in the stir zone due to excessive heating and high strain rate deformation experienced by metals during FSW. The tool pin exerts high strain rate deformation on the metals and makes it soft and easily flowable. The temperature attained in this region is generally above the recrystallization temperature of the lower melting point material (Al). Therefore, dynamic recrystallization of Al is observed in this zone. The tool pin is generally kept offset towards softer metal, and the softer metal itself is kept towards the retreating side. Therefore, the shape of the weld nugget or stir zone is asymmetrical and is shifted more towards the Al side. The zone adjacent to the stir zone also experiences the thermal and mechanical effect of the stirring. The grains of the parent metal adjacent to the weld nugget gets deformed/pulled/aligned in a particular direction. This zone is generally very narrow and termed as thermo-mechanically affected zone (TMAZ). Very thin zone of steel plate is being stirred compared to the Al plate; therefore, the TMAZ region of the steel side is smaller than the Al side, as shown in Fig. 10.6.

To achieve a sound joint in Al-steel dissimilar welding, the appropriate amount of heat generation is highly important. A high heat generation may lead to excessive softening and melting of the metal with a lower melting point (i.e. Al here), leading to a defective joint. Similarly, a low heat input may not soften the metal enough to be joined to facilitate flowability. Thus, it is vital to optimize the heat generation to have adequate heating during friction stir welding of dissimilar metal systems. As discussed above, heat generation depends on various process parameters of FSW. In addition to the above-discussed parameters, tool offset is also an important parameter to be optimized in the case of dissimilar metal systems.

10.5 Process Parameters Affecting Joining in Butt Joining

10.5.1 Tool Rotational Speed

Tool rotational speed is the first and foremost parameter to influence the joining in Al-steel butt joining. The heat generation is directly proportional to the tool's rotational speed. As the tool rotational speed increases, the total heat available to the

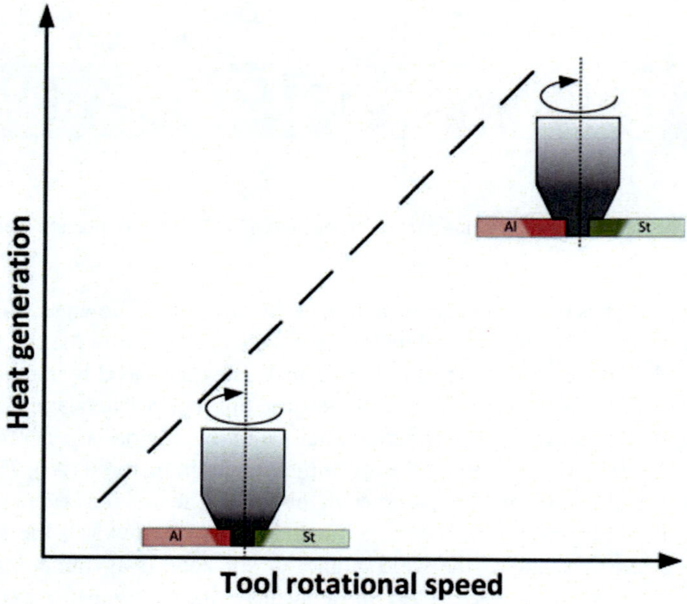

Fig. 10.7 Schematic diagram showing relationship between heat generation and tool rotational speed

joint also increases, which in turn will increase the width of the thermo-mechanically affected zone (TMAZ), as shown in Fig. 10.7. Heat generation will be lower at lower tool rotational speed, and therefore, the TMAZ region will be very narrow. Higher rotational speeds generate higher heat, which spreads over a larger area, as shown in Fig. 10.7. The metal having higher heat conductivity will be affected up to a larger area than the lower conductivity metal. So, Al side will show a broader TMAZ than steel.

There are added problems in Al-steel dissimilar systems associated with heat generation. The most prominent are differential expansion, contraction and thick intermetallic compound layer formation. At higher tool rotational speed, increased heat generation due to the differential thermal expansion coefficient of Al and steel leads to a higher difference in expansion of Al compared to steel. A large difference in the expansion and contraction of both metals leads to the development of residual stresses in the joint.

Excessive heat generation may also be correlated to a higher interdiffusion of both the metals into each other, forming a thick intermetallic compound (IMC) layer. IMC layer of Fe and Al is reported to be highly brittle in nature. Thus, IMC layer of thickness greater than 10 μm is reported detrimental to the joint strength. As the tool rotational speed increases, the thickness of the IMC layer also follows the same trend, as shown in Fig. 10.8a. Relating the strength with the IMC layer, Kundu et al. (2013) reported a decreasing trend in the strength of the joint with an increase in tool rotational speed, as shown in Fig. 10.8b

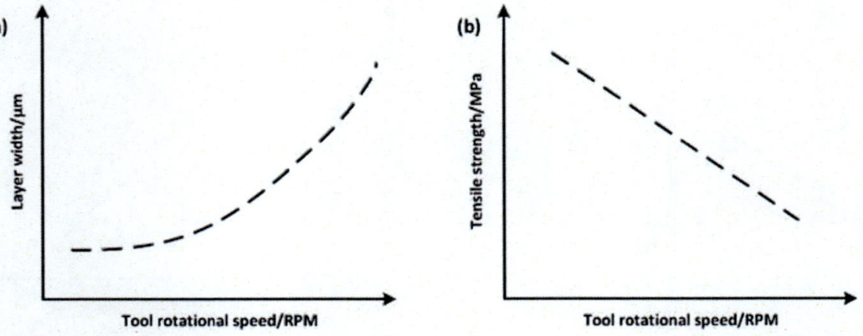

Fig. 10.8 Schematic diagram showing relation between **a** intermetallic layer thickness and tool rotational speed **b** weld strength and tool rotational speed (Kundu et al. 2013)

The formation of a sound joint using optimum heating is found to be beneficial for joint strength. Therefore, the tool rotational speed should be established suitably in case of dissimilar Al-steel FSW to have optimum heat generation, narrow TMAZ and a thin IMC layer. While keeping the tool rotational speed to a minimum, it must be considered to have at least some minimum rotational speed to account for the thermal softening of the metals while reducing the chances of IMC layer formation.

10.5.2 Tool Traverse Speed

Tool traverse speed is the speed of the tool with respect to the weld plate along the joint line. In some cases, when the tool is rotating at a fixed axis and the table holding the workpiece plate is movable, the speed of the table with respect to the tool is known as traverse speed or welding speed. Traverse speed is a crucial parameter in estimating the net heat input. In friction stir welding, the traverse speed determines the number of effective revolutions of the tool at a location or unit weld length say per mm. When traverse speed is low, the number of revolutions per unit length of the joint line increases, increasing the net heat input for welding.

The effect of tool traverse speed on net heat input is opposite to that of the tool rotational speed. As the traverse speed or welding speed increases, net heat input decreases as shown in Fig. 10.9. A suitable combination of tool rotational speed and welding speed is required to have an optimum level of heat input in Al-steel friction stir welding. Therefore, optimization of the process parameters is required to have a sound welding joint.

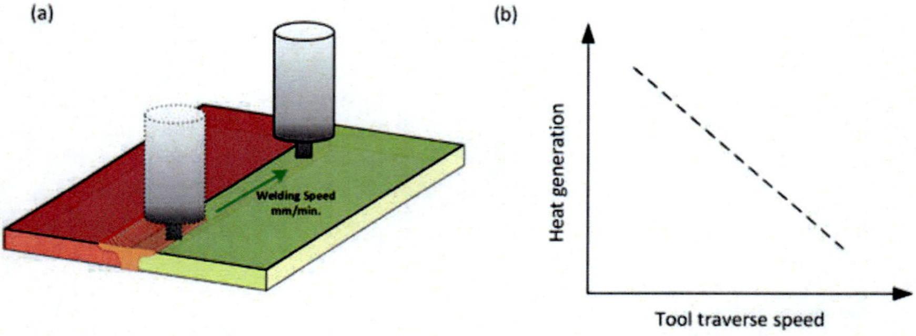

Fig. 10.9 Schematic diagram showing representation of welding traverse speed (in mm/min) and its effect on heat generation

10.5.3 Tool Pin Offset

Tool offset is an essential parameter in friction stir welding. It is more critical in the case of dissimilar welding having metals with different physical and mechanical properties. In Al-steel FSW, offsetting the tool towards one metal may change the overall morphology of the weld nugget. There can be three different types of offsetting possible in Al-steel FSW. The usual offset position as in similar welding is keeping the tool axis over the joint line, as shown in Fig. 10.10b. The axis of the tool pin is offset towards the Al side or steel side when extra stirring of a specific metal is to be carried out, as shown in Fig. 10.10a and c, respectively.

Keeping the tool pin entirely offset towards Al may only stir the Al metal, and the steel plate does not participate in joint formation. It could result in the formation of joints with very low strength. Conversely, keeping the tool pin entirely on the steel side would lead to excessive heating, and Al metal may melt leading to no weld formation. Generally, in Al-steel friction stir welding, the large part of the tool pin is kept offset towards the Al side to have more stirring of the Al metal. The Al metal becomes soft, flows under the stirring action of the tool pin and provides the forging action. Only a small part of the tool pin is kept towards the steel side to rub the steel plate surface to activate the surface for bonding, as shown in Fig. 10.11a. The

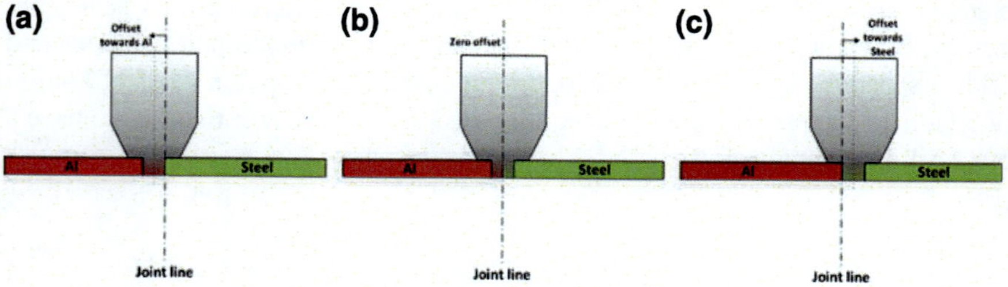

Fig. 10.10 Schematic diagram showing various tool offset positions in Al-steel FSW: **a** towards Al, **b** zero offset and **c** towards steel side

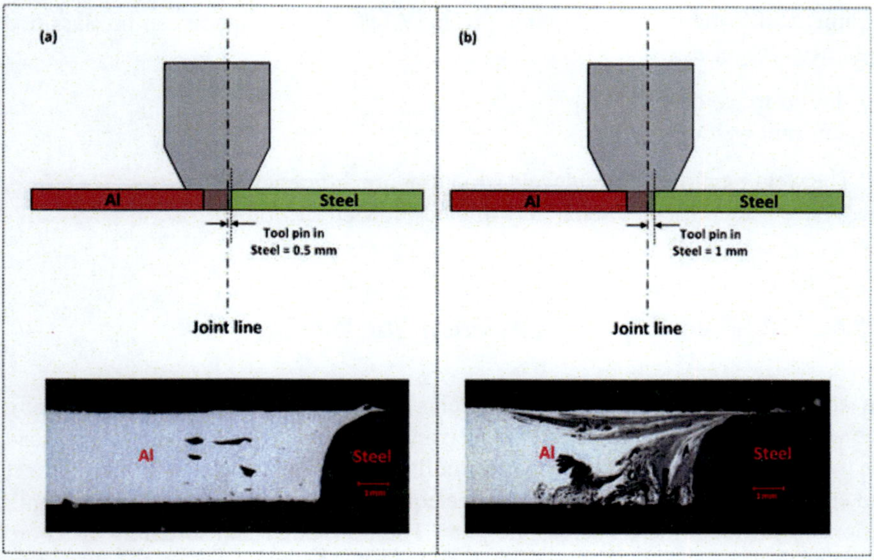

Fig. 10.11 Schematic diagram and actual image of **a** 0.5 mm tool in steel **b** 1 mm tool in steel

chipping away of metal from the steel during (steel-Al) FSW leads to entrapment of the steel fragments in weld nugget due to rubbing of the tool pin with the steel surface. Increased interaction of the tool pin with the steel causes a large number of steel fragments in weld nugget and deteriorates the weld joint strength because of insufficient filling up of trails of large steel fragments, as shown in Fig. 10.11b. The rubbing of the tool pin with the steel plate generates a large amount of heat. Excessive heating may also lead to poor joint properties in terms of strength and ductility. Furthermore, it may lead to the formation of voids and cracks in joints due to tensile residual stresses.

Therefore, considering the detrimental effects of steel fragments and heat generation, it is recommended that the tool pin is offset towards the Al side, keeping a small part of the pin on the steel side.

10.6 Al-Steel Lap Joint

Lap joining of thin sheet materials is widely used in industries, specifically in automobile industries. Lap joining of Al-steel is also of significant importance in the hybrid structure of thin sheets. Lap joints are generally fabricated if butt joint is either difficult to make or highly unreliable due to inherent stress raisers. In friction stir lap joining, one thin sheet is placed over another, and both the sheets are joined together. Lap joining of Al-steel experiences the similar issues as butt joining. However, there are several challenges encountered in lap joining due to different geometrical features. The lap joining of the Al-steel combination can be classified into two categories.

1. lap seam welding,
2. lap spot welding.

The welding procedure and weld structures are different in both cases. Therefore, both have been presented separately in the following section.

10.6.1 Al-Steel Lap Seam Friction Stir Welding

In lap seam friction stir welding of Al-steel, a continuous lap weld is developed along the weld line. The softer material is placed on the top, and the hard material is placed on the bottom. Generally, in Al-steel dissimilar welding, the thickness of the Al sheet is kept more than the steel sheet to balance/equalize the load-bearing capacity at both sides of the parent metals across the joint. The flowability and softening of Al are more as compared to steel; therefore, it is generally placed on the top to achieve the adequate stirring. Placement of hard material on the top causes excessive heating and improper joint formation. The tool pin is plunged into Al sheet, and only the bottom part of the pin slightly interacts with the steel plate.

There are three steps in the lap seam FSW of Al-steel, as shown in Fig. 10.12.

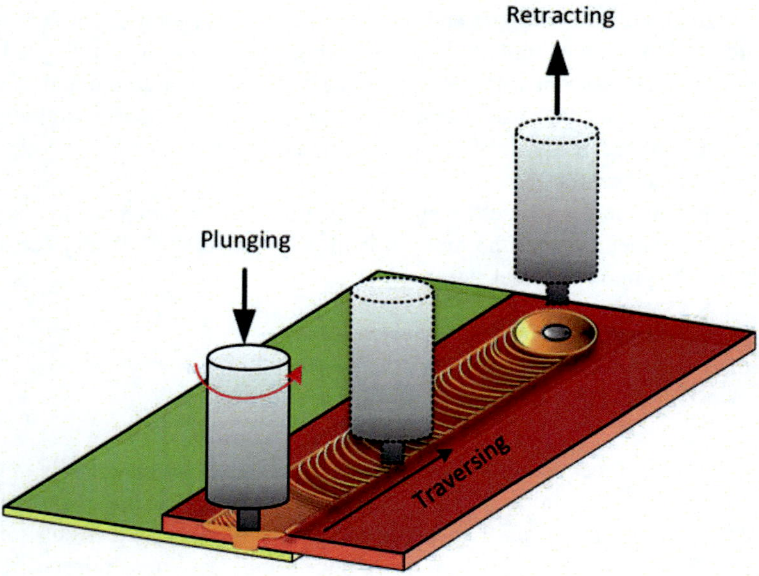

Fig. 10.12 Schematic of welding steps used in lap seam FSW

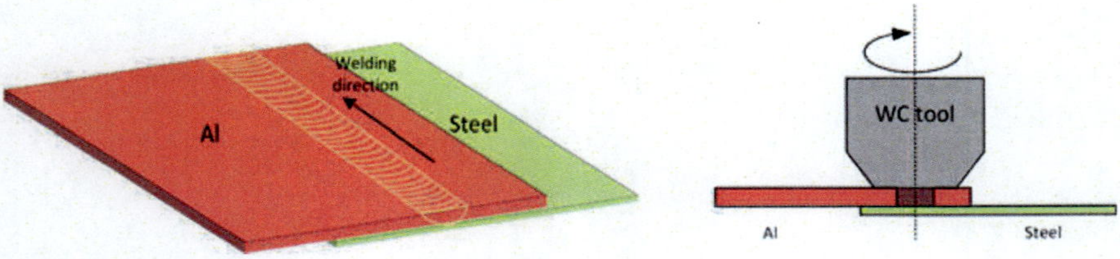

Fig. 10.13 Schematic of the weld joint and placement of the tool and workpiece plates

1. Plunging: In the initial step, the tool is rotated and plunged into the plates up to the required depth.
2. Traversing: After plunging, the tool or worktable is traversed longitudinally along the weld line.
3. Retracting: On completion of the weld up to the required length, the tool is retracted back to its home position.

The schematic of the joint formed and the placement of the tool with respect to plates is shown in Fig. 10.13.

The placement of the Al plate on the top enables a significant stirring in the Al plate compared to steel. The TMAZ is formed in an Al sheet adjacent to the stir zone. The steel sheet placed at the bottom has a very narrow heat-affected region underneath the rotating tool pin, as shown in Fig. 10.14. The weld formed in Al-steel lap welding has a very distinct morphology. On stirring action of the tool pin, the steel adjacent to the tool pin flows upwards and forms two unique features in the weld joint. These upward-flowing steel shows finger-like structures known as "hooks". The hooks near the periphery of the rotating tool pin distinguish the stir zone from the TMAZ region.

These hooks are very critical in joint performance and load-bearing capacity. These hooks in the weld structure indicate proper plunging and the active participation of steel plates in joint formation. The absence of these hooks signifies that only the upper plate is being stirred, and the bottom plate is not involved in joining. The size and shape of these hooks dictate the fracture location in the component under external loading. The absence of hooks generally leads to an interfacial failure of the joint. The presence of an oversized hook decreases the effective load-bearing cross-sectional area of the Al plate, which in turn leads to an overload failure of Al from the top of the hook.

The absence of these hooks, or very small size as well as oversized hooks, both are detrimental to the joint's strength (Kaushik and Dwivedi 2022b), as shown in

Fig. 10.14 Schematic of the cross-sectional view of the joint formed on Al-steel lap FSW

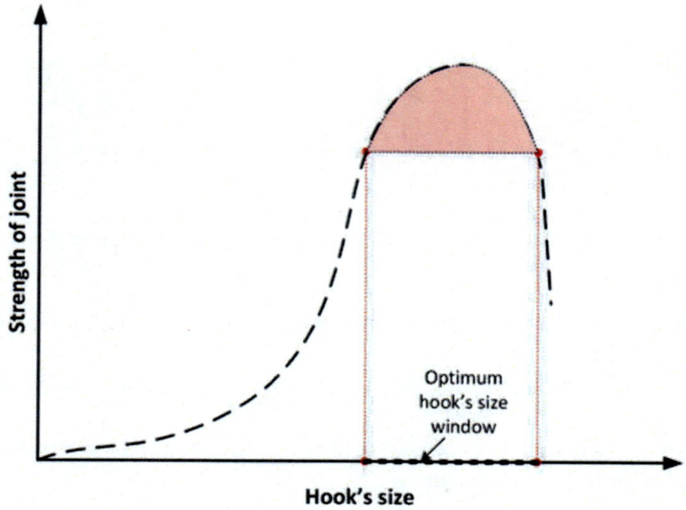

Fig. 10.15 Schematic diagram showing relationship of joint strength versus hook size

Fig. 10.15. A joint with an optimum-sized hook, high strength and ductility before fracture performs better in elongation. Thus to have an optimum-sized hook, the plunging of the tool should be optimized.

10.6.2 Al-Steel Lap Spot Friction Stir Welding

The second variant of lap joints in FSW is the spot welding. This method is used for joining thin sheets at specific locations. These joints are used where a continuous seam and leak-proof joints are not required. These joints are relatively easier to fabricate, and the processing time is significantly less. Spot joining is vastly used in automobile thin sheet joining. Easy automation, less processing time and high reliability make this process suitable for industrial applications.

Friction stir lap spot welding is a suitable alternative for resistance spot welding, especially for dissimilar metal joining. In resistance spot welding of dissimilar Al-steel, the large difference in electrical and thermal conductivity of Al and steel metal poses many problems. Such issues are not present in FS spot joining, where flow properties and interlocking phenomena play important role in the mechanisms for joint formation. In friction stir spot welding (FSSW), both plates are placed in overlapping position, and the tool stirs the material to form the spot joint, as shown in Fig. 10.16.

The steps used in Al-steel FSSW are.

1. plunging,
2. dwell,
3. retracting.

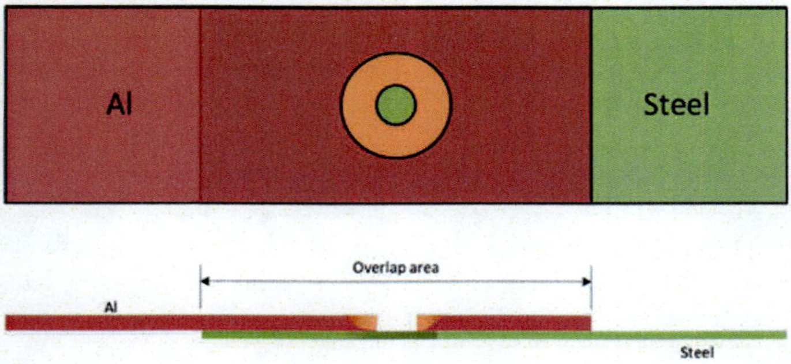

Fig. 10.16 Schematic of friction stir lap spot joint

In the first step, the rotating tool pin is plunged into the upper sheet and is allowed to slightly penetrate up to the bottom sheet. Then in the dwell step, the tool rotates at its position for a certain period to ensure proper stirring action for good interlocking. After desirable stirring action, the tool is retracted back to its home position. The complete cycle takes place in less than 30 s. Important to note is that there is no traversing of tool/workpiece.

The cross section of Al-steel stir spot weld is shown in Fig. 10.17. The Al metal is placed on top, and the steel sheet is placed at the bottom. The most prominent feature of the weld is the formation of the exit hole due to the retraction of the tool pin. The conventional tool pin generally leaves an unfilled hole in the friction stir spot weld joint. The present hole is the weakest area in FSSW joint, and the weld generally fails from the exit hole. Moreover, the unfilled cavity deteriorates the aesthetics of the welded sheets, which is usually of utmost importance in automobile manufacturing industries.

There are multiple approaches in trend to evade the problem of exit holes in friction stir spot weld joint. The most common one is to re-stir the already formed joint using a pin-less tool having shoulder only. In this method, the pin-less tool stirs the Al metal surrounding the exit hole and fills up the hole to a significant level. Another technique is to use a refill friction stir spot welding setup (Schmal and Meschut 2020) with a specially designed tool with a sleeve and clamping ring, as shown in Fig. 10.18.

The refilled stir joint is found to be superior in load-bearing properties with improved aesthetics, as shown in Fig. 10.19.

Fig. 10.17 Cross-sectional view of Al-steel friction stir spot weld joint

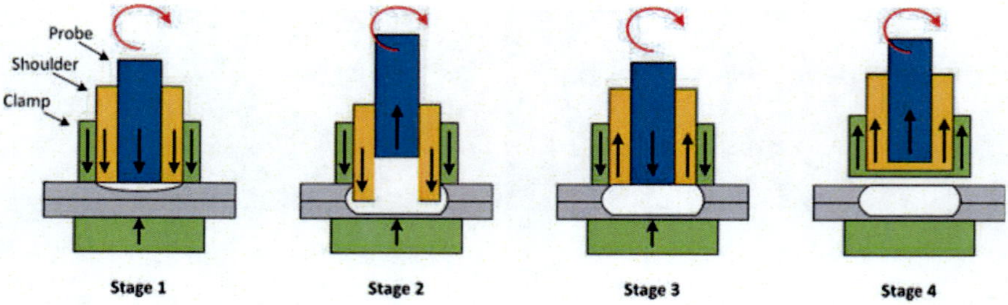

Fig. 10.18 Schematic representation of refill friction stir spot welding process

Fig. 10.19 Image of refill friction stir spot weld joint

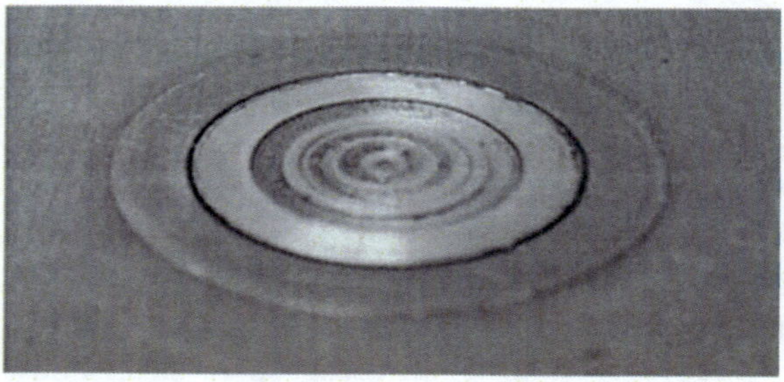

10.7 Summary

Joining the Al-steel combination using the friction stir welding process helps avoid the melting and solidification of parent metals and, in turn, evades the difficulties associated with the joining of this dissimilar metal combination. The amount of heat generation in the FSW process is dependent on the process parameters used during welding, and those can be controlled accordingly for optimum heating. Fabrication of sound and defect-free weld joint is dependent on the process parameter optimization in Al-steel butt joining. The discussion of process parameters and their effects on weld properties is helpful for a better understanding of the mechanism of the joint formation. Classification of joints depending upon the configuration of metal plates is essential to discuss them discretely. The lap joining of Al-steel poses a different set of challenges than butt joining. Therefore, optimizing plunging and hook size is critical in lap joining, whereas refilling the exit hole is of utmost importance in lap spot welding of the Al-steel.

References

Kaushik P, Dwivedi DK (2020) Effect of tool geometry in dissimilar Al-Steel friction stir welding. J Manuf Process. https://doi.org/10.1016/j.jmapro.2020.08.007

Kaushik P, Dwivedi DK (2021) Induction preheating in FSW of Al-Steel combination. Mater Today Proc 1–5. https://doi.org/10.1016/j.matpr.2021.01.438

Kaushik P, Dwivedi DK (2022a) Al-steel dissimilar joining: challenges and opportunities. Mater Today Proc. https://doi.org/10.1016/j.matpr.2022.05.211

Kaushik P, Dwivedi DK (2022b) Influence of hook geometry in failure mechanism of Al6061-Galvanized Steel dissimilar FSW lap joint. Arch Civ Mech Eng

Kundu S, Roy D, Bhola R, Bhattacharjee D, Mishra B, Chatterjee S (2013) Microstructure and tensile strength of friction stir welded joints between interstitial free steel and commercially pure aluminium. Mater Des 50:370–375. https://doi.org/10.1016/j.matdes.2013.02.017

Kusuda Y (2013) Honda develops robotized FSW technology to weld steel and aluminum and applied it to a mass-production vehicle. Ind Rob 40:208–212. https://doi.org/10.1108/01439911311309889

Schmal C, Meschut G (2020) Refill friction stir spot and resistance spot welding of aluminium joints with large total sheet thicknesses (III-1965-19). Weld World 64:1471–1480. https://doi.org/10.1007/s40194-020-00922-2

Schmidt H, Hattel J, Wert J (2004) An analytical model for the heat generation in friction stir welding. Model Simul Mater Sci Eng 12:143–157. https://doi.org/10.1088/0965-0393/12/1/013

Chapter 11
Adhesive Joining of Dissimilar Metals

This chapter presents the fundamentals of adhesive joining, applications, advantages and limitation of adhesive joints and factors affecting the joint performance. The role of different types of adhesives, joint design and procedural steps of adhesive joining like surface preparation and curing on adhesive joint strength have been elaborated using suitable schematics. A section has been dedicated to the adhesive joining with specific reference to dissimilar metal joints.

11.1 Introduction

The adhesive joining is a very commonly used technique for joining of variety of materials including similar and dissimilar material in automotive, construction, electrical appliances, etc. The joint is developed by putting in a non-metallic adhesive material at interface between components to be joined called adherends which on hardening either at room temperature or little high temperature after curing produces reasonably good strength (Fig. 11.1). The adhesive joining is a non-metallurgical joining. Similar to brazing and soldering, there is no melting of the faying surfaces during adhesive joining, though little heat may be applied to facilitate fast curing and is sometimes used to ensure sound and defect-free joint.

11.2 Developing Good Adhesive Joint

The development of a sound adhesive joint needs (a) good wetting of adhesive with faying surfaces of adherends followed by curing/hardening, (b) good interaction of adhesive with faying surfaces and good strength of adhesive itself for desired mechanical performance, (c) very clean surfaces free from dust dirt, oil grease, paint, (d) joint free from gaseous packets/bubble from the interface and (e) properly designed

© The Author(s), under exclusive license to Springer Nature Singapore Pte Ltd. 2023
D. K. Dwivedi, *Dissimilar Metal Joining*,
https://doi.org/10.1007/978-981-99-1897-3_11

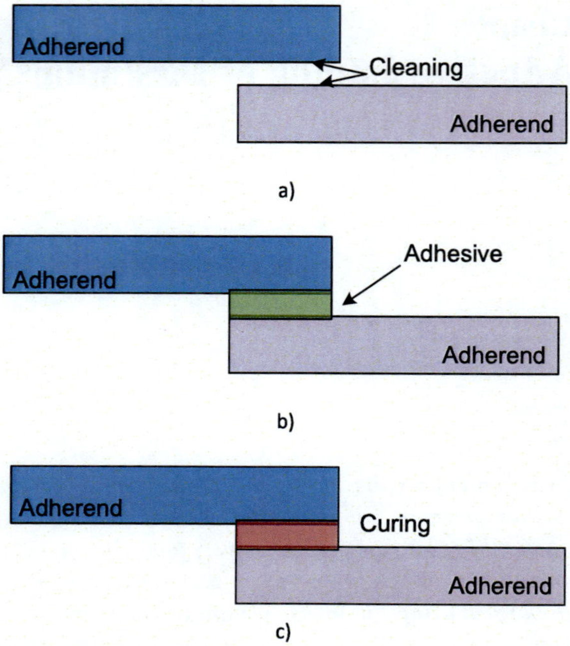

Fig. 11.1 Schematic showing sequential steps of the adhesive joining namely **a** surface cleaning, **b** adhesive application at the faying surfaces and **c** curing

joint with reasonably high bonding area and minimum tendency of failure under peeling and cleavage load conditions.

11.3 Wetting in Adhesive Joining

Success of the adhesive joining significantly depends on the proper wetting and spreading of the adhesive at the faying surfaces, which in turn depends on relative free surface energies of the parent metals and adhesive. The surface energy of faying surfaces of metals to be joined must be greater than polymeric adhesive for proper wetting (Fig. 11.2). Surface energy of metals, in general, reduces in presence of adsorbed contaminants and impurities; therefore, metallic surfaces must be properly cleaned using mechanical and chemical methods. Further, roughening of adherends may deteriorate the wetting and flow of adhesive but can facilitate the mechanical interlocking to improve the joint strength (Fig. 11.3). The low viscosity adhesive spreads better, improves the wetting with adherends and lowers the gas pocket and bubble entrapment tendency.

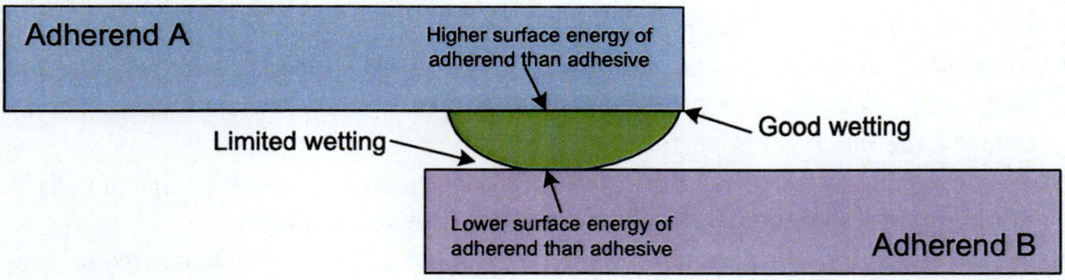

Fig. 11.2 Schematic showing differential wetting behaviour due to higher free surface engineering of adhesive than that of adherend B and lower than that adherend A

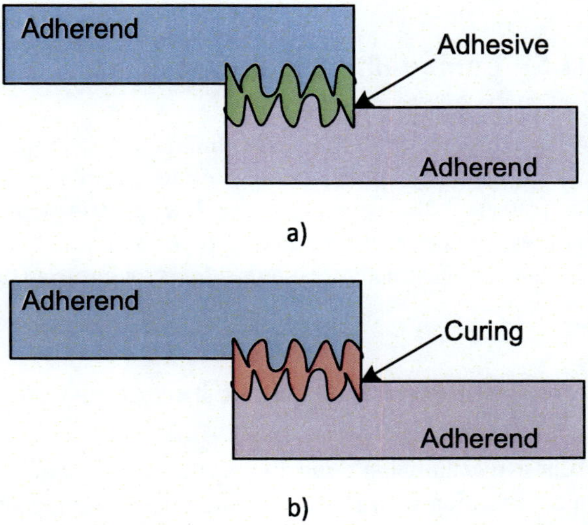

Fig. 11.3 Schematic showing surface roughness leading to **a** filling in of adhesive into the valleys during adhesive application and **b** mechanical interlocking due to roughness on curing

11.4 Adhesive Joining Offers Multiple Advantages Over Metallic Joining

- Adhesive joining allows joining of both dissimilar metals and metal and non-metal combination.
- Electrical isolation by non-conducting adhesive in these joints reduces the galvanic corrosion tendency of dissimilar metal joints. Thermal and electrical insulations are useful for electronic and electrical appliances. Even electrical conducting adhesive joints can be developed using carbon fillers.
- Joining of thin sheets and multiple layers of thin sheets for fabricating sandwich panels.
- Low heating temperature (65–176 °C) for curing of adhesive joints than soldering/brazing (> 220 °C) and in many cases even room temperature curing is performed. Adhesive joints are capable to withstand at moderately high temperature (76–150 °C) depending upon the type of adhesives.

- Leak proof/water-tight joining is offered by adhesion bonding.
- Adhesive joining offers comparatively more uniform stress distribution over the large area coupled with reduced stress, and less stress concentration in turn improve the mechanical performance.
- As compared to weld joints, and riveted joints, adhesive offers advantage of weight saving which is of special interest for automotive applications.
- Sound and vibration damping characteristics of adhesive allow improved life, fatigue performance and comfort.
- Adhesive remains between the adherends, and therefore, it does not affect appearance.

11.5 Limitations of Adhesive Joining

Adhesive joints suffer from many limitations like low joint strength against peel and cleavage loading, limited service life under hostile service condition of high humidity (> 70%) and little high temperature, low operating temperature (max. 260 °C), needs fixtures, furnace and long curing time to realize the desired strength, absence of suitable non-destructive testing methods for ensuring desired quality and soundness.

11.6 Adhesive and Joint Characteristics

Adhesive comprising a suitable type of resin (thermos-setting and thermos-plastic) in combination with elastomers and fillers used for joining determines mechanical performance and thermal stability of the adhesive joint. Thermo-setting and thermos-plastics show significantly different mechanical behaviour against external load in terms of the load carrying capacity and toughness (Fig. 11.4a). Thermo-setting (TS) resins harden/cure through chemical reactions. These reactions under the influence of heat (as per type of resin at moderate temperature) and radiation are accelerated. The hardening of the TS resin is a unidirectional and irreversible process, therefore; these cannot be further hardened/softened via heating/cooling. Thermos-plastics (TP) are long chain molecular compounds, which do not go through chemical reactions during heating, and cooling therefore these can be hardened/softened through controlled heating and cooling as per need multiple times. However, thermo-plastics decompose and oxidize at high temperature. Curing of the thermo-setting involves chemical reaction and cross-linking while that of thermo-plastic occurs in form of change of liquid- to solid-phase transformation.

Thermo-plastics resins also get softened in organic solvents and get hardened on evaporation of the solvents. Both thermo-plastics and thermo-setting resins can also be mixed suitability to improve the mechanical performance like peel strength of the joints. An addition of elastomers increases resilience, peel strength and resistance

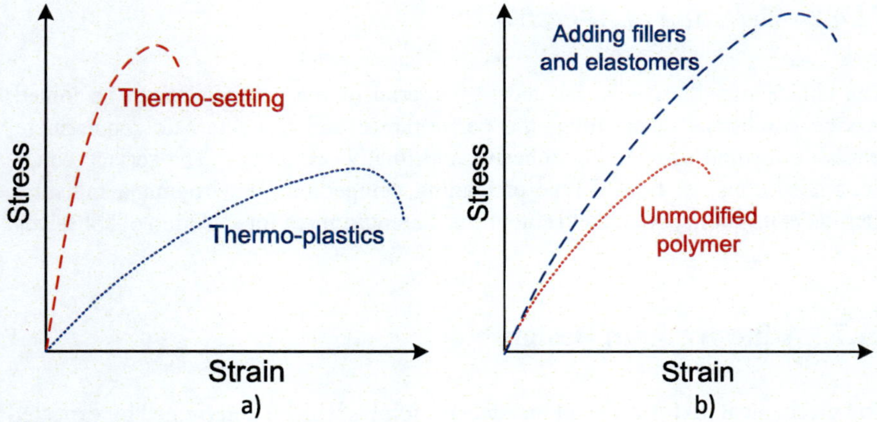

Fig. 11.4 Schematic showing **a** differential tensile behaviour of the thermos-setting and thermos-plastic in form of stress–strain diagram and **b** effect of adding fillers and elastomers

to shock and vibrations while fillers (suitable inorganic compounds) reduce thermal expansion coefficient and modulus of elasticity (Fig. 11.4b).

11.6.1 Adhesive

The adhesives are also categorized based on physical form (tape, liquid), chemical group (silicate, phenolic and epoxy), base bonding material (paper, plastic, metal) and method of application (spraying, hot melt). Thermos-plastics (polyester, polyamide, ethylene and vinyl acetate) are applied in hot molten state at the faying surfaces of components to be joined, and later on, cooling forms a joint but these are found unsuitable for structural applications. Pressure-sensitive adhesive suits for low load conditions only. These gain strengths on applying pressure and are found in form of the transfer tape, i.e. single/double-side tape. Chemical reaction-type adhesives are basically thermo-setting resins which are found suitable for structural applications. These are available in the form of solid, liquid, tape and films. Thermo-setting resins (epoxy, phenolic, acrylic and anaerobic) are activated by hardener and heat and offer strength around 20–30 MPa. The following factors need to consider while choosing suitable adhesive (a) availability of adhesive in desired form, (b) desired method of application of adhesive, (c) joint design, (d) desired production rate and (e) requirement of tooling, fixtures and equipment. Liquid adhesive can be applied using roller, brushes, spraying and dipping while solid adhesive uses heat to melt the adhesive in form of rods and powder. Films are applied using rollers.

11.6.2 Selection of Adhesive

The choice of adhesive is dictated by material of the components to be joined, service conditions determining the performance and characteristic requirement, available method of applying adhesive and finally economics. The service conditions need consideration of type of loading, temperature, environment, moisture, thermal expansion, toxicity and colour match requirement for selection of adhesives.

11.7 Adhesive Joint Design

The mechanical performance of an adhesive joint is heavily determined by expected failure mode as per the weak link location in connection/joint (Fig. 11.5). The failure can occur from adhesive, adhesive-adherend interface(s) accordingly, and these are termed as a cohesive and adhesive failure, respectively. Additionally, failures of adhesive joints can also occur from the weak adherends or parent materials (in dissimilar materials joining), or failure can be combined type wherein cracks follow zigzag path between adhesive, adhesive-adherend interface, and parent materials. In case of cohesive failure, the strength of adhesive itself will dictate the joint strength under optimal joining and curing conditions. The strength of adhesive joint in case of adhesive-adherend interface failure depends on surface roughness, cleanliness/cleaning methods, curing time and temperature, and interaction between the adhesive and adherends in forms of chemical and covalent bond formation, if any.

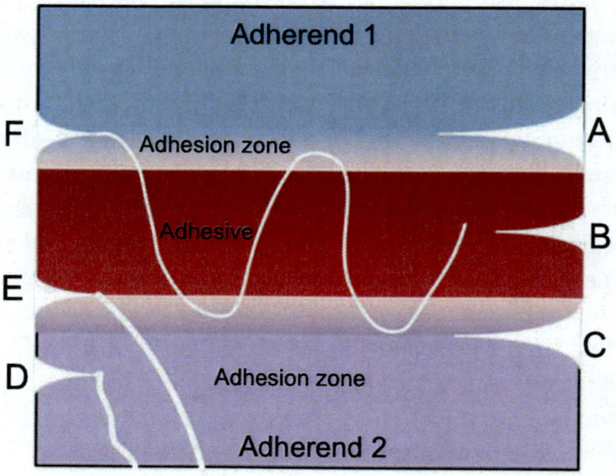

Fig. 11.5 Schematic showing various modes of failure namely adhesive failure (A, C), cohesive failure (B), adherend failure (D), failure triggered from adhesive-adherend interface then crack propagates through adherend (E), failure triggered from adhesive-adherend interface then crack propagates in zigzag manner through adhesive

11.7.1 Common Type of Loading on Adhesive Joints

The primary focus of adhesive joint design is to increase the joint strength by controlling the adhesive failure; accordingly, overlap length, surface conditions and curing for adhesive joining are optimized. The adhesive joint design significantly affects the load carrying capacity (under tensile, shear, peeling and cleavage loading) and allowable plastic strain (Fig. 11.6). Adhesive may experience plastic strain/creep under external load/fatigue conditions. A good adhesive joint design helps to achieve the desired load carrying capacity economically using suitable formulation of adhesive, and simplified procedures needing minimum quality checks to ensure desired quality and performance.

11.7.2 Overlap Length

Adhesive joints offer higher load carrying capacity against tensile and shear than peeling and cleavage loading conditions. Therefore, adhesive joint is designed in such a way that the service load causes tensile/shear loading. The adhesive joint area (attained using suitable overlap length) must be large enough to ensure that the (a) static service load/stress remains within the plastic strain limit of the adhesive and (b) fatigue service load/stress remains within the limit to minimize the creep of adhesives.

The load carrying capacity of adhesive joint is significantly affected by overlap length of adhesive joint, besides type of adhesive and type of load (Fig. 11.7). An increase in overlap length increases the area over which load is distributed; however, increasing overlap length compromises with uniformity of stress distribution. An increasing overlap length initially increases the tensile shear load carrying capacity of the joint; then, after reaching maxima it stabilizes. The stabilization of load carrying capacity despite of increasing overlap length and bonding area is attributed to increased possibility of peeling and cleavage loading due to high stress concentrations. The joint efficiency, therefore, depends on the parent metal properties, adhesive strength, thickness of adhesive, type of loading and service environment (Faseeulla and Dwivedi 2012a, b; Mittal and Dwivedi 2012).

11.7.3 Overlap Length and Stress Distribution in Adhesive Lap Joints

A short overlap length of an adhesive joint causes high shear stress so the plastic straining of adhesive results in low joint strength as failure is triggered from the adhesive near edge of the joint. An increase of overlap length decreases the plastic strain and improves the joint strength. Presence of high tensile/shear stress at the end

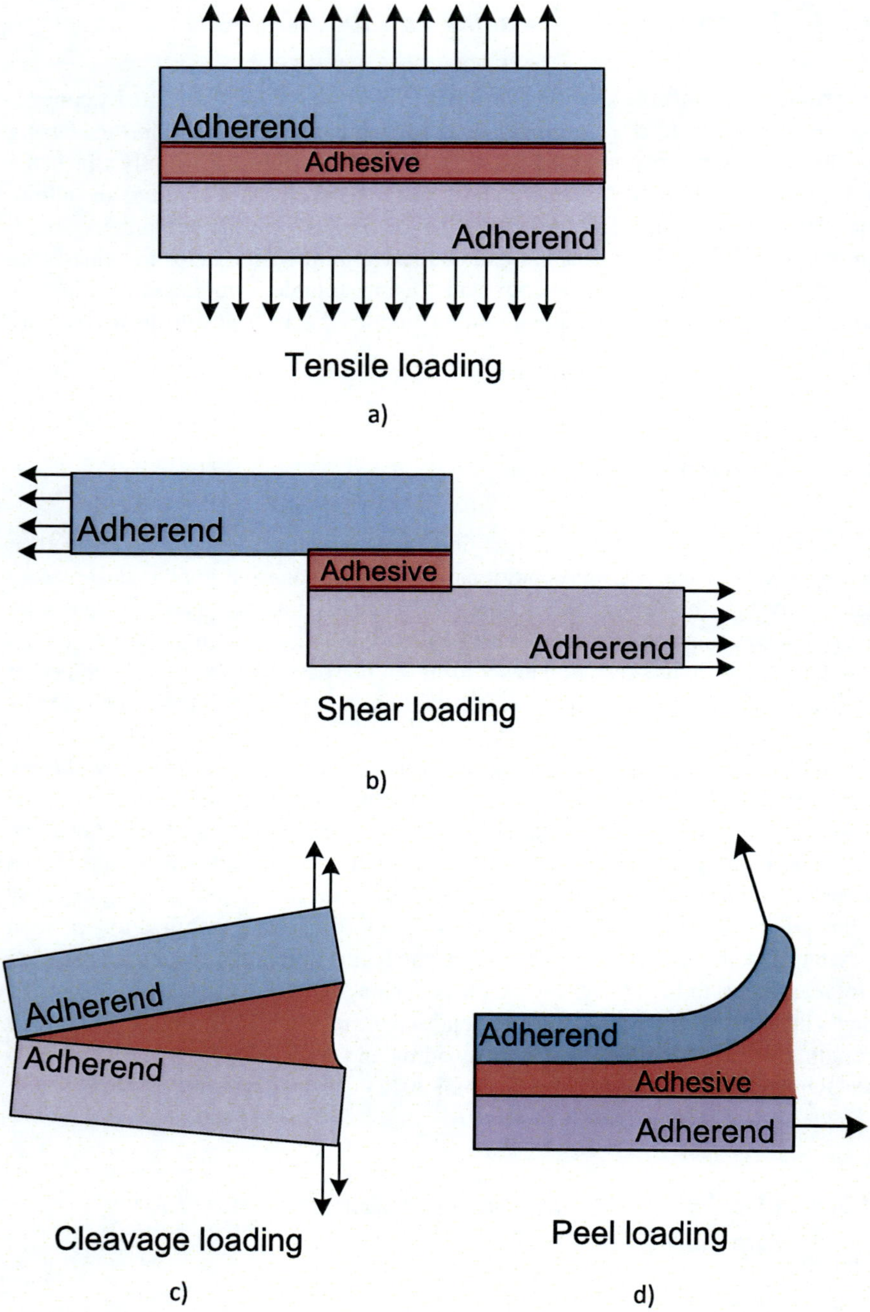

Fig. 11.6 Schematic showing common types of loading experienced by adhesive joints **a** tensile loading, **b** shear loading, **c** cleavage loading and **d** peel loading

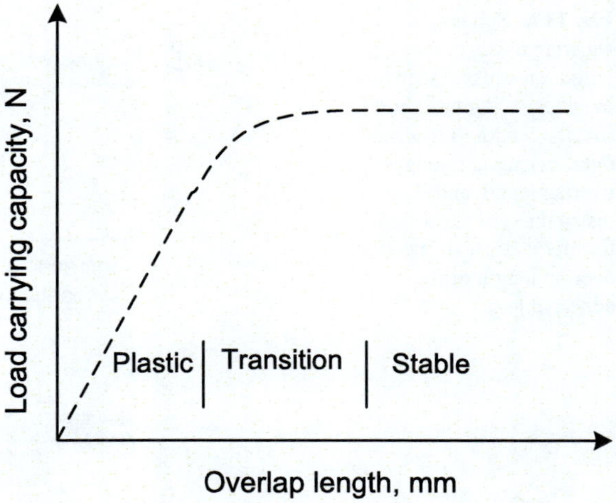

Fig. 11.7 Schematic showing effect of overlap length on load carrying capacity of adhesive joint

of lap joints can trigger the failure; therefore, joint design must take care of such stress distribution (Fig. 11.8).

In real life, adhesive joints hardly experience the single type of designed load (as per design) as they frequently come across the peeling and cleavage loading which complicates and deteriorates the joint performance as peeling/cleavage load carrying capacity of adhesive joints is only a fraction of their tensile/shear load carrying capacity. For example, a simple lap adhesive joint under tensile shear loading tends to get aligned by bending of the joints which imposes the peeling and cleavage loading at edges of the joint (Fig. 11.9).

11.7.4 Common Design Joint Designs

Many adhesive joint designs have been developed with consideration of reducing stress concentration at the edges of the joint, reducing the bending of joint and increasing bonding area to enhance joint strength. Schematics of common adhesive joint designs along with expected performance in the light of stress concentration and load carrying capacity have been mentioned as poor, fair, good, very good and excellent (Fig. 11.10).

11.8 Mechanisms Responsible for Mechanical Performance of Adhesive Joints

Many attractive forces (namely adsorptive force, electrostatic force, diffusion force besides mechanical interlocking) acting at adhesive and adherends interface govern

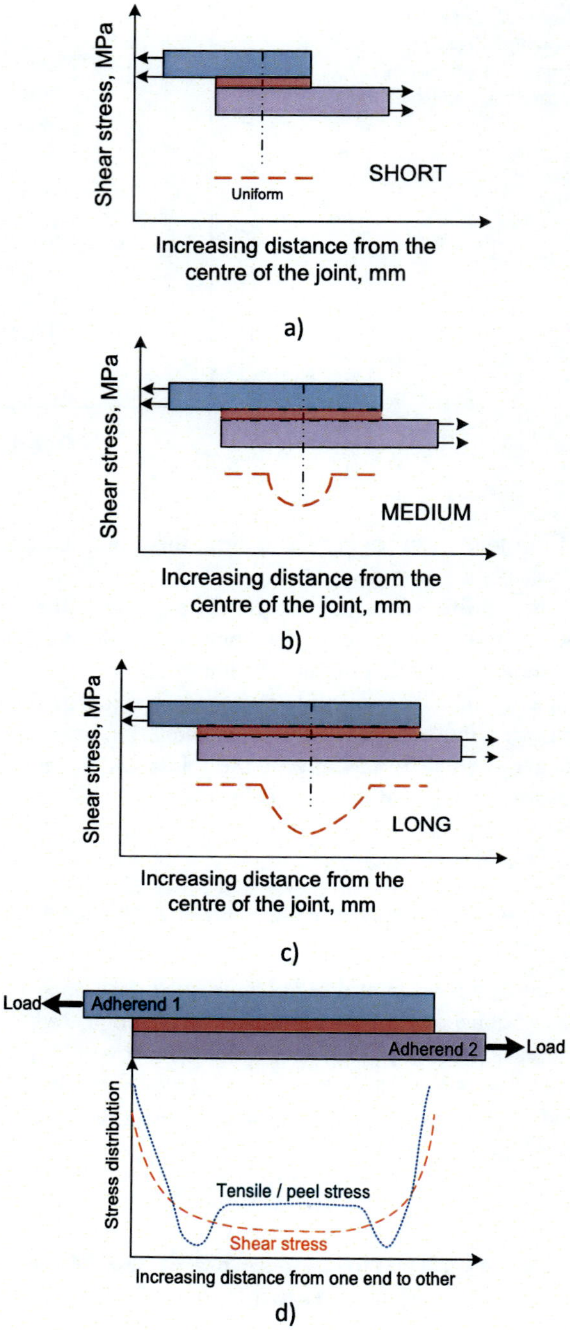

Fig. 11.8 Schematic showing effect of overlap length on stress distribution in adhesive joints **a** short overlap length, **b** medium overlap length, **c** long overlap length and **d** distribution of shear and tensile/peel stress in entire overlap length of the adhesive joint

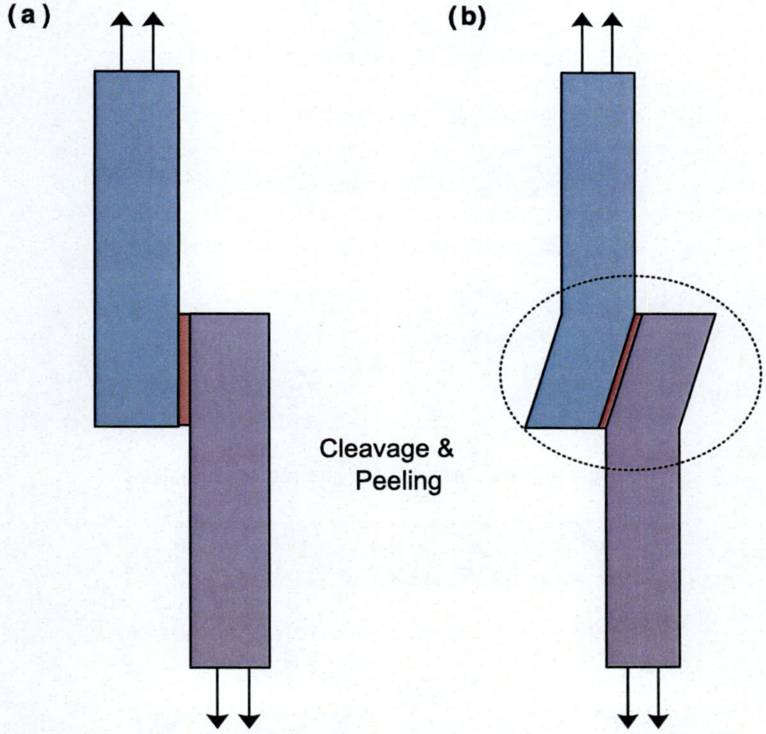

Fig. 11.9 Schematic showing effect of tensile loading on single lap adhesive joint **a** shear loading and **b** peel and cleavage loading owing to rotational bending of the joint

the load carrying capacity of the adhesive joints. Adsorptive forces act due to very close intimate contact between adhesive and substrate surfaces in form of van der Waals force, covalent bond formation and dipolar force. The electrostatic force occurs due to attraction between oppositely charged molecular and ionic bonding. The diffusion force observed due to molecular chain entanglement between adhesive and substrate as adhesive diffuses across the interface (Fig. 11.11). The presence of macro- and micro-level surface irregularities allow entry of adhesive in valleys, which on hardening effectively resists the shear loading through mechanical interlocking. Therefore, roughening of surfaces, etching and chemical cleaning help to improve the joint efficiency.

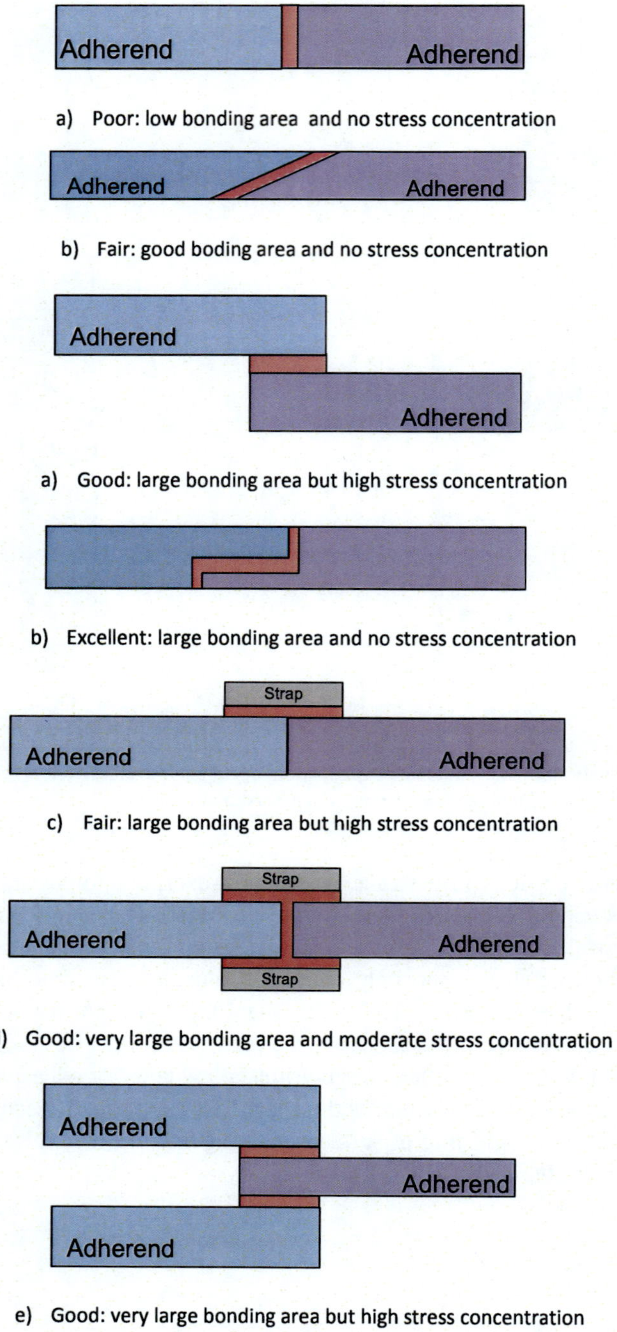

Fig. 11.10 Schematic of common designs of adhesive joint **a** butt joint, **b** scarf joint, **c** single lap joint, **d** double butt single lap joint, **e** single strap lap joint, **f** double strap lap joint **g** double lap joint, **h** bevelled lap joint, **i** bevelled double strap lap joint, **j** joggle lap joint, **k** tongue joint, **l** tongue and scarf and **m** scarf joint

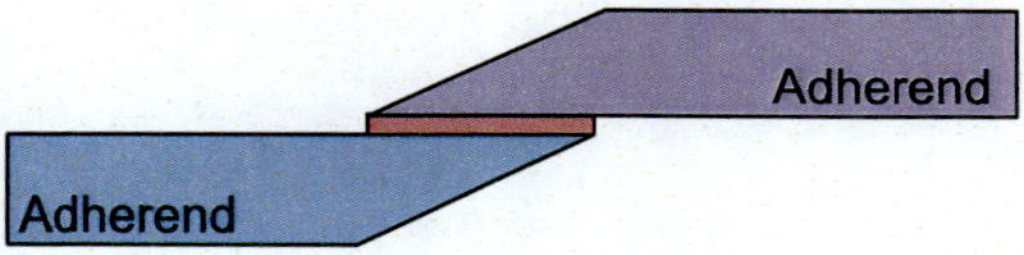

f) Very good: very large bonding area and low stress concentration

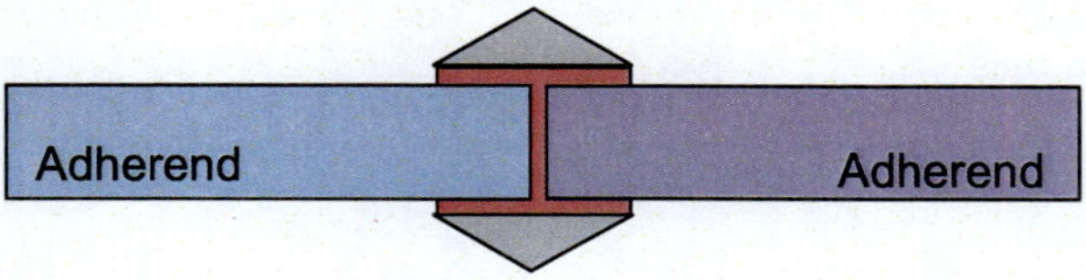

g) Excellent: very large bonding area and very less stress concentration

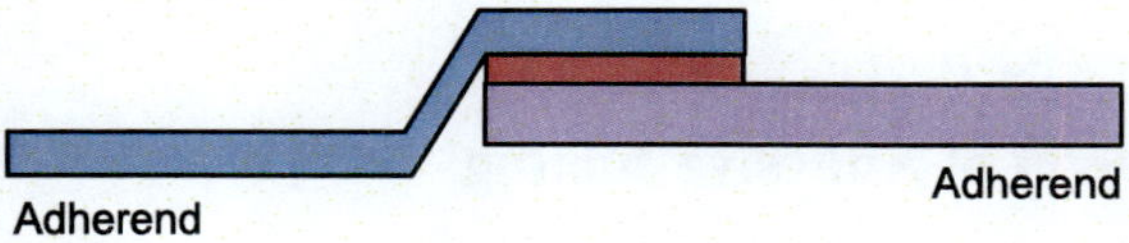

h) Good: large bonding area and moderate stress concentration

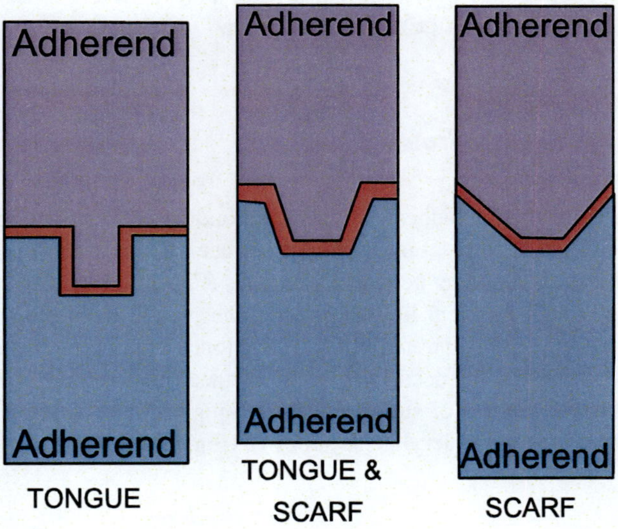

k) l)_ m)

(large bonding area and no stress concentration with k, l, m)

Fig. 11.10 (continued)

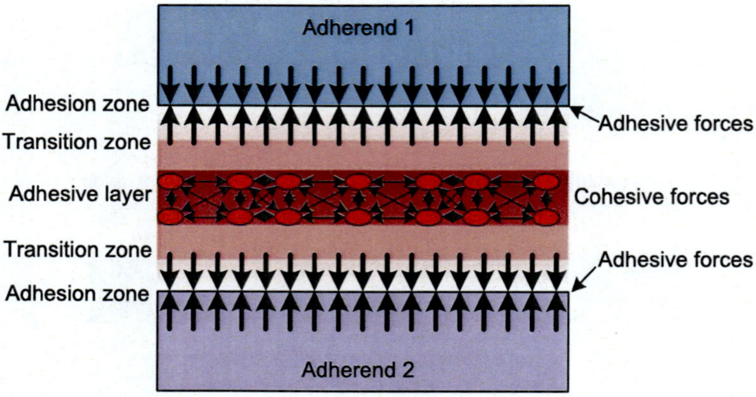

Fig. 11.11 Schematic showing different zones formed in typical adhesive joint and common active bonding forces

11.9 Parameter of Adhesive Joining

The strength of adhesive joints is influenced by many parameters of adhesive joining including surface preparation (cleanliness, methods of cleaning and surface roughness), type of adhesive, thickness of adhesive at joint interface, soundness, curing (time and temperature condition) and joint design.

11.9.1 Surface Cleanliness

Strength of adhesive joint (adhesive-adherend bonding) depends upon the quality of adhesive-adherends interaction. A clean surface of adherends results in good interaction and so the bonding between adherends (Fig. 11.12). Clean surface helps in developing strong chemical links between adhesive/adherent and so good joint strength. For example, shear strength of adhesive joints of polypropylene-reinforced glass fibre is strongly influenced by type of surface treatment as maximum joint strength is obtained in case of surface cleaned using trichloroethylene plus primer treatment while using cyanoacrylate as adhesive (Fig. 11.12b) (Reis et al. 2012).

11.9.2 Surface Roughness

The surface roughness of the adherends is another important aspect, which needs to be optimized as it affects the joint strength due to two factors (a) mechanical interlocking between adhesive and adherends and (b) possibility of formation of unbonded zones (Fig. 11.13. A little rough surface of adherends results in good

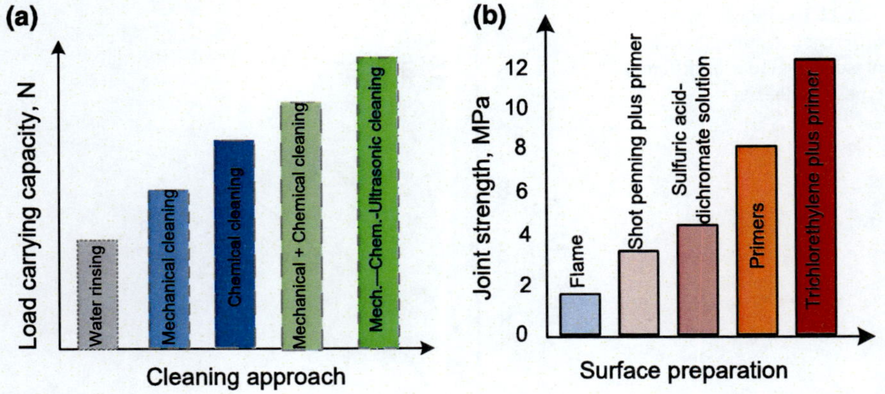

Fig. 11.12 Schematic showing effect of cleaning method on load carrying capacity of the adhesive **a** adhesive joining of dissimilar metals and **b** adhesive joining of polymers

joint strength due to dominance of first factor, i.e. mechanical interlocking. Too high surface roughness makes wetting of adherends and filling all asperities (peaks and valley) by adhesive difficult, which in turn leads to formation of unbonded regions.

Fig. 11.13 Schematic showing effect of surface roughness of adherends on **a** load carrying capacity of the adhesive joints and **b** possibility of unbonded adhesive zone formation in case of high surface roughness

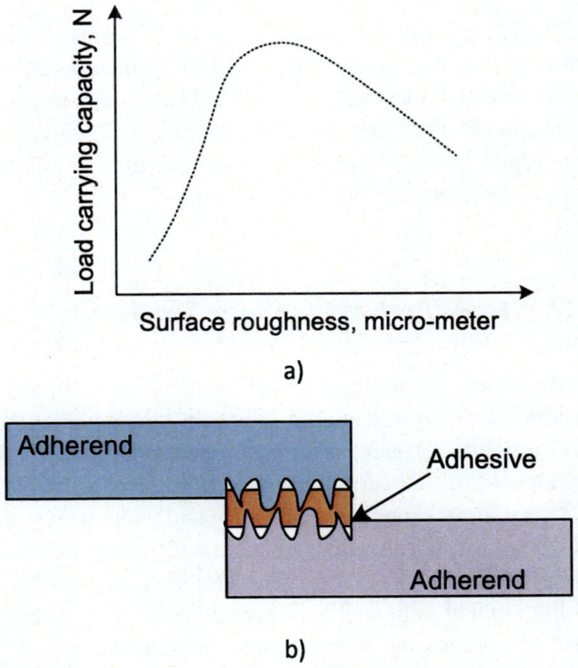

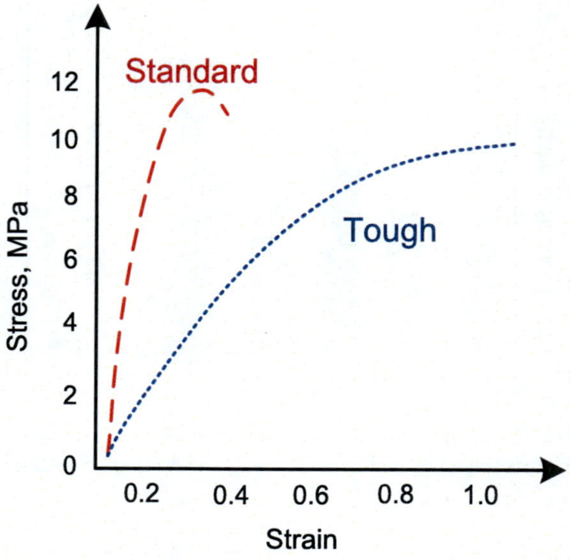

Fig. 11.14 Plot showing difference in strength and plastic behaviour of different types of adhesives

11.9.3 Type of Adhesive

Effect of type of adhesive on joint strength depends on multiple factors like the formation of chemical/covalent bond, diffusion/reaction layer due to interaction between adhesive and adherends, and related hardening, physical and chemical changes. As described earlier, thermo-setting adhesives offer higher strength than thermo-plastics in terms of mechanical properties (on curing), i.e. allowed strength and plastic strain prior to the failure (Fig. 11.14).

11.9.4 Adhesive Bond Line Thickness

Thickness of adhesive layer significantly affects the failure mode (adhesive/cohesive/mixed mode) which in turn governs the joint strength appreciably. A very thick adhesive layer encourages cohesive failure; therefore, joint strength in such cases usually corresponds to strength of adhesive only. However, the presence of very limited amount of adhesive and under heavy pressure conditions may result in very thin adhesive layer coupled with many unbonded regions leading to reduced joint strength (Fig. 11.15a). Therefore, an optimum thickness of adhesive layer with fully bonded joint results in high adhesive joint strength. Further, in case of sound adhesive joints, an increasing thickness of adhesive (beyond optimum level) reduces the joint strength due to increasing tendency of the cohesive failure of adhesive (Fig. 11.15b).

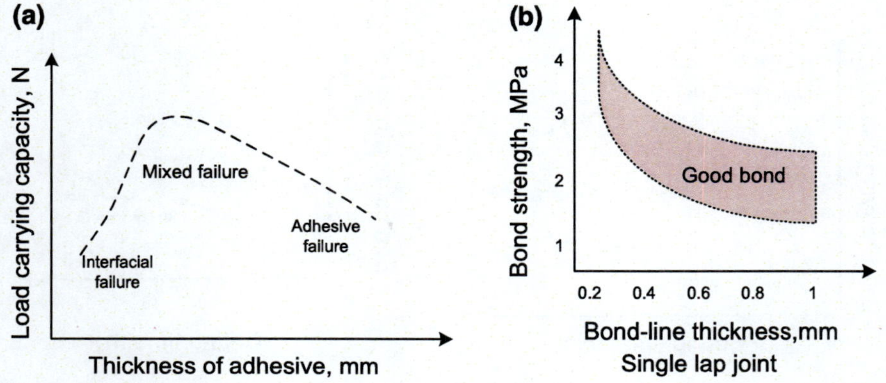

Fig. 11.15 Schematic showing effect of adhesive thickness layer on joint **a** load carrying capacity due to varying failure modes and **b** adhesive joint strength as function of bond line adhesive thickness when there is no unbonded region

11.9.5 Curing Time and Temperature

Adhesive joint gains its strength after hardening of adhesive. Hardening of the adhesive occurs during curing. The curing of thermo-setting adhesives occurs differently from thermos-plastic adhesives. Curing of the thermo-plastics (TP) primarily involves phase transformation from liquid to solid without any chemical reactions while that thermo-setting occurs through chemical reaction and cross-linking. Curing is temperature and time-dependent phenomenon. The high temperature accelerates the curing. The high-temperature exposure for curing must be done for an optimum time to avoid under-curing or overcuring (Fig. 11.16a, b). Under-curing results in limited hardening while overcuring causes thermal damage of adhesive. Therefore, under/overcuring both compromise the adhesive joint strength. Curing temperature and time therefore considering adhesives and adherends should be established as a part of adhesive joining procedure (Fig. 11.16c) (Mittal and Dwivedi 2012, 2013; Faseeulla Khan et al.; Faseeulla Khan et al. 2010).

11.10 Dissimilar Metal Joining

The adhesive joining results in a non-metallurgical joint between two similar/dissimilar metals or metal and non-metals as there is no metallurgical interaction between the components to be joined. Therefore, ease of adhesive joining of dissimilar metals may not be much different from similar metal joining except the difference in terms of joint strength and minor residual stress development. The adhesive joint strength might be affected by the parent materials of the component to the extent it affects (a) the covalent bond formation between surface of component to

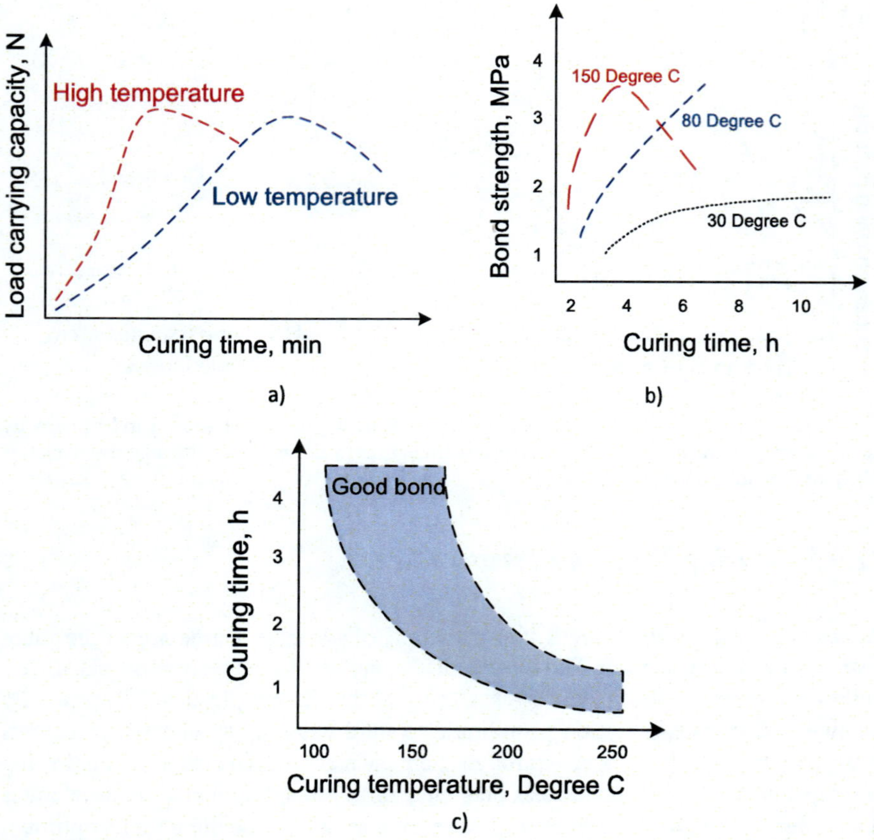

Fig. 11.16 Schematic showing effect of curing conditions **a** and **b** load carrying capacity of adhesive joint as function of curing temperature and time and **c** need of identifying window of curing temperature and time for good adhesive joint strength

be joined and adhesive and (b) diffusion of elements across the interfaces. The development of internal/residual stress is also affected by parent materials depending upon the mismatch/difference in thermal expansion coefficient of adhesive, parent material 1 and parent material 2 (Fig. 11.17). Residual stress may arise from contraction of the liquid adhesive during curing either due to liquid- to solid-phase transformation or differential expansion and contraction of adherends during artificial heating/cooling applied for curing (Fig. 11.18).

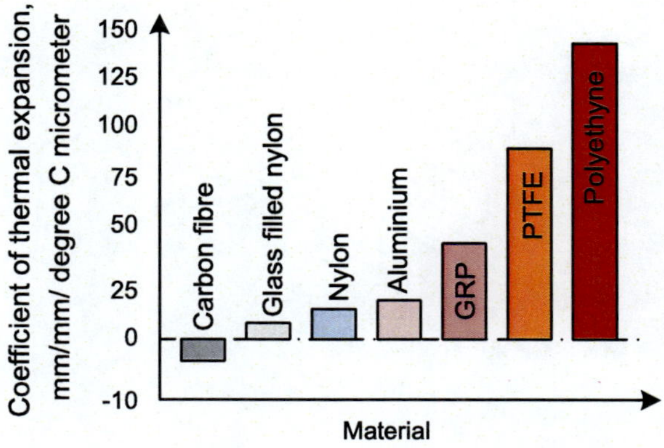

Fig. 11.17 Bar chart showing thermal expansion coefficient of the common engineering adherends subjected to adhesive joining

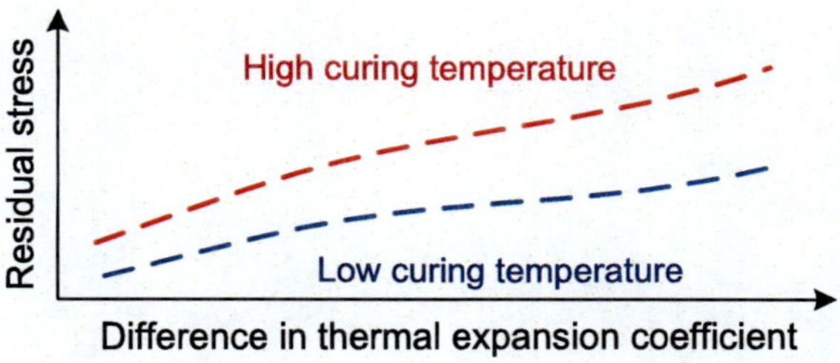

Fig. 11.18 Schematic showing effect of difference in thermal expansion coefficient of adherends subjected to adhesive joining and curing temperature on residual stress at joint interface

References

Faseeulla Khan MD, Dwivedi DK (2012a) Development of response surface model for tensile shear strength of weld-bonds of aluminium alloy 6061 T651. Mater Design 34:673–678

Faseeulla Khan MD, Dwivedi DK (2012b) Mechanical and metallurgical behaviour of weld-bonds of 6061 aluminium alloy. Mater Manuf Process 27(6):670–675

Faseeulla Khan MD, Dwivedi DK, Ghosh PK (2010) Studies on the effect of process parameters on the shear performance of weld-bonds of aluminium alloy. In: Proceedings of 36th international MATADOR conference held in Manchester 14th–16th July 2010

Faseeulla Khan MD, Sharma G, Dwivedi DK, Weld-bonding of 6062 Aluminium Alloy. Int J Adv Manuf Technol 78(5–8):863–873

Mittal M, Dwivedi DK (2012) Statistical analysis of influence of input process parameters on characteristics of weld-bonds of Al 5052 H32 alloy using Box-Behnken Design (BBD). Proc Instit Mech Eng Part B, J Eng Manuf 226(6):1001–1017

Mittal M, Dwivedi DK (2013) Studies on fatigue behavior of weld-bonds of Al–Mn–Mg alloys. In: Proceedings of international conference on manufacturing research (ICMR 2013) held at Cranfield University, Bedford, UK during 19–20 Sep 2013

Reis PNB, Ferreira JM, Richardson MOW (2012) Lap joints strength effect of the surface preparation on PP reinforced glass fiber adhesive. J Thermopl Compos Mater 25:1–8

Chapter 12
Residual Stress and Thermal Treatment of Dissimilar Metal Joints

This chapter presents the fundamental of residual stress development in dissimilar metal joints developed by fusion and solid-state joining processes. Mechanisms, causes and issues related of residual stresses in dissimilar metal joints have been explained using suitable schematics. The factors related to joining and parent metals affecting the residual stress have been elaborated. Remedial measures like joint design, preheat and post-joining treatment of dissimilar metal joints have been discussed.

12.1 Residual Stress

The residual stresses primarily arise due to differential volumetric changes in different zones of the same metal or different components connected rigidly/metallurgically. Differential volumetric change can occur due to many thermal, mechanical and metallurgical events during manufacturing and fabrication. These events include (a) localized plastic deformation (like in contour rolling, shot peening, burnishing), (b) localized heating (like in fusion and even in a few solid-state joining, case hardening, thermal spray coating), (c) differential cooling rate experienced by different zones (e.g. surface and subsurface cooling during casting/welding) and (d) metallurgical transformation (like austenite to martensite and other phases). An increase in differential volumetric change simply increases the residual stress. The nature of residual stress induced can be tensile/compressive. However, the balancing counter stress is also induced in the neighbouring regions for sake of the equilibrium.

© The Author(s), under exclusive license to Springer Nature Singapore Pte Ltd. 2023
D. K. Dwivedi, *Dissimilar Metal Joining*,
https://doi.org/10.1007/978-981-99-1897-3_12

12.1.1 Effect of Residual Stress

The magnitude of residual stress at the most can be equal to the yield strength of metal as any induced stress more than yield strength may cause either plastic deformation (distortion) in ductile metals or even cracking in hard and brittle metals. Further, locked-in residual strain as per modulus of elasticity of metal in a joint induces residual stress. The presence of residual stress can be good or bad for the performance of dissimilar metals joint as per loading conditions. The presence of residual stress at the surface has more effect on the mechanical performance of joint in terms of tensile strength, fatigue resistance and stress corrosion cracking than those present in the subsurface region. Residual stress of the opposite type (tensile/compressive) than externally applied stress generally improves mechanical performance, load carrying capacity and life of the joint.

12.1.2 Residual Stress in Similar and Dissimilar Metal Joint

Residual stress in metallic joint significantly depends on the approach of joining (fusion welding, solid-state joining, solid/liquid joining). However, increasing differences in thermo-physical, mechanical properties of both parent metals, buttering layer and filler metal properties coupled with dilution as per the joining process increases the severity and further complicates the development of residual stress in dissimilar metal joints (Fig. 12.1).

12.2 Factors Affecting Residual Stress

Residual stress in dissimilar metal joints depends upon the thermo-physical, mechanical and metallurgical characteristics of both the parent metals and filler metal/interlayer if any, besides procedural steps of joining like preheat, process parameters affecting heat input, post-weld treatment (stress relieving, heat treatment), etc. Approach of joining like fusion or solid-state joining, autogenous or heterogeneous joining (with filler/interlayer), heat input/generated, if any, the difference in yield strength and ductility of both the parent metals and filler metal are probably the most important aspects determining the residual stress followed by other factors like size and shape of weld/nugget/joint geometry and morphology (Fig. 12.2).

Considering two predominant causes of residual stress in dissimilar metal joining, namely (a) differential thermal expansion/contraction due to thermal cycle imposed during joining owing to localized heating either applied or generated under in-situ condition followed by rapid cooling and (b) localized yielding of metal at/near joint interfaces, the following section describes the development of residual stresses in joints.

Fig. 12.1 Schematic of residual stress distribution in transverse section of the weld joint **a** similar metal weld joint and **b** dissimilar metal joint with flat/straight joint interface

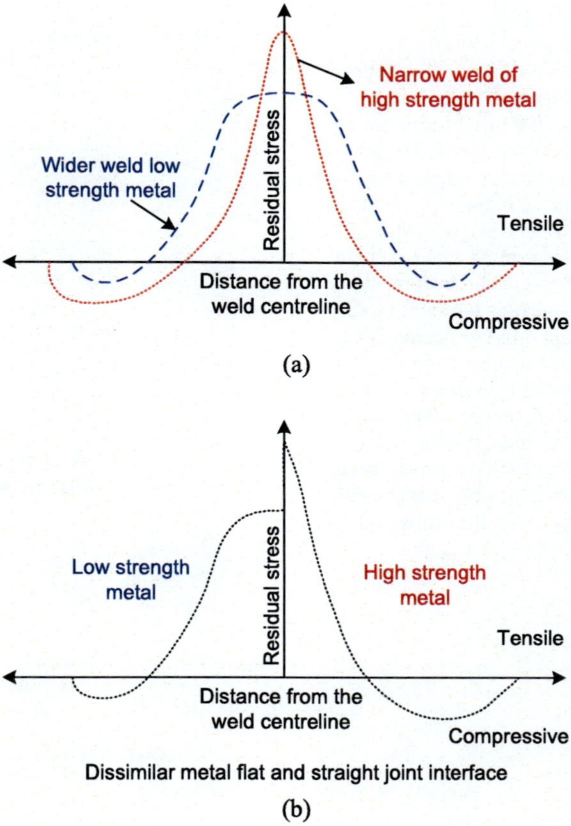

12.2.1 Thermal Stress

The typical thermal cycle of a high heating rate followed by a relatively slow cooling rate causes thermal expansion during heating and contraction during cooling (Fig. 12.3). However, the contraction is not the same as the expansion because expansion is facilitated by the weakening of metals during the heating regime as per $(l.\alpha.\Delta T)$ thermal expansion coefficient (α), temperature rise (ΔT) and region (l) of parent metals being heated during joining. While metals gain strength during the cooling regime, therefore, even after contraction, the original dimensions are not restored. This in turn results in some locked-in tensile strain in the weld/joint interface and the neighbouring regions. The locked-in tensile strain causes residual tensile stress in joints made by fusion-based joining or high heat input solid-state joining processes. Moreover, balancing residual compressive stresses are also induced in nearby regions. The locked-in thermally induced strain, therefore, depends on heat input, thermal expansion coefficient and rise in temperature as per the metals system. An increase in all above three aspects increases locked-in strain and so residual stress.

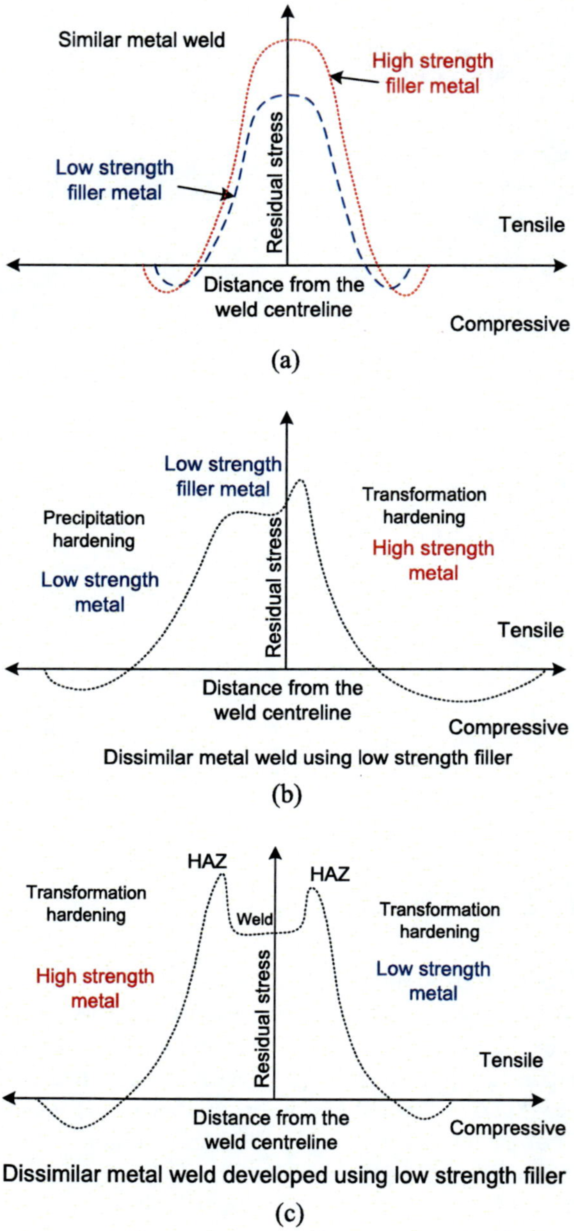

Fig. 12.2 Schematic showing longitudinal residual stress distribution in transverse section of **a** similar metal fusion weld joint developed using low- and high-strength filler metal, **b** dissimilar metal fusion weld joint wherein one parent metal is of high strength of transformation hardening type and another one is of low strength and precipitation hardenable type and **c** dissimilar metal fusion weld joint developed using low-strength filler metal where in both parent metals are transformation hardening type with little difference in their yield strength

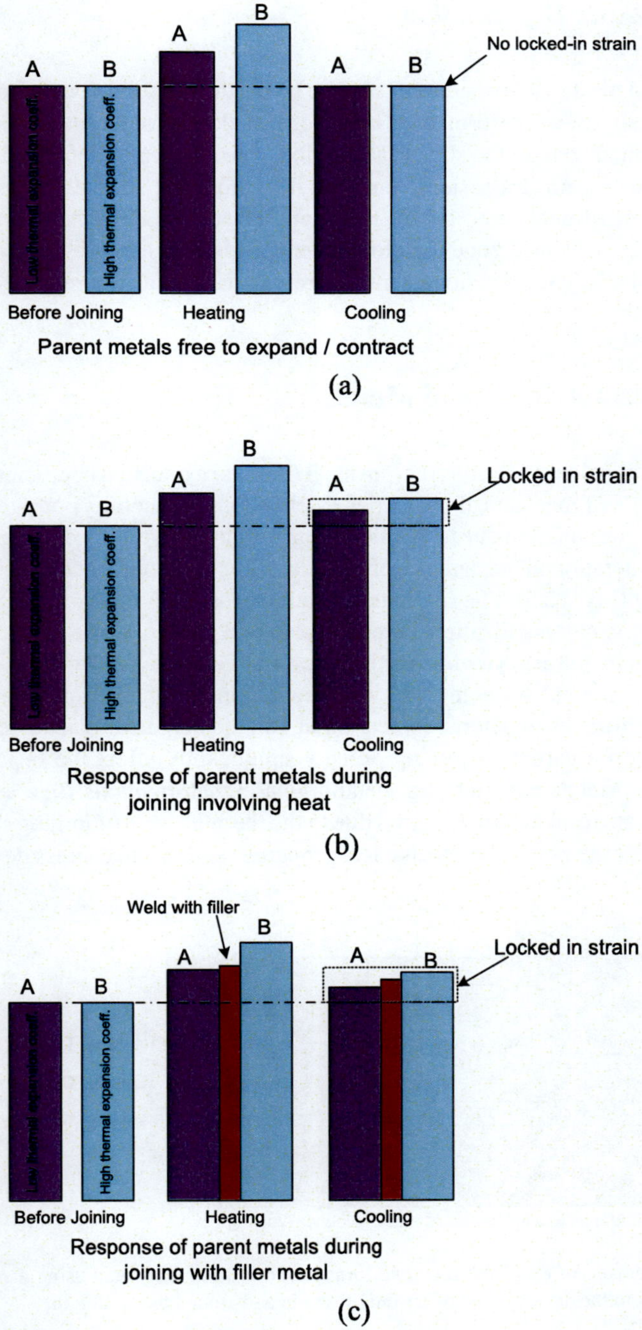

Fig. 12.3 Schematic showing thermal expansion and contraction due to thermal cycle imposed **a** both parent metals of dissimilar combination are free the expand/contract, **b** autogenous dissimilar metal joining and **c** dissimilar metal joining using suitable filler

12.2.2 Plastic Deformation

The localized plastic deformation of surface layers during solid-state joining mostly accompanied by plastic deformation (in the form of elongation) leaves behind an elastically deformed zone in the subsurface region. The elastically deformed/elongated zone tends to regain dimensions, but that does not happen due to metallurgical connectivity with unaffected parent metal and plastically deformed surface layer. The elastically deformed zone therefore develops residual compressive stress at the surface and the balancing residual tensile stress in the subsurface region (Fig. 12.4).

12.3 Thermal Stress and Metals

Further, the variations in the thermo-physical (thermal expansion coefficient, thermal conductivity) and mechanical properties (yield strength, ductility) of parent metals of dissimilar combination due to the rise of temperature and plastic deformation need careful consideration in the analysis of residual stress development during dissimilar metal joining (Fig. 12.5). The rise in temperature causes thermal softening of metals due to recovery, recrystallization, reversion and gain growth, while plastic deformation causes an increase in yield strength at the cost of ductility due to strain hardening (Fig. 12.6a). Thermal softening and work/transformation hardening have opposite effects. Therefore, heat generation/heat input and the degree of plastic deformation associated with a particular joining process significantly affect the residual stress development. Metals experiencing metallurgical transformations (like austenite to martensitic transformation hardening) due to the thermal cycle of joining should also be analysed for their effect on mechanical properties and residual stress development (Fig. 12.6b).

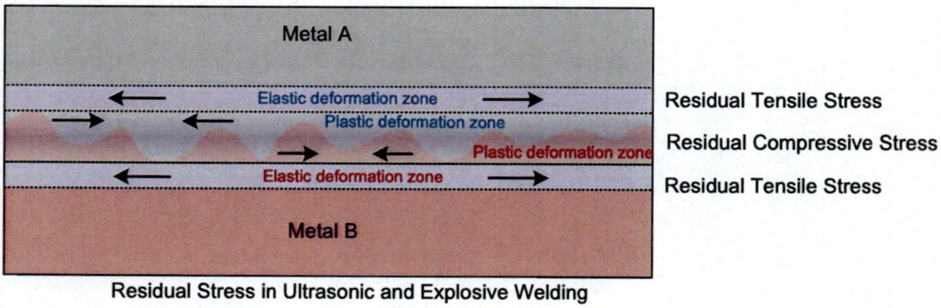

Fig. 12.4 Schematic of elastic/plastic zone formation in a weld fabricated using a solid joining process leading to the development of residual stress in and around the weld joint

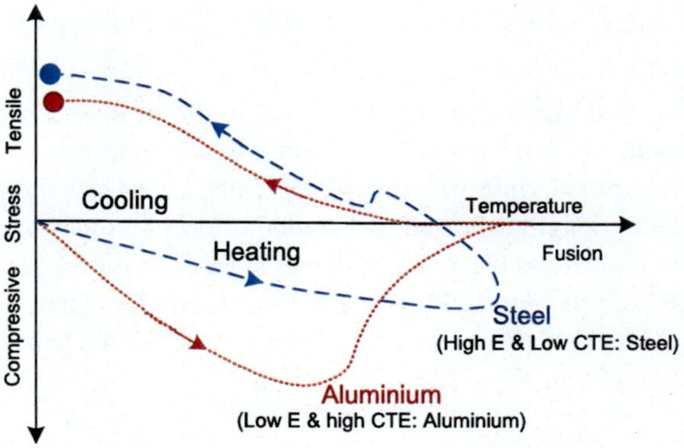

Fig. 12.5 Schematic showing the effect of thermal cycle on residual stress development in two different metals having varying thermal expansion coefficient and modulus of elasticity

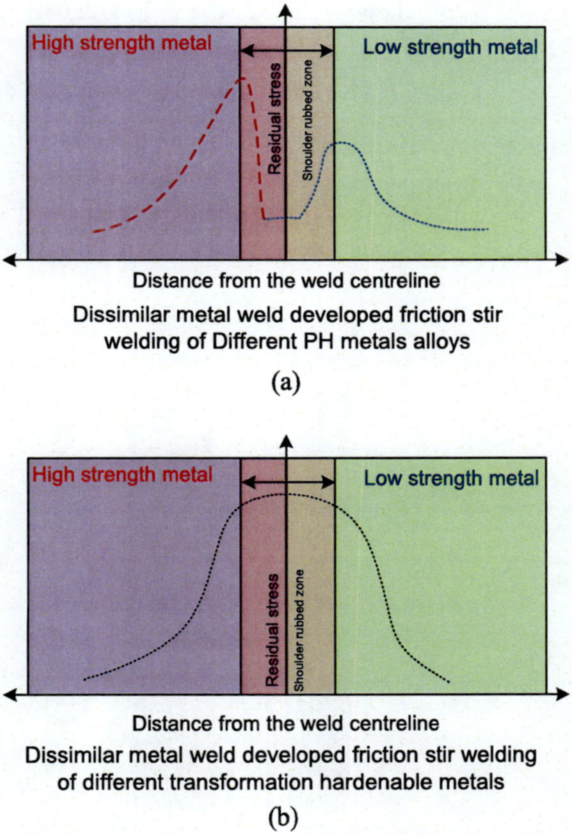

Fig. 12.6 Schematic showing longitudinal residual stress distribution across the dissimilar metal weld joint developed by FSW where in both parent metals are **a** precipitate hardened and **b** transformation hardened

12.3.1 Thermal Stress in the Dissimilar Metal Weld

The residual stress developed due to the heat generated/applied and plastic deformation can be understood easily by considering each of the parent metals and weld/nugget/interface separately if any. Let us say, if the edge of metal A (mild steel) of unit length is heated and cooled in a similar way to the joining, then longitudinal residual stress (f.l.α.ΔT.E) due to locked-in strain under simplified conditions would be 2856fE (Eq. 12.1). While that for metal B (austenitic stainless steel) and metal C (aluminium alloy) would be 3474fE and 1079fE (Eqs. 12.2 and 12.3), respectively, as shown in Fig. 12.7.

$$\text{Metal A : RS due to locked-in strain considering fraction } (f) \text{ of plastic strain : } f \times (1 \times 1400)2.04\text{E : 2856fE} \quad (12.1)$$

$$\text{Metal B : RS due to locked-in strain considering fraction } (f) \text{ of plastic strain : } f \times (1.5 \times 1200)1.93\text{E : 3474fE} \quad (12.2)$$

$$\text{Metal C : RS due to locked-in strain considering fraction } (f) \text{ of plastic strain : } f \times (2.0 \times 650)0.83\text{E:1079fE} \quad (12.3)$$

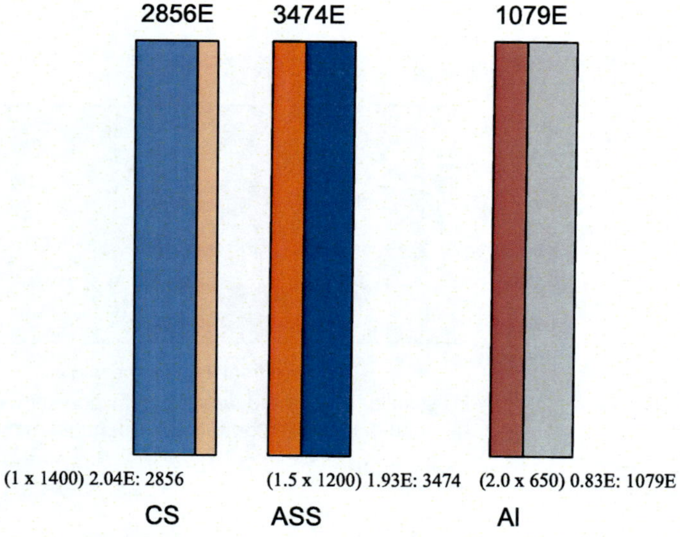

Fig. 12.7 Schematic showing variation in residual stress in different metals when the edge of the plate is heated and cooled

Since parent metals are metallurgically joined and rigidly connected during and after joining, therefore, locked-in strain that remains a fraction (f) of plastic strain would be approximately the same in both the members in similar as well as dissimilar metal joining, which in turn will be leading to undue localization and overstraining of weak metal of dissimilar combination (Fig. 12.8). This can cause more damage (in the form of distortion, cracking and fracture) to a weak member than the strong member of the dissimilar metal combination.

Conversely, the peak longitudinal residual tensile stress experienced on metal A and metal B sides of the dissimilar metal joint (with flat/straight joint interface) would be 2856fE and 3474fE. The peak longitudinal residual tensile stress in the case of metal (a) A and metal C combination would be 2856fE and 1079fE and (b) metal B and metal C combination would be 3474fE and 1079fE, respectively (Fig. 12.8).

The residual stress development due to the thermal cycle of the joining process is significantly affected by variations in mechanical properties as a function of temperature and solid-state phase transformation if any occurring during cooling as observed in the case of steel and aluminium alloys. The pattern of residual stress development in the transformation of hardenable steel depends along on the thermal cycle experienced by weld metal, CGHAZ and HGHAZ (Fig. 12.9a–c).

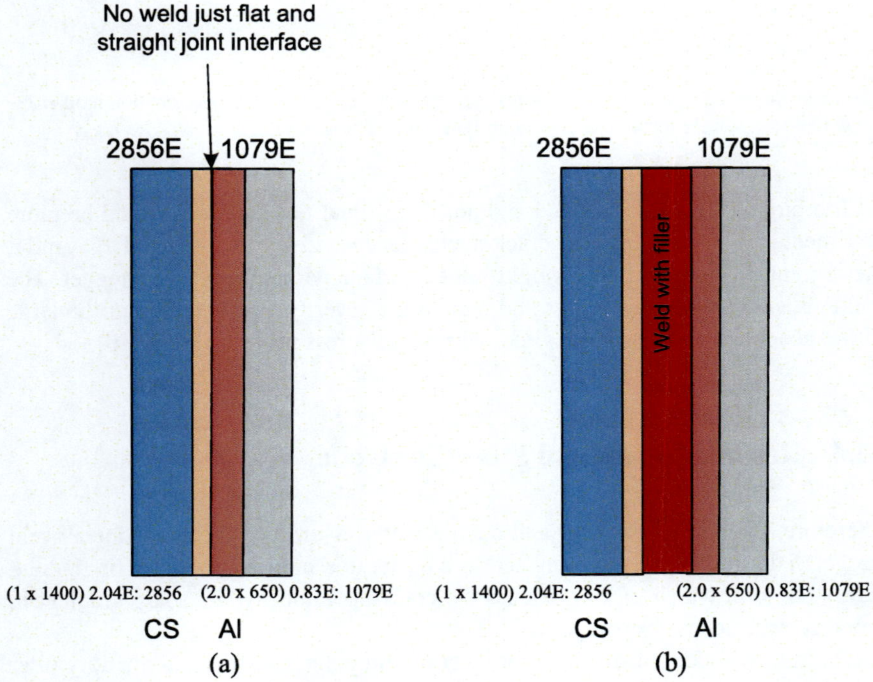

Fig. 12.8 Schematic showing the possibility of differential residual stress development in dissimilar metal (carbon steel and aluminium) joining **a** with a flat and straight interface like in FSW joint and **b** typical fusion weld joint with filler

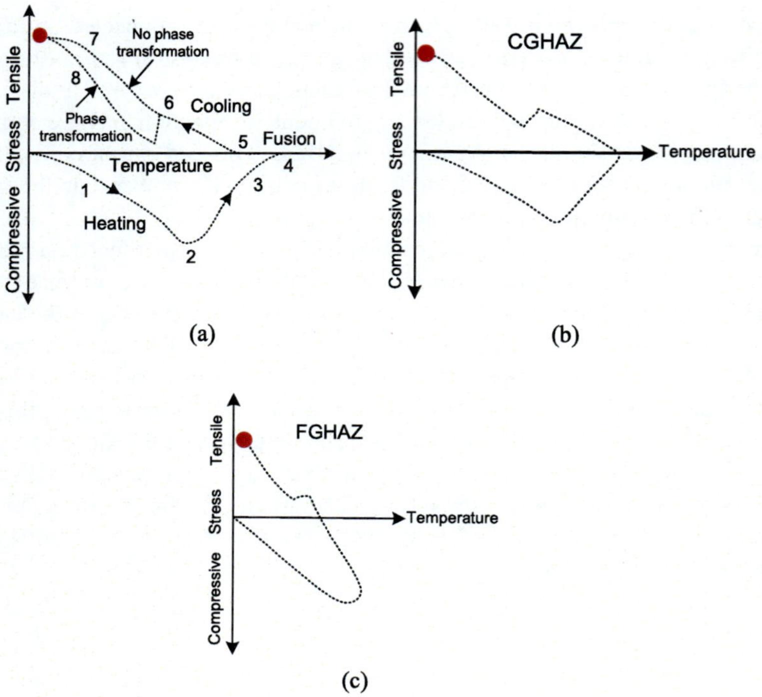

Fig. 12.9 Schematic showing states of residual stress development in transformation hardenable metal like steel **a** weld metal, **b** coarse-grain HAZ and **c** fine-grain HAZ

Too large difference in peak longitudinal residual tensile stress would be more detrimental to the soundness, cracking tendency and performance of dissimilar metal joint, if there is a flat straight joint interface without any weld/nugget. The weld/nugget can act as a buffer and reduce the severity as evident from the large difference in transverse and longitudinal residual stress profile (Fig. 12.10).

12.4 Residual Stress and Filler/Electrode

Therefore, the presence of soft and ductile butter layer/interlayer/weld metal/weld nugget in dissimilar metal joining acts as a buffer to establish some kind of gradient across the joint which in turn reduces the cracking tendency and sharp gradient in property variation across the joint.

The fusion welding using a butter layer of comparatively soft and ductility filler metal/electrode (austenitic stainless steel, Ni alloy) is considered somewhat better for dissimilar metal joining than autogenous fusion welding as these not only reduce the issues related to residual stress but also help to deal with the problem related to metallurgical incompatibility (Fig. 12.11).

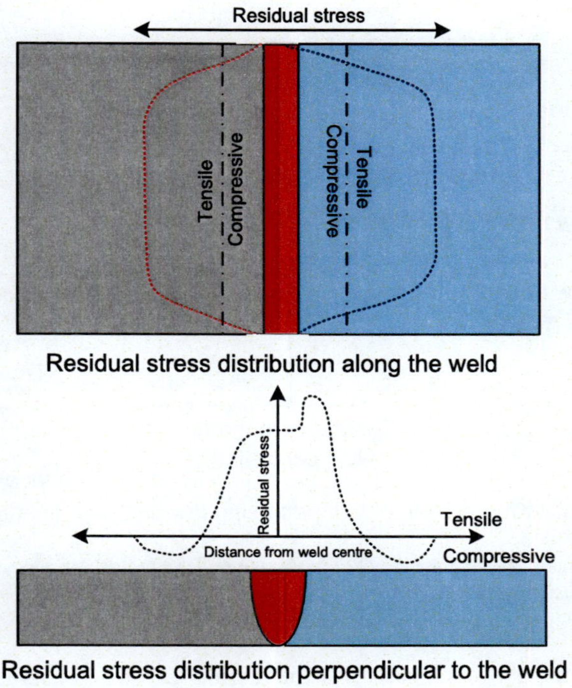

Fig. 12.10 Schematic showing residual stress distribution in a dissimilar metal joint

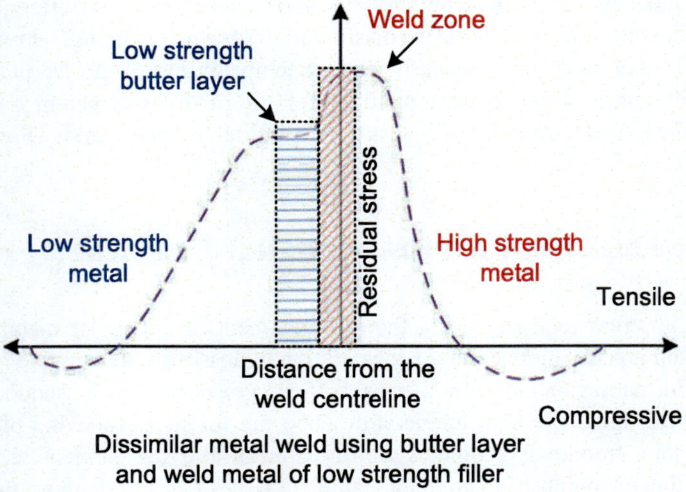

Fig. 12.11 Schematic showing residual stress distribution in dissimilar fusion weld joint developed with low-strength filler metal

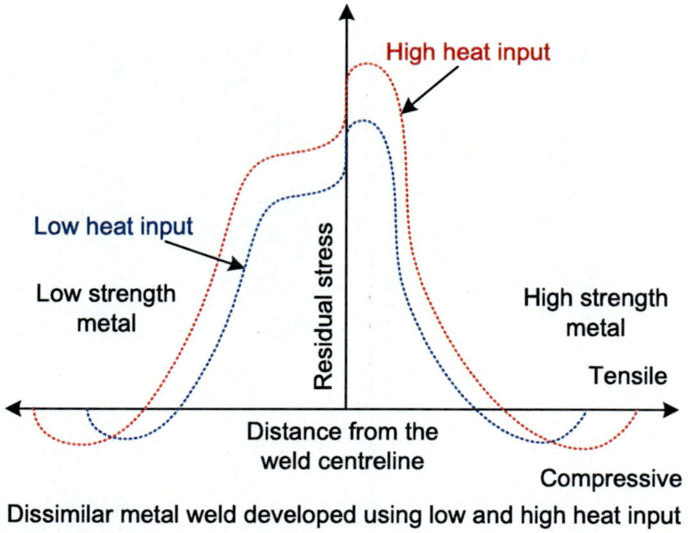

Fig. 12.12 Schematic showing the effect of heat input on longitudinal residual stress distribution in the dissimilar fusion weld joint

Analysis of the above suggests that the thermal expansion coefficient (α), heat input affecting temperature rise (ΔT) and metal property (E) are the most important factors controlling the distribution of residual stresses and their localization in dissimilar metals joining. The length of section (l) heated in one go during joining is same for both the parent metals during dissimilar/similar metal joining. Increasing the difference of thermal expansion coefficient (α), temperature rise (ΔT) as per thermal conductivity and heat generated/applied and elastic modulus of parent metals (E) increases the issues related to residual stress in dissimilar metal joining (Fig. 12.12).

12.5 Residual Stress and Post Weld Heat Treatment

The residual stress relieving using thermal treatment of dissimilar metal joint is complex, and it needs more careful consideration of possible unfavourable metallurgical transformation and different thermal softening response and subsequent weakening of weld joint. The high-temperature exposure for stress relieving of dissimilar metal joint frequently promotes the diffusion of alloying elements across the weld joint due to composition gradient leading to issues like carbon migration, IMC formation, reversion, coarsening, formation of unfavourable micro-constituents, etc.

Stress reliving through thermal treatment is primarily based on the principle of locked-in strain relaxation through thermal softening of weld, HAZ and parent metal due to a rise in temperature, which can also cause recovery, recrystallization and even

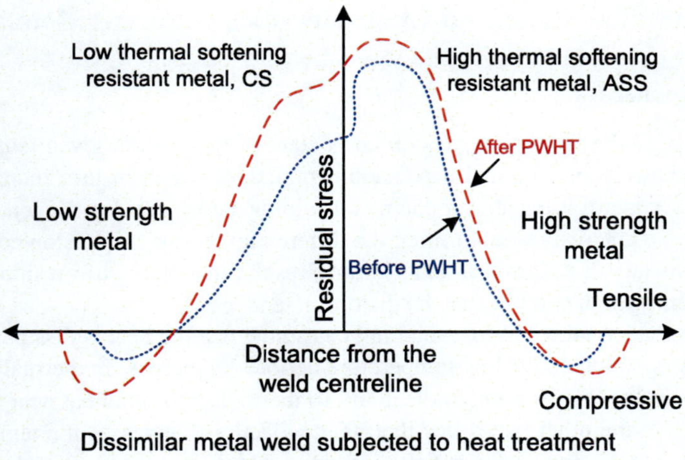

Fig. 12.13 Schematic showing the effect of post weld heat treatment on longitudinal residual stress distribution in dissimilar fusion weld joint

metallurgical transformations. However, different metal shows different thermal softening behaviour as few metals (carbon steel, aluminium) get softened more significantly than others (stainless steel, cobalt and nickel alloys) with the rise in temperature during thermal treatment.

Therefore, in dissimilar metal joints, the parent metal showing high thermal softening tendency results in more relaxation of the locked-in strain and so greater reduction in the residual stress as compared to those (other parent metal/weld metal) exhibiting low thermal softening tendency. Accordingly, the dissimilar metal weld joint of carbon steel and stainless steel subjected on post weld heat treatment leads to greater stress relief on the carbon steel side than austenitic stainless steel side (Fig. 12.13).

Post weld heat treatment (PWHT) of austenitic (AISI 316/304) and martensitic steel combination (P 22/91/92) is expected to cause tempering of untempered martensite of weld zone, overtempering of HAZ and formation of fine-grain zone heated to the intercritical temperature range in martensite steel side. Additionally, PWHT can also result in carbon migration from martensitic steel side to austenitic stainless steel side leading to type IV cracking, knife line cracking and sensitization and weld decay issues. Similarly, dissimilar weld joints of carbon/alloy steel and precipitation hardenable aluminium alloy respond differently with regard to metallurgical changes in both the parent metals in the form of thickening of IMC layer at joint interface/weld, tempering of HAZ on steel side and reversion on the aluminium side. These metallurgical changes in such dissimilar metal joints on PWHT can degrade the mechanical performance significantly. Therefore, PWHT of dissimilar metal joints even for stress relieving must be designed carefully; otherwise, despite relieving residual stresses PWHT of the dissimilar metal joint can be counter-productive.

12.6 Residual Stress and Characteristics of Parent Metals

Physical properties

An increase of the thermal expansion coefficient of metal for a given temperature variation causes more thermal expansion/contraction, which in turn increases the locked-in strained during dissimilar metal joining and increases the tendency of residual stress and distortion. Further, the differential thermal expansion coefficient of parent metals in dissimilar metal joining results in asymmetric residual stress development, which complicates the distortion tendency.

The thermal conductivity of metal indicates how fast the heat is dissipated from the high-temperature to the low-temperature region. An increase in thermal conductivity reduces the temperature gradient and so thermal strain gradient near the weld centreline. On the other hand, low thermal conductivity increases the temperature gradient near the fusion boundary/joint interface, which in turn increases the locked-in strain and so residual stress. Conversely, high thermal conductivity results in more uniform thermal expansion/contraction during the welding/joining. Similar to the thermal expansion coefficient, the increasing difference in thermal conductivity of both parent metals in dissimilar metal joining promotes the asymmetric residual stress and distortion (Fig. 12.14a, b).

Mechanical properties

An increase in both yield strength and elastic modulus of parent metals and filler metals in general increases the residual stress in the dissimilar metal joint (Fig. 12.14c). Increase in yield strength results in higher locked-in strain in the dissimilar metal joint (caused by thermal stress and metallurgical transformation) in the absence of any local strain relaxation. The locked-in strain as per the modulus of elasticity of metal (E) in consideration determines the development of residual stress. The modulus of elasticity of metal indicates the ability to get strained due to thermal/mechanical stress without plastic deformation. An increase in the modulus of elasticity of metal reduces the distortion tendency. Differential mechanical properties of dissimilar metal lead to differential locked-in strain in the dissimilar metal joint (Table 12.1).

12.7 Residual Stress and Component Geometry

The type of residual stresses developed in plate and pipe joints of dissimilar metals differs significantly. The residual stresses in dissimilar metal joint primarily arise due to locked-in shrinkage strain leading to the tensile residual stress in the weld/joint interface and its vicinity. The locked-in strain is a strain by which contraction of the metal should have taken place but could not occur due to metallic continuity with unaffected parent metals. All the factors affecting the thermal expansion and contraction directly or indirectly affect the residual stresses as well.

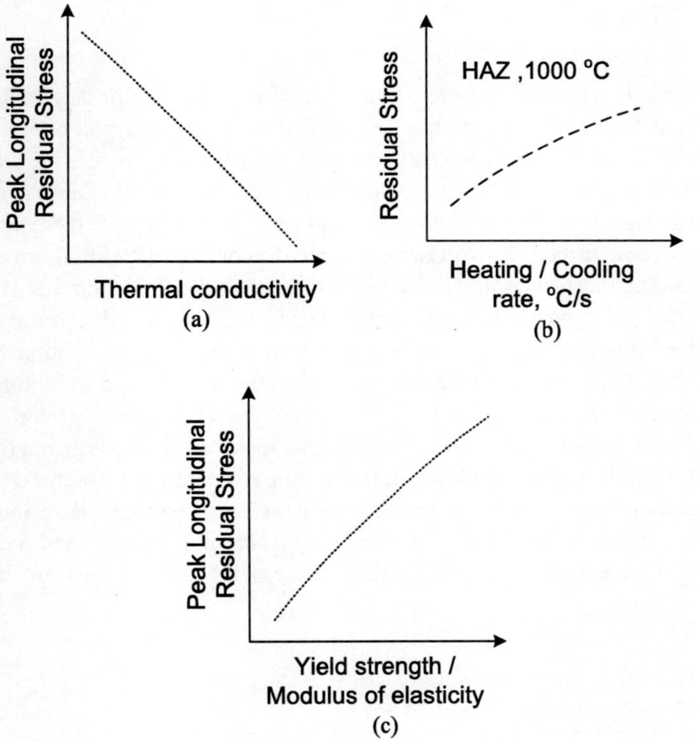

Fig. 12.14 Schematic showing the effect of **a** physical properties, **b** heating/cooling rate and **c** mechanical properties

Table 12.1 Representive mechanical and physical properties of a few common metals

Metal	Thermal conductivity , W/cm K	Thermal expansion coeff, 10^{-6} (°C)$^{-1}$	Yield strength, MPa	Elastic modulus, GPa
Stainless steel	0.15	18	500	180
Carbon steel	0.5	12	380	200
Aluminium	1.9	22	200	70
Copper	3.9	16	280	120

The direction of locked-in shrinkage strain depends on the orientation and configuration of the weld/joint. Like in the case of joining of plates, mostly the linear joint is developed, while in the case of pipe joining, it can be a circular ring or oval shape. Accordingly, locked-in shrinkage strain is of different types, which is leading to the development of different types of residual stresses.

12.7.1 Plate Joining

In the case of plate joining, the stress in the direction of the weld/joint is called longitudinal stress and that in the direction perpendicular to the weld/joint is termed transverse stress. The distribution of longitudinal stress can be seen across the weld/joint, while that of transverse stress can be observed along the weld. The magnitude of longitudinal and transverse stress is maximum and tensile type at the centre of the weld joint in both directions with respect to weld/joint (Fig. 12.15). Thermal expansion coefficient, thermal conductivity, yield strength and elastic modulus of all three metals (both parent metals and filler metal) besides heat input and restraint are some of the factors affecting the peak tensile residual stress and their distribution. Since the length of the weld/joint is significantly larger than the width of the weld/joint, and so metal experiencing locked-in strain in direction of the weld is much greater than that in the direction perpendicular to the weld (which includes weld width and both sides of HAZs). Therefore, the peak longitudinal tensile residual stress (in direction of the weld) was found to be much higher than the peak transverse tensile residual stress (in the direction perpendicular to the weld). Still, thermo-physical and mechanical properties of parent metals and filler dictate the peak stress in each of the parent metal(s).

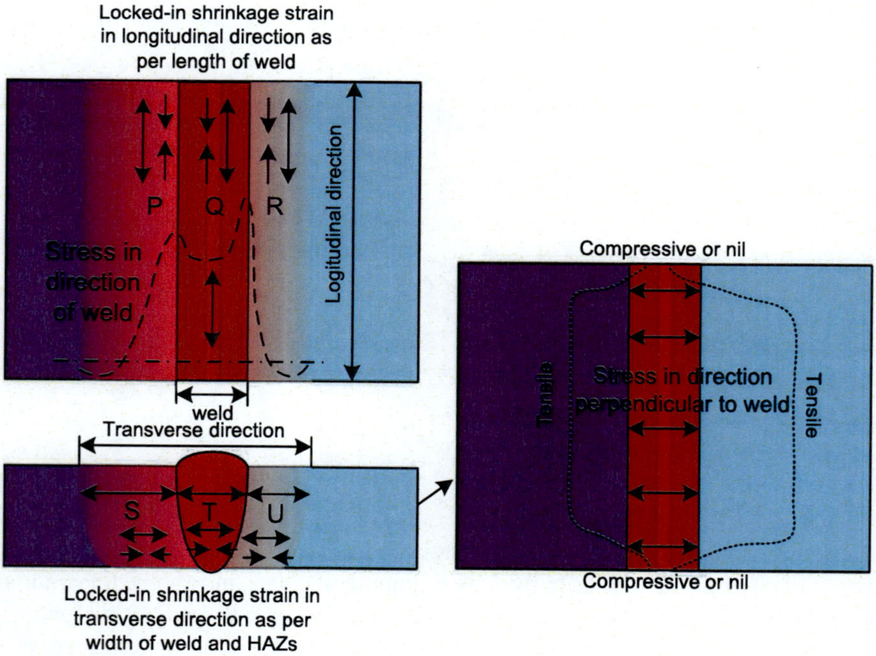

Fig. 12.15 Schematic of **a** longitudinal residual stress distribution, **b** transverse residual stress distribution and **c** schematic of varying locked-in strain in the transverse direction of joint

12.7.2 Pipe Joining

In the case of pipe joining, three types of residual stress may arise, namely hoop stress, axial stress and radial stress. The hoop stress and axial stress need proper consideration due to their significant effect on the performance of pipe joints. Hoop stress occurs tangentially along the circumference of the weld joint, while axial stress is observed across the weld joint in the direction parallel to the axis of the pipe. Residual stress occurring in the direction perpendicular to the axis of the pipe and through the thickness of the weld joint and HAZs is termed radial stress. The typical residual stress distribution on the transverse section of the pipe weld/joint is shown in Fig. 12.16a–c. In general, residual stress in weld joint and HAZ especially at the outer periphery is tensile in nature and is the maximum at the centre of the weld joint in case of similar metal welding using identical filler metal. The inner periphery of the pipe joint may show tensile/compressive stress residual stress. In the case of the dissimilar metal weld joint of pipe, the maximum residual tensile stress at the outer periphery is noticed at the fusion boundary of both the parent metals primarily due to the application of the preferred low-strength filler metal for dissimilar metal joining.

Apart from the composition of dissimilar metals, diameter of the pipe, thermo-physical properties (thermal expansion coefficient, thermal conductivity), mechanical properties (yield strength and elastic modulus of all three, i.e. both parent metals, filler metal), root gap, heat input and restraint are some of the factors affecting the peak tensile residual stress and their distribution in the pipe joint. Residual tensile stress and axial stress at the outer periphery of the pipe weld joint have more effect on mechanical performance, cracking and fracture tendency than radial stress and those present at the inner periphery of the joint.

12.8 Residual Stress and Performance of Dissimilar Metal Joint

The performance of dissimilar metal joints especially with regard to residual stress affects mechanical properties (tensile, fatigue), corrosion resistance (galvanic corrosion, stress corrosion cracking), soundness (cracking) and distortion (angular, transverse, longitudinal distortion). In general, the effect of residual stress in the dissimilar metal joint is more severe on the weaker metal than the stronger metal of dissimilar combination in the form of increased cracking and distortion tendency. For example, the joining of steel and aluminium frequently results in cracking of weld joint from weld/joint interface primarily due to localization of residual stress. Increasing residual stress, in general, reduces tensile load/fatigue load carrying capacity, especially of weak metal in dissimilar metal joining (Fig. 12.17).

The presence of high tensile residual stress in high stress concentration areas like weld toe and root decreases the tensile and fatigue strength of dissimilar metal joints.

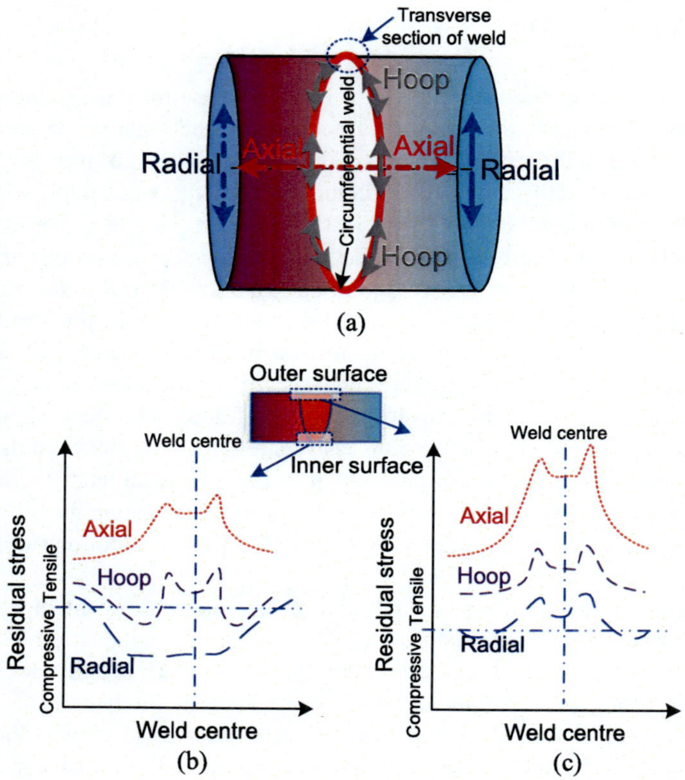

Fig. 12.16 Schematic of residual stress is pipe joining different dissimilar metals **a** types of residual stress in pipe joint, **b** residual stress distribution at the outer periphery and **c** residual stress distribution at the inner periphery

Fig. 12.17 Schematic showing the effect of residual stress on load/stress-carrying capacity

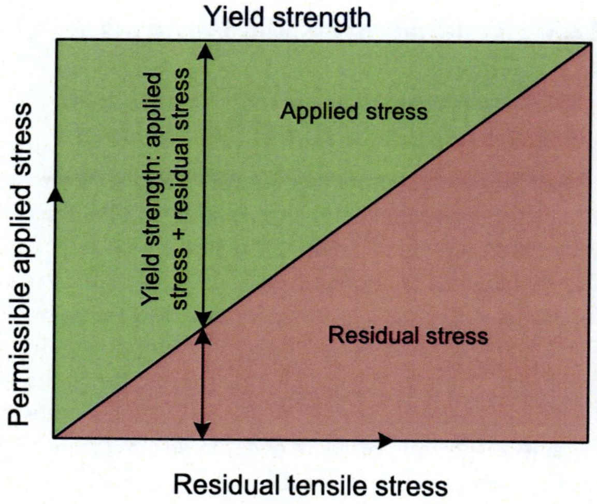

Similarly, metals of dissimilar combination, which are widely spaced in electronegativity series, encourage galvanic corrosion. The presence of tensile residual stress coupled with corrosion-sensitive environment makes dissimilar metal joints prone to stress corrosion cracking. The stress corrosion cracking in presence of tensile residual stress significantly decreases the tensile load carrying capacity and increases the fracture tendency of dissimilar metal joints. Cracking of dissimilar metal joint is common in presence of residual tensile stress coupled with the formation of hard and brittle intermetallic compounds and unfavourable micro-constituents. Solidification cracking in the weld metal and HAZ cracking (in metallurgically incompatible dissimilar metal combinations) due to embrittlement are primarily caused by residual tensile stress. The differential thermal expansion and contraction of HAZ of both the parent metals and weld metal/joint interface causes distortion. Longitudinal stress results in shrinkage in the length of assembly dissimilar metal joint roughly 1 mm/m, while the width of weld, groove geometry and the volume of weld metal per unit length marginally affect the transverse shrinkage. These shrinkages therefore must be kept in mind while designing a dissimilar metal joint, else shorter and smaller assembly than the design requirement would be obtained at the end.

Selection of joint design, the fixture for restraint, low-strength filler metal, the appropriate sequence of welding, suitable groove geometry, welding/joining process for controlling the heat input and post-weld/joining treatment for reliving residual stress, all help in reducing issues related to residual stress of dissimilar metal joints.

12.9 Thermal Treatment in Dissimilar Metal Joining

The joining of dissimilar metals using solid-state joining, fusion joining and brazing approach may need heating of parent metals in the form of preheating and post-joining treatment to develop a sound dissimilar metal joint with desired mechanical and metallurgical characteristics. The purpose of thermal treatment in the form of preheating and post-heat treatment may vary significantly depending upon the composition and physical metallurgy of both the parent metals and filler metal if any, approach of joining (solid state, fusion joining and brazing), process, the tendency of cracking, residual stress and distortion.

12.9.1 Preheating

The preheating of the parent metals in the solid-state joining is applied for increasing ductility and lowering the yield strength to facilitate the localized/bulk metal plastic flow at the interface. On the contrary, preheating in the fusion metal joining approach is primarily applied to increase the penetration and produce symmetric weld, reduce the cooling rate and reduce the cracking tendency, residual stress and distortion. Preheating can be applied either selectively on one side only or on both sides of parent metals in dissimilar metal joining. However, the difference in thermo-physical properties (thermal conductivity, melting temperature) and physical metallurgy (temperature of recovery, recrystallization, reversion, transformation, etc.) of both metals can complicate the selection of preheating temperature. In such cases, the choice of preheat temperature should be based on consideration of preheating requirement of weak metal of dissimilar combination (steel and Al alloy) to apply uniform preheating (temperature) of both the members to be joined. However, this may not be optimum or enough for other metals of dissimilar metal combination.

12.9.2 Preheat and Metal Strengthening Mechanism

The metal strengthening mechanisms of both the metals of dissimilar metal combination to be joined must be considered to minimize the adverse effect on the structure and properties of dissimilar metal joints. The common metal strengthening mechanisms are solid solution strengthening, grain refinement, work/strain, precipitation, dispersion and transformation hardening either singly or in the combination of the above. The influence of preheating on the structure and properties of metals strengthened by solid solution formation and dispersion hardening is marginal, while the metals strengthened by grain refinement, work hardening, precipitation hardening and transformation (Q & T condition) hardening are significantly affected by application of external/in-situ heat during dissimilar metal joining (Fig. 12.18). Therefore, the selection of the preheat temperature needs to be done carefully. Preheating to high temperature is expected to increase the softening of HAZ regions especially due to recovery, recrystallization, reversion, grain growth and overtempering as per the strengthening mechanism of metal involved in dissimilar metal joining (Fig. 12.19). For example, PH aluminium alloys show reversion of hardening precipitates (GP I, GP II) on heating above 150 °C, and steel in Q & T condition (tempered say at 350 °C) will get overtempered on heating above tempering temperature (350 °C). Hence, the same preheating temperature for both the metals of dissimilar metal combination cannot be optimum/enough.

12.9.3 Preheat and Thermal Cycle

Preheating primarily alters the thermal cycle experienced by two parent metals during dissimilar metal joining. However, even under identical conditions of heat input, the thermal cycle experienced by two parent metals may not be the same due to differences in thermo-physical properties (Fig. 12.20). Variation in thermal cycles can be observed in terms of heating rate, peak temperature, high-temperature retention period and the cooling rate at a particular location away from the fusion boundary/joint interface on both the sides of parent metals during dissimilar metal joining.

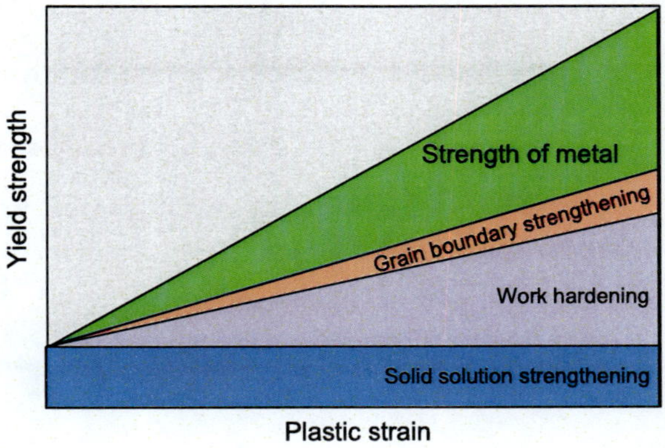

Fig. 12.18 Schematic gives a general idea of the relative contribution of different metal strengthening mechanisms on yield strength

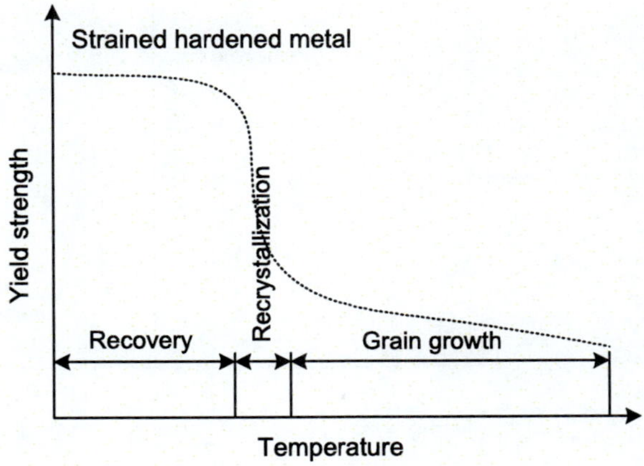

Fig. 12.19 Schematic showing the effect of temperature on yield strength due to different mechanisms

The finalization of preheat temperature must be based on consideration of purpose, e.g. reducing cooling rate, thermal softening of a specific metal (of dissimilar metal combination) in addition to possible external heat input application/in-situ heat generation. Efforts should always be made to apply minimum possible heat as unnecessarily high preheat temperature simply increases the efforts (power, heat, cost, time) needed and difficulty in handling hot components during dissimilar metal joining.

12.10 Post-dissimilar Metal Joining Thermal Treatment

The post-dissimilar metal joining thermal treatment changes the metallurgical and mechanical properties and residual stress state. These changes might be favourable or unfavourable as well in some of the aspects. Therefore, the feasibility of thermal treatment of dissimilar metal joints must be evaluated carefully for the potential benefits vis-a-vis the compromises in mechanical, metallurgical and corrosion behaviour.

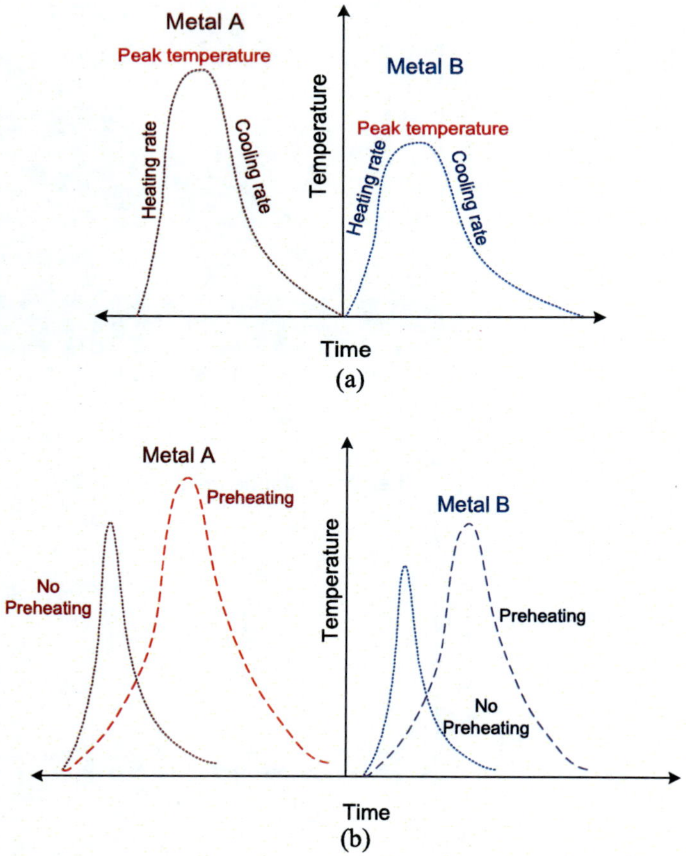

Fig. 12.20 Schematic showing weld thermal cycle in vicinity of fusion boundary/joint interface of **a** two different parent metals of dissimilar metal combination and **b** dissimilar metal joining with and without preheating

12.10.1 Obstacles in Thermal Treatment

The heat treatment cycle commonly performed on carbon and alloys steels includes heating 30–50 °C above the upper critical temperature (> 730 °C) holding time for 30–60 min for austenitization/homogenization and followed by cooling. The cooling rate can vary from 2–5 °C/sec, 10–20 °C/sec and 60–100 °C/sec for annealing, normalizing and quenching treatment depending upon the carbon equivalent and hardenability of steel in consideration (Fig. 12.21). On the other hand, thermal treatment of the common non-ferrous metal like PH Al alloys includes T6 and T4 treatment involving solutionizing (450–510 °C), quenching followed by artificial ageing (140–200 °C) or natural ageing, respectively. The significant difference in thermal cycle requirements for heat treatment of Al and steel imposes difficulties in thermal treatment of dissimilar metal joint of Al-steel combination. Thermal treatment designed for steel shall cause melting of Al member of the dissimilar metal joint of Al-steel combination, while thermal treatment designed for Al alloy shall not cause any major metallurgical transformation in steel member except in case of tempering/overtempering of Q & T steel and reducing residual stress. Moreover, low-temperature thermal treatment (150–200 °C) primarily designed for residual stress reliving of Al-steel dissimilar metal joint shall release residual stress primarily from the Al side and not much from the steel side (Dwivedi 2000a; Kaushik and Dwivedi 2021).

12.11 Metallurgical Transformation(s) During Thermal Treatment

The design of thermal treatment in terms of peak temperature and soaking/holding time followed by controlled cooling rate determines metallurgical transformation in dissimilar metal joints. The difference in physical metallurgy of two parent metals and weld metal leads to differential response to the thermal treatment of the entire assembly of the joint. The thermal treatment brings in favourable metallurgical changes in one of the sections between the two parent metals and weld metal/weld nugget, while the remaining section may suffer badly or remain unaffected depending upon the thermal cycle applied and their physical metallurgy. Therefore, the design of thermal treatment in case of dissimilar metal joining is more complicated especially in fusion welding with filler metal, which is different from both the parent metals.

12.11.1 Physical Metallurgy of Ferrous Metals

The physical metallurgy of ferrous and non-ferrous metals is completely different. The physical metallurgy of ferrous metals needs an understanding of the Fe–C diagram, TTT diagram, CCT diagram and Schaeffler diagram, besides the concept of hardenability, carbon equivalent, chromium and nickel equivalent, etc. The family of ferrous metals includes wrought iron, cast iron, and steel with increasing carbon content along with the minor concentration of alloying elements (as traces) predominantly affecting the ease of joining. The ferrous metals are primarily transformation hardened with limited contribution from solid solution strengthening and grain refinement. An increase of carbon and other alloying elements generally increases hardness and hardenability, which in turn encourages the tendency of the cracking and embrittlement, thereby making the joining of ferrous metals more and more difficult due to increased requirement of efforts and resources in the form of better-quality filler metals, higher preheat temperature and need of post weld heat treatment. However, metallurgical transformation in ferrous metals usually occurs on heating

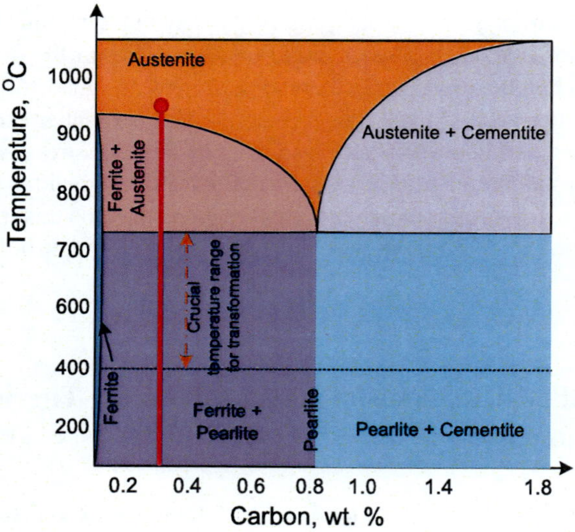

(a)

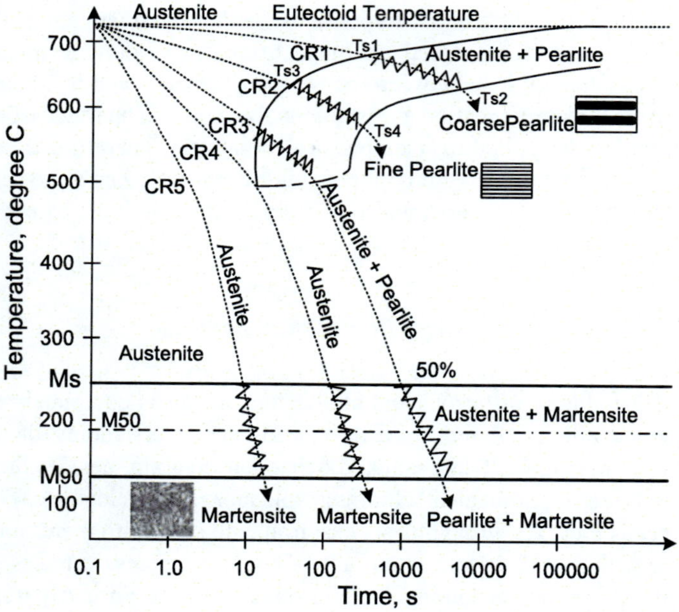

(b)

Fig. 12.21 Two important diagrams **a** a part of the Fe–C phase diagram and **b** a continuous cooling transformation diagram used to determine the peak heating temperature and cooling rate to design a post-joining thermal treatment, respectively

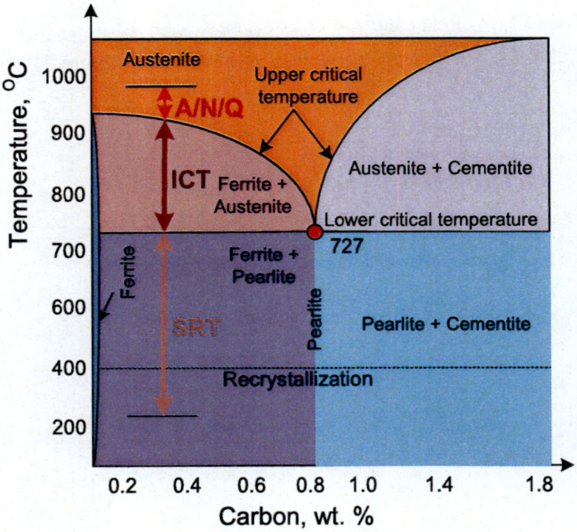

Fig. 12.22 Fe–C phase diagram showing peak heating temperature for different post-joining thermal treatments

to a temperature more than lower critical temperature (> 730 °C) except in the case of Q & T steel (Fig. 12.22). The post-joining thermal treatment (even in the range of 200–730 °C) of dissimilar metal joints involving ferrous metal in dissimilar metal combination can result in residual stress relieving and a variety of embrittlement. Conversely, ferrous metals are not much affected in respect of metallurgical transformation if post-joining thermal treatment is carried out in subcritical temperature regions except in tempering, overtempering, stress relieving and a few types of embrittlement (Dwivedi 2000a).

12.11.2 Physical Metallurgy of Non-ferrous Metals

On the other hand, the physical metallurgy of non-ferrous metals like Al, Mg, Cu, Ti, etc. can be entirely different with respect to solid solution strengthening, grain refinement, work hardening and precipitation hardening. The metallurgical transformation in the form of recrystallization, recovery, reversion and grain growth can occur at relatively lower temperatures starting from 150 °C for Al alloy to high temperatures as per the type of non-ferrous metal system (Table 12.2). The recrystallization temperature of metals is around 0.4–0.5 T_m (where T_m melting temperature is in K), and it decreases with plastic deformation. Similarly, recovery and reversion (dissolution of precipitates) in the work hardened and precipitation hardened non-ferrous metals (Al, Mg, Cu alloys) take place at low temperatures during post-joining thermal treatments (Sharma and Dwivedi 2005; Dwivedi et al. 2006; Shah et al. 2004, 2007; Sharma and Dwivedi 2007; Haro et al. 2009; Singh et al. 2010; Sharma et al. 2013, 2014a, b, 2015; Dhiman et al. 2014; Dwivedi 2000b, 2002; Ramirez et al. 2007).

Table 12.2 Recrystallization and melting temperature of the common metals

Metal	Recrystallization temperature (°C)	Melting temperature (°C)
Sn	− 4	232
Pb	− 4	327
Zn	10	420
Al	150	660
Mg	200	650
Ag	200	962
Cu	200	1085
Fe	450	1540
Ni	600	1453
Mo	900	2610
W	1200	3410

12.11.3 Mechanical Behaviour

The response of thermal treatment to the mechanical performance (hardness, tensile, toughness, fatigue and creep) of the dissimilar metal joint depends on the strengthening mechanisms of metals, metallurgical compatibility and metallurgical transformation occurring in both the parent metals, weld/joint interface and heat-affected zone. Additionally, the way by residual stress affected, i.e. either reduced or induced due to differential thermal expansion and contraction due to thermal treatment also influences the mechanical performance significantly.

The post-joining thermal treatment of dissimilar metal joints (based on different strengthening mechanisms) responds differently. For example, dissimilar combinations of (a) ferritic/martensitic steel and austenitic stainless steel, (b) PH aluminium and transformation hardening steel, (c) PH aluminium alloy and copper alloys, (d) copper alloy and austenitic stainless steel and (e) Ti alloy and austenitic stainless steel are commonly used in industries to fabricate engineering system. The post-joining thermal treatment of dissimilar metal joints of such combinations must be designed carefully as per need and purpose. Thermal treatment of dissimilar metal joints may not always be beneficial for mechanical and metallurgical performance. For example, intermetallic compounds formed during dissimilar metal joining may grow during post-joining thermal treatment and deteriorate the mechanical and corrosion performance of dissimilar metal joints.

The post-joining thermal treatment (simple high-temperature exposure for some time) of metal strengthened by work hardening, precipitation hardening, grain refinement and Q & T may show weakening and soften due to recovery, recrystallization, grain growth, overtempering and even coarsening of IMCs coupled with relaxation of residual stresses. Designing thermal treatment of either one of the parent metals in dissimilar metal joint or weld metal might be considered if other metal is expected to remain unaffected. For example, T6 or T4 treatment to PH aluminium alloy member

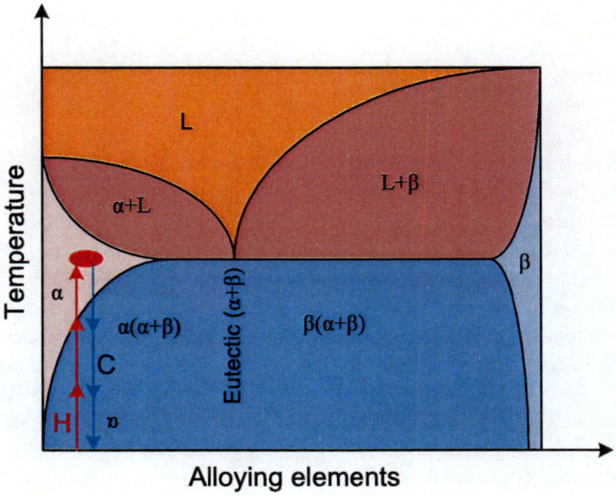

Fig. 12.23 Schematic of the binary phase diagram of an alloy having complete solubility in liquid state but partial solubility in solid state showing identification of solutionizing temperature for post-joining thermal treatments like T6 and T4

joined with steel can be given thermal treatment using solutionizing (450–500 °C), followed by quenching and then artificial ageing (140–200 °C) or natural ageing (room temperature) as shown in Fig. 12.23. However, coarsening of IMCs and their adverse effect on mechanical and corrosion performance must be considered while designing such thermal treatments. In such a case, steel may not be affected much by thermal treatment except for residual stress relaxation.

12.11.4 Residual Stress

Thermal treatment designed carefully to relieve residual stress certainly improves the mechanical performance of the dissimilar metal joints. Poorly designed stress relieving thermal treatment of dissimilar metal joint can also induce a new set of residual stress leading to the reduced mechanical performance in terms of tensile strength, stress corrosion cracking, etc.

References

Dhiman M, Dwivedi DK, Sehgal R, Bhat IK (2014) Effect of iron on wear behavior of as cast and heat treated hyper-eutectic Al-18Si-4Cu-0.5Mg alloy: a taguchi approach, proceedings of the institution of mechanical engineers, Part L. J Mater: Des Appl 228(1):2–16

Dwivedi DK (2000a) Effect of cutting parameters and heat treatment on specific—power consumption in machining of En-31. Trans Indian Instit Metals 54(4–5):539–543

Dwivedi DK (2000b) Heat treatment of cast Al–Si base alloys for improved mechanical properties. Aluminium India 28(4):21–24

Dwivedi DK (2002) Study the effect of cutting parameters and heat treatment on machining behavior of spheroidized steel En-31. Instit Eng (india) 82:57–60

Dwivedi DK, Sharma R, Kumar A (2006) Influence of silicon content and heat treatment on the mechanical properties of cast Al–Si–Mg alloys. Int J Cast Metal Res 19(5):275–281

Haro S, Ramírez J, Dwivedi DK, Martínez E (2009) Influence of solutionizing and ageing temperatures on microstructure and mechanical properties of cast Al–Si–Cu alloy. Mater Sci Technol 25(7):886–890

Kaushik P, Dwivedi DK (2021) Induction preheating in FSW of Al-Steel combination. Mater Today: Proc 46(1091–1095):2021

Ramirez J, Martinez E, Dwivedi DK, Haro S (2007) Solution and aging heat treatment of cast Al–Si–Cu alloys. In: Proceeding of international materials research congress on materials characterization held in Cancun, Mexico, Oct 28 Nov 1 2007

Shah KB, Kumar S, Dwivedi DK (2004) Influence of heat treatment parameters on abrasive wear behaviour of two cast Al–Si alloys. Trans Indian Instit Metals 57(5):28

Shah KB, Kumar S, Dwivedi DK (2007) Aging temperature and abrasive wear behaviour of cast Al-(4, 12, 20%)Si-0.3%Mg alloys. Mater Des 28(6):1968–1974

Sharma R, Dwivedi DK (2005) Influence of silicon (wt.%) and heat treatment on abrasive wear behaviour of cast Al–Si–Mg alloys. Mater Sci Eng A 408(1–2): 274–280

Sharma R, Dwivedi DK (2007) Solutionizing temperature and abrasive wear behaviour of cast Al–Si–Mg alloys. Mater Des 28(6):1975–1981

Sharma C, Dwivedi DK, Kumar P (2013) Effect of post weld heat treatments on microstructure and tensile properties of friction stir welded joints of Al–Zn–Mg alloy AA7039. Mater Des 43:134–143

Sharma C, Dwivedi DK, Kumar P (2014a) Fatigue behaviour of friction stir weld joints of Al–Zn–Mg alloy AA7039 developed using base metal in different temper condition. Mater Des 64:334–344

Sharma C, Dwivedi DK, Kumar P (2014b) Investigating the microstructure and mechanical properties of friction stir weld joints of solution hardening aluminum alloy AA5086. Indian Weld J 47(4):65–73

Sharma C, Dwivedi DK, Kumar P (2015) Influence of pre-weld temper conditions of base metal on microstructure and mechanical properties of friction stir weld joints of Al–Zn–Mg alloy AA7039. Mater Sci Eng A 620:107–119

Singh RR, Sharma C, Dwivedi DK, Mehta NK, Kumar P (2010) The microstructure and mechanical properties of friction stir welded Al–Zn–Mg alloy in as welded and heat treated conditions. Mater Des 32:682–87

Index

© The Editor(s) (if applicable) and The Author(s), under exclusive license to Springer Nature Singapore Pte Ltd. 2023
D. K. Dwivedi, *Dissimilar Metal Joining*,
https://doi.org/10.1007/978-981-99-1897-3

J

K

L

M

N

O

P

R

S